Concrete and Concrete Structures

Concrete and Concrete Structures

Edited by
Jasper Bates

WILLFORD **P**RESS

www.willfordpress.com

Published by Willford Press,
118-35 Queens Blvd., Suite 400,
Forest Hills, NY 11375, USA

ISBN: 978-1-64728-341-4

Cataloging-in-Publication Data

Concrete and concrete structures / edited by Jasper Bates.
p. cm.
Includes bibliographical references and index.
ISBN 978-1-64728-341-4
1. Concrete. 2. Concrete construction. 3. Building materials. 4. Structural engineering. I. Bates, Jasper.
TA439 .C66 2022
620.136--dc23

For information on all Willford Press publications
visit our website at www.willfordpress.com

WILLFORD PRESS

Contents

Preface

Over the recent decade, advancements and applications have progressed exponentially. This has led to the increased interest in this field and projects are being conducted to enhance knowledge. The main objective of this book is to present some of the critical challenges and provide insights into possible solutions. This book will answer the varied questions that arise in the field and also provide an increased scope for furthering studies.

Concrete is a composite material that is composed of coarse and fine construction aggregate bonded together with fluid cement that hardens over time. The construction aggregate is mixed with water and dry cement, to form fluid slurry which is easy to pour and mold into shapes. The cement forms a hard matrix by reacting with water and other ingredients. This matrix binds together the materials into a durable stone-like material which has many uses. Additives like pozzolans can be added in the mixture in order to improve its physical properties. Concrete is the most frequently used building material. It must be produced carefully since it is time sensitive. This book elucidates the concepts and innovative models around prospective developments with respect to concrete and concrete structures. The readers would gain knowledge that would broaden their perspective about this field. Coherent flow of topics, student-friendly language and extensive use of examples make this book an invaluable source of knowledge.

I hope that this book, with its visionary approach, will be a valuable addition and will promote interest among readers. Each of the authors has provided their extraordinary competence in their specific fields by providing different perspectives as they come from diverse nations and regions. I thank them for their contributions.

Editor

An Experimental Investigation on the Mechanical Properties of Gangue Concrete as a Roadside Support Body Material for Backfilling Gob-Side Entry Retaining

Peng Gong [iD],[1] Zhanguo Ma [iD],[1] Xiaoyan Ni [iD],[1] and Ray Ruichong Zhang[2]

[1]*State Key Laboratory for Geomechanics and Deep Underground Engineering, School of Mechanics and Civil Engineering, China University of Mining and Technology, Xuzhou, Jiangsu 221116, China*
[2]*Department of Mechanical Engineering, Colorado School of Mines, Golden, CO 80401, USA*

Correspondence should be addressed to Zhanguo Ma; zgma@cumt.edu.cn

Academic Editor: Peter Majewski

Development of a safe and economical roadside support body (RSB) material is the key to successful backfilling gob-side entry retaining (GER). By means of laboratory tests, this paper studied the effects of the water-cement ratio, aggregate content, and age on the contractibility and resistance increasing speed, compressive strength, and postpeak carrying capacity of the concrete with gangues as an aggregate. It also discussed the rationality and adaptability of gangue concrete as a RSB material for backfilling GER. The experimental results show that the compressive strength of gangue concrete increases with age, and that the strength of gangue concrete demonstrates a nonlinear decreasing trend with the increase of the cementing material's water-cement ratio. The water-cement ratio in the range of 0.46–0.60 has the most significant regulation effect on the strength of gangue concrete. Mixing with a certain amount of coal gangue enhances the postpeak carrying capacity of concrete, preventing the sample from impact failure. The field experimental results report that as a RSB material, gangue concrete can meet the design and application requirements of GER with gangue backfilling mining. A RSB material featuring high safety, high waste utilization rate, fast construction speed, and low costs is provided.

1. Introduction

The gangue backfilling GER (Figure 1(b)) is an innovative mining technology based on gob-backfilled adaptive to multiple complex mining geological conditions without a coal pillar [1]. Based on the present theoretical and technological research on GER in thin and medium thick coal seams under simple conditions, we basically mastered the mine pressure law of the GER with roof caving (Figure 1(a)). In addition, applications under single complex mining geological condition, such as deep mine or the working faces of fully mechanized top coal caving, were explored [2–4]. But to date, GER for working faces with high mining height is still faced with difficulties under deep, heavily stressed, and multiple complex geological conditions, limiting the GER technology development [5]. Since the RSB generally needs to withstand a high vertical stress in the GER process in which roofs are managed using the caving method, the hanging roof at gob side is sheared down to reduce additional load, ensuring the late-stage movement stability of the roof [6–8]. However, under multiple complex geological conditions, the mechanical properties of the traditional RSB materials cannot adapt to the roadway pressure and deformation under the influence of mining, which results in a high supporting cost and a lower entry retaining success rate.

In recent years, as a safe, environmentally friendly mining method, gob-backfilled mining has been applied in China's eastern and central regions. Many scholars have conducted creative research on backfilling mining [9–12], the results of which show that it can reduce the working face pressure and avoid main roof breaking [13–16]. The combination of the backfilling mining technology and the GER technology enables mining without using a coal pillar under multiple complex geological conditions. Because of differences in roof

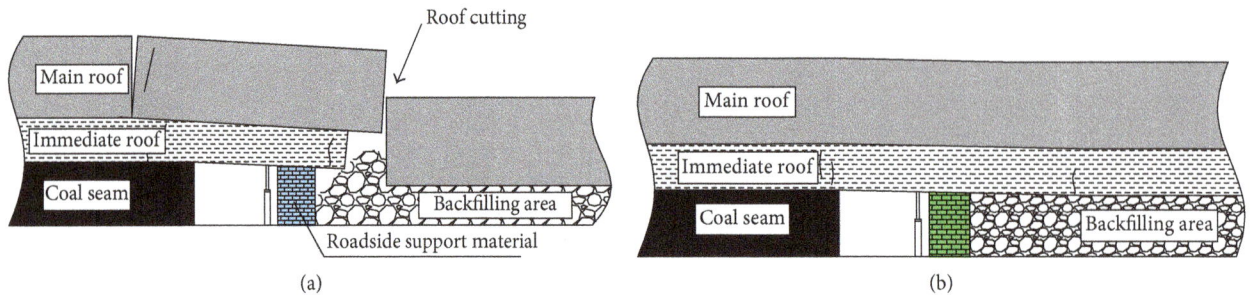

FIGURE 1: GER of gob roof managed by different methods. (a) GER with roof managed by caving method. (b) GER with roof managed by backfilling method.

"large structure" and the roadside support principle [17–19], there are significant differences in roadside support material requirements between the GER with fully mechanized gangue backfilling mining and the GER in which roofs are managed using the caving method. Consequently, the key to the gangue backfilling GER lies in the development of a safe and economical RSB material which is adaptable to the law of mine pressure in the gob-backfilled GER process.

In the GER development, a variety of RSB materials have been tried by engineers and scholars to maintain roadway stability, such as timber cribs, dense pillar, gangue piling, and masonry walls [20, 21]. However, due to issues including low support force, large deformation, poor gob isolation, and poor roof connection, the GER stability and safety have been greatly affected. The high water-material backfilling technology that came out in recent years has been applied in GER engineering in which roofs are managed using the caving method, but the high costs limit its application on backfilling GER working faces. However, for working faces with gangue backfilling, in case a large amount of granular gangues can be used as the aggregate to produce the RSB material in entry retaining, then the conveying and back-filling systems can be shared, greatly reducing the roadside support costs, which is favorable to the promotion of the backfilling GER. At the same time, the use of gangues as a RSB material can help remove gangues on the ground, turning the waste into a benefit as well as alleviating the ground environmental problems [22–24].

Coal gangue is a concrete material utilizing gangues as a coarse aggregate and cement as a cementing material, mixed with a certain amount of additives. Recently, scholars have studied the performance of gangue concrete for different application fields: Wang et al. [25] predominately discussed the permeability of the concretes composed of coal gangue and fly ash when applied in farmland drainage ditches. The results indicate that the concrete using waste coal gangue and fly ash had better water permeability than ordinary concrete. And it had a high subsurface drainage modulus, making it feasible for application in coal mining subsidence areas with high groundwater levels. Li et al. [26] developed finite element analysis models of rebar-penetrated connection between gangue concrete filled steel tubular column and reinforced gangue concrete beam using software ABAQUS 6.10 and revealed cyclic behavior of rebar-penetrated connection between gangue concrete filled steel tubular column and

reinforced gangue concrete beam. Wu et al. [27] designed a numerical model to predict the development of suction in a cemented gangue backfill mixture containing fly ash.

Through laboratory tests, this paper focuses on the study of the influence of the water-cement ratio, aggregate content, and age on the early stage contractibility and the transitional stage resistance increasing speed, the late-stage compressive strength, and the postpeak carrying capacity of gangue concrete. Through this process, the sensitive control ranges of the various factors affecting gangue concrete performance were obtained. The adaptability of gangue concrete as a RSB material was discussed by taking into account the behavior law of mine pressure of GER with fully mechanized gangue backfilling mining. Based on the analysis of the test results, field tests were carried out in China's Shandong mining area, the results of which suggest that the gangue concrete material designed according to the ratio obtained in this paper can meet the RSB load-deformation requirements in different stages during the entire backfilling GER process. A safe, environmentally friendly and economical RSB material is provided for the development and promotion of GER with fully mechanized gangue backfilling mining.

2. Test Method and Scheme Design

2.1. Test Materials. Cement clinker (p.c32.5) was used, and granular gangues of 0–25 mm in diameter were used as the aggregate in the test. Considering the economy and rationality of GER engineering, the gangues were obtained directly from the gangue piles in the experimental mining area. In order to give a reasonable explanation of gangue concrete strength evolution, the particle size distribution and mechanical properties of the aggregate were measured first.

(1) Particle size distribution characteristics of crushed gangues

The initial crushed gangue particles obtained from the field are continuously graded granular materials. A Talbot formula was used to quantitatively describe the particle size distribution characteristics of gangues:

$$P_x = 100\left[\frac{d}{D}\right]^n (\%),\qquad(1)$$

where P_x is the passing percentage of the d aggregate, D is the maximum particle size of the granular gangues, d is the

FIGURE 2: Gangue strength and particle size distribution.

current size of the granular gangues, and n is the coefficient of the Talbot formula.

The analysis results show that the aggregate basically satisfies the continuously graded distribution characteristics with a Talbot coefficient ($n = 0.6$). Figure 2 shows the comparison of the particle size distribution characteristics between the gangue aggregate and when $n = 0.6$.

(2) Mechanical properties of the complete gangue samples

To obtain the strength characteristics of the gangue aggregate for analyzing the coordination mechanism of the aggregate and the cementing material, uniaxial compression tests were carried out on multiple complete gangue samples ($\Phi 50$ mm $\times 100$ mm) in this paper, and the average unconfined uniaxial compressive strength obtained is 9.83 MPa. The compressive strength results of the standard gangue samples are shown in Table 1. From the uniaxial compression characteristic curve of gangues shown in Figure 2, it can be seen that the gangues have a certain postpeak carrying capacity. Moreover, a number of shear cracks occur during the crushing process, suggesting that the gangues had piled in the open air and experienced long-term weathering and soaking. Their initial fissures are relatively developed, and the strength of

the granular gangue aggregate is lower than the ordinary concrete aggregate.

2.2. Test System and Sample Preparation. Samples were prepared according to the concrete sample forming and curing standards with a design size of 150 mm $\times$ 150 mm $\times$ 150 mm. To guarantee the reliability of the data, 3 samples were prepared for the same ratio and age. The test used the MTS 816 test system (Figure 3) that ensures highly accurate and high sensitivity results.

2.3. Experimental Scheme Design. Walker et al. [28] found that the concrete strength is mainly determined by the following factors: (1) the strength of the cement matrix; (2) the bonding strength of the mortar and the coarse aggregate; (3) the strength, stiffness, and grading of the coarse aggregate; and (4) the maximum size of the aggregate particles. After determining the strength, stiffness, grading, and the maximum size of the coarse aggregate in the field, the test used the orthogonal test design method (Table 2) to design pure cement (nonaggregate) samples with a same water-cement ratio for each gangue concrete group to provide a comparison. The following factors influencing gangue concrete properties were mainly considered:

TABLE 1: Unconfined uniaxial compression test results for the standard cylindrical gangue samples.

Gangue no.	Sample size (mm)		Compressive strength (MPa)	Elastic modulus (GPa)	Poisson's ratio
	Diameter	Height			
G-01	49.8	98.3	9.17	5.11	0.21
G-02	49.8	97.7	10.33	4.32	0.26
G-03	49.9	102.3	7.32	6.22	0.18
G-04	49.8	101.7	11.82	4.83	0.22
G-05	49.9	98.6	10.52	5.08	0.22

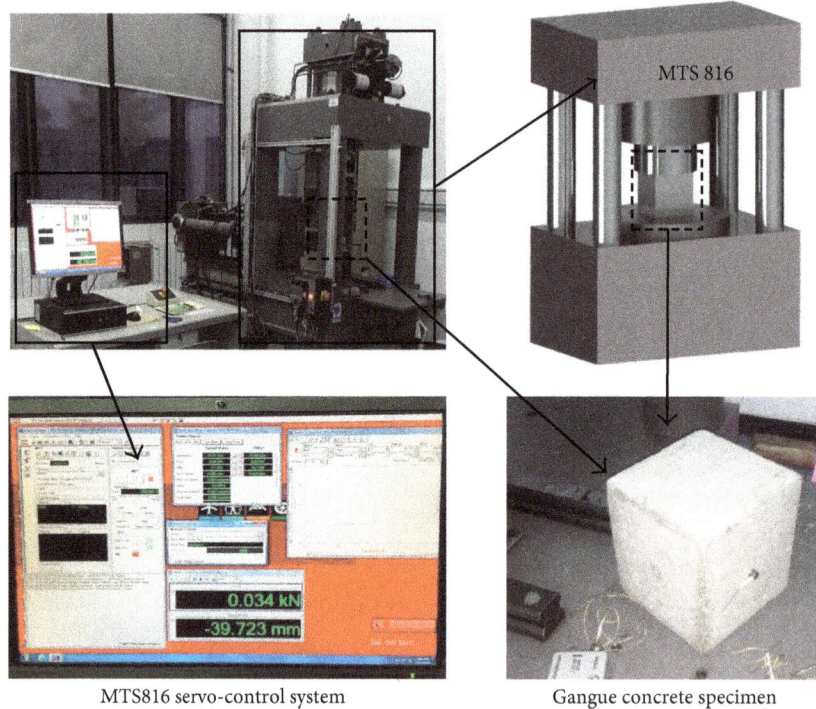

MTS816 servo-control system   Gangue concrete specimen

FIGURE 3: Test system and formed samples.

(1) Water-cement ratio: water-cement ratio is an important factor affecting the strength of the cementing material. The water-cement ratios of 0.40, 0.43, 0.46, 0.50, 0.55, and 0.6 were considered in this test.

(2) Aggregate content: four aggregate content levels were considered, respectively, which include 45%, 50%, 55%, and 60%.

(3) Curing age: for gangue concrete samples and pure cement samples with each water-cement ratio and gangue content level, three ages were considered, that is, 7 days, 14 days, and 28 days, for studying the evolution rules of the compressive strength properties of gangue concrete in different hardening stages.

3. Test Results and Discussion

3.1. Effect of the Water-Cement Ratio on the Mechanical Properties of Gangue Concrete. The average concrete strength with a 50% gangue content level under different water-cement ratios can be obtained through uniaxial compression tests of the gangue concrete sample groups 1–6, as noted in Figure 4. With the increase of the water-cement ratio of the cementing material, the gangue concrete strength demonstrates a nonlinear decreasing trend. When the ratio is greater than 0.46 and less than 0.60, the gangue concrete strength becomes more sensitive to the ratio. When the ratio is less than 0.46, the decrease of the ratio has little effect on the late stage strength of the gangue concrete. Within the sensitivity range of the water-cement ratio, the strength of the 28-day gangue concrete is adjustable at 4.97–10.18 MPa.

The average strength of the cementing material under different water-cement ratios can be ascertained through uniaxial compression tests of pure cement sample groups 7–12, as depicted in Figure 5. The raising of the water-cement ratio increases the free water in the cementing material during the cement hardening process, and the average compressive strength of the pure cement samples monotonously decreases with the increase of the water-cement ratio.

TABLE 2: Sample gangue concrete ratios (mass ratios).

No.	Gangue content (%)	Water-cement ratio	Water content (%)	Cement content (%)	Age (day)
1		0.40	14.29	35.71	7, 14, 28
2		0.43	15.03	34.97	7, 14, 28
3	50	0.46	15.75	34.25	7, 14, 28
4		0.50	16.67	33.33	7, 14, 28
5		0.55	17.74	32.26	7, 14, 28
6		0.60	18.75	31.25	7, 14, 28
7		0.40	14.29	35.71	7, 14, 28
8		0.43	15.03	34.97	7, 14, 28
9	0	0.46	15.75	34.25	7, 14, 28
10		0.50	16.67	33.33	7, 14, 28
11		0.55	17.74	32.26	7, 14, 28
12		0.60	18.75	31.25	7, 14, 28
13	45		14.29	35.71	7, 14, 28
14	55	0.4	14.29	35.71	7, 14, 28
15	60		14.29	35.71	7, 14, 28

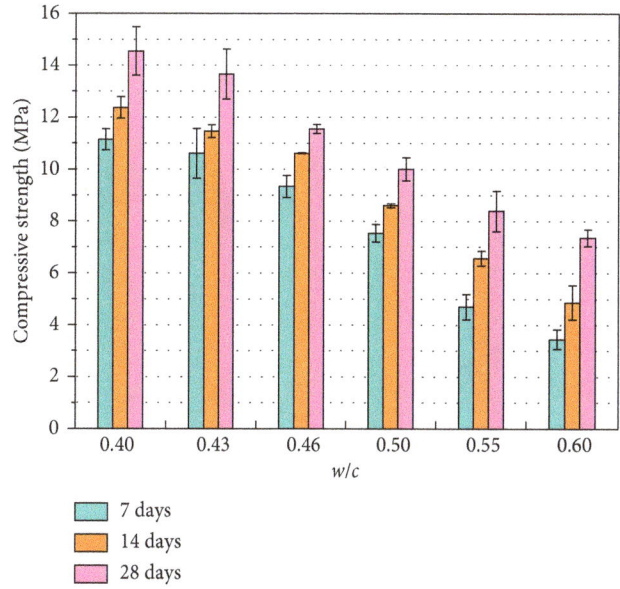

FIGURE 5: Average strength and error of the pure cement samples with different water-cement ratios.

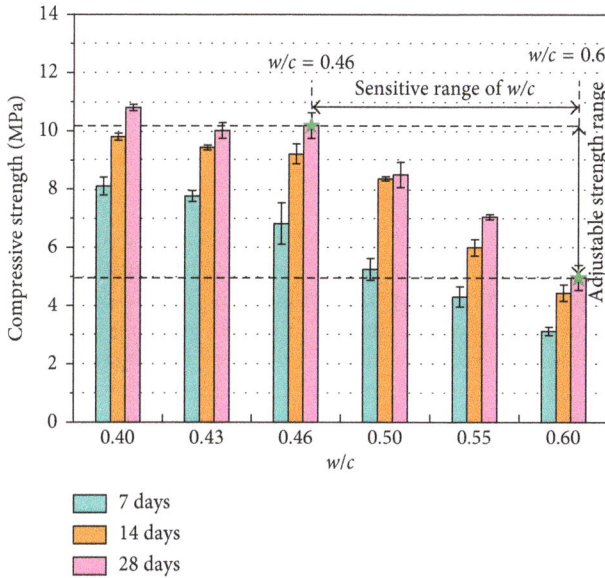

FIGURE 4: Average strength and error of the concrete with different water-cement ratios and a gangue content of 50%.

With a comparison between Figures 4 and 5, it can be assumed that the strength of the gangue concrete is jointly determined by the aggregate, the cementing material, and their bonding surface. Damage and failure are first present in the weakest one of them, resulting in the overall instability of the concrete samples. The strong water absorption of the gangues during the mixing of the concrete mix causes a local low water-cement ratio on the surface of the gangue granules and thus increases the compactness of cement stone near the surface of the gangue aggregate. Concurrently, because the surface of the gangue particles is rough with micropores, the bonding surface between gangues and the cementing material

is enlarged, and the binding power between gangues and the cement stone is improved. Hence, the bonding strength at the interface between the aggregate and the cement paste in the gangue concrete is higher than that of the ordinary concrete. When the water-cement ratio lies within the range of 0.46–0.60, the cementing material's strength in the gangue concrete is lower than the strength of the aggregate. Therefore, the cementing material's strength determines the overall strength of the gangue concrete, and changing the water-cement ratio has a relatively significant influence on the gangue concrete's strength. When the water-cement ratio is within the range of 0.40–0.46, the strength of the cementing material is greater than that of the aggregate. When the vertical stress is greater than 10 MPa during the loading process, damages appear inside the aggregate, eventually leading to the instability of the samples. Therefore, increasing the strength of the cementing material within the above range has no notable effects on the strength of the gangue concrete.

Therefore, within the water-cement ratio range described in this test, to allow a full harmonization between the strengths of the aggregate and the cementing material, it can be expected that the sensitivity range of the water-cement ratio lies in 0.46–0.60 for controlling the strength of gangue concrete. In the application of backfilling GER engineering, the RSB load requirements may be satisfied plus a full use of the carrying capacity of the granular gangues through adjusting the water-cement ratio of the cementing material within the sensitivity range based on understanding the roof load law and by considering the strength characteristics of gangues.

3.2. Effect of the Aggregate on the Mechanical Properties of Gangue Concrete. The average concrete strength with dissimilar gangue contents can be obtained through uniaxial

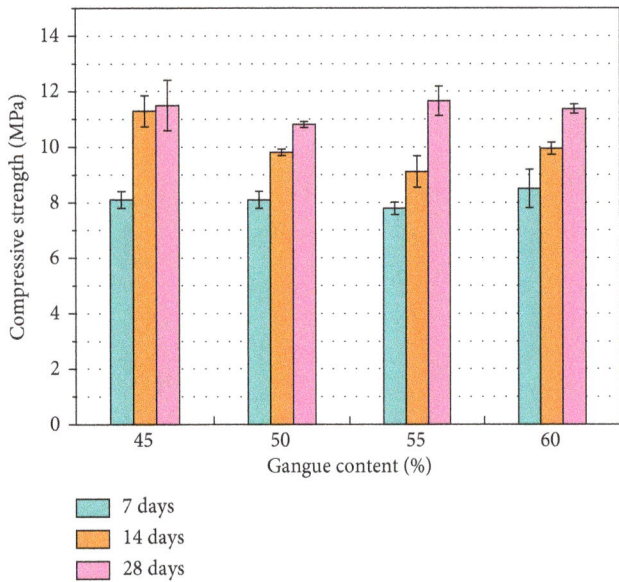

FIGURE 6: Concrete strengths at various ages with different gangue contents.

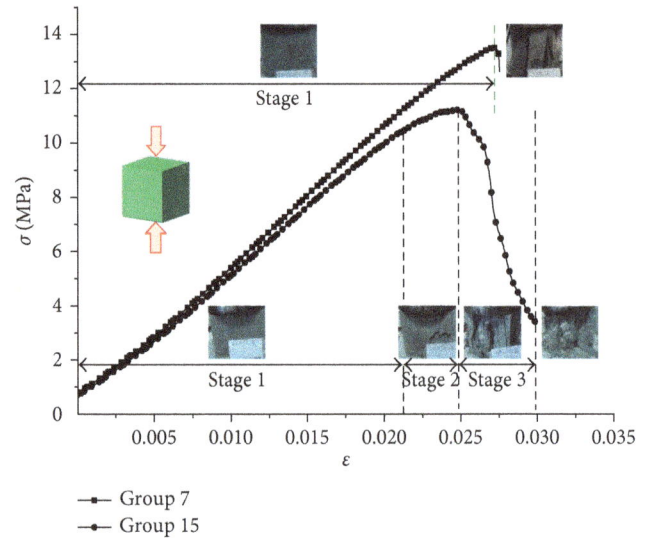

FIGURE 7: Destruction characteristics of the gangue concrete and the pure cement samples with the same water-cement ratio.

compression tests of the gangue concrete sample groups 1, 13, 14, and 15, as noted in Figure 6. It can be seen from the figure that with the gangue content falling in the range of 45% to 60%, the strength of the gangue concrete with a 28-day age reaches 10.8 to 11.6 MPa. Since the granular gangues have many fissures, the strength of the coal gangue aggregate is lower than that of the complete gangue samples. However, the rough and water-absorbing surface of gangues allows both a larger contact area between the aggregate and the cementing material and a reduced local water-cement ratio on the aggregate surface. A hard cement stone shell forms around the gangues, which constrains the lateral deformation of the aggregate. It puts the gangues in a three-dimensional stress state in the concrete, improving the ultimate strength of the fractured granular gangues. Therefore, the uniaxial compressive strength of the gangue concrete samples is higher than that of the complete gangue samples. Concurrently, the aggregate's own damage is a key factor leading to the overall concrete destruction.

With the water-cement ratio at 0.40, the increase of the gangue content has little effect on the concrete strength. Therefore, it can be deduced that when the water-cement ratio and the aggregate strength are kept unchanged, the gangue content within a certain range is not a notable factor affecting the strength of gangue concrete. However, by comparing the uniaxial compression characteristics of the sample groups 7 and 15 (Figure 7), it can be seen that as an aggregate, coal gangue has a vital effect on the destruction mode and the postpeak carrying characteristics of concrete.

As shown in Figure 7, the entire gangue concrete compression process can be divided into three stages according to the destruction characteristics. Stage 1 is the elastic stage, in which the stress and the strain maintain a linear relation following the initial compaction and the elasticity modulus remains constant. There are no obvious cracks on the gangue concrete surface in this stage. Stage 2 is the stiffness attenuation

stage, which begins with the appearance of nonlinear stress-strain growth and ends when the peak stress is achieved. With the increase of the strain, the stiffness of the samples is reducing continuously, but the stress is still rising. In this stage, fine cracks appear on the surface of the samples, which gradually evolve into local surface spalling. Stage 3 is the postpeak stage. With the increase of the strain, the stress demonstrates a decreasing trend, and the crack development is relatively slow until finally the samples are destructed due to excessive deformation. The samples have a certain postpeak carrying capacity. Such features allow the adaptability to the large deformation requirement of the RSB and a slow release of the internal elastic energy in the RSB, avoiding the impact damage caused by a sudden release of the internal elastic energy. In the destruction of the nonaggregate samples with the same water-cement ratio, only the characteristics in stage 1 (the elastic phase) are present throughout the compression process, and once at the strength limit, the samples will experience a brittle failure. The destruction is mainly in a form of tensioning and splitting to a relatively great degree of damage with a certain impact. The postpeak carrying capacity of the samples is poor with a small yield. Obviously, in terms of safety and economy, the gangue concrete material mixed with aggregate at a certain proportion has obvious advantages in GER engineering practices.

3.3. Effect of the Age on the Mechanical Properties of Gangue Concrete

3.3.1. Effect of the Age on the Strength of Gangue Concrete. The strength limit of the gangue concrete was obtained for various ages with various water-cement ratios and a gangue content of 50% through uniaxial compression tests of the gangue concrete groups 1 to 6, which is shown in Figure 8. Regardless of the water-cement ratio, the compressive strength of the gangue concrete demonstrates an

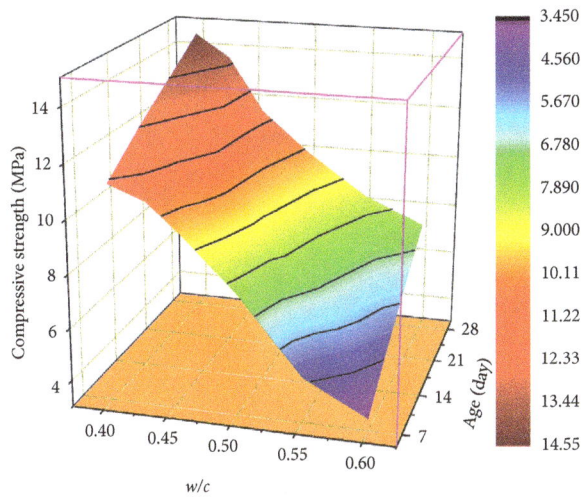

FIGURE 8: Average compressive strength of the gangue concrete at different ages with different water-cement ratios (gangue content: 50%).

TABLE 3: Strength development degree of the early age gangue concrete.

Water-cement ratio	f_{7d} (%)		f_{14d} (%)	
	Without gangue	50% gangue content	Without gangue	50% gangue content
0.40	76.60	75.00	85.01	90.74
0.43	77.64	77.49	83.93	94.23
0.46	80.89	66.95	91.96	90.48
0.50	75.27	61.81	85.94	98.32
0.55	56.05	61.24	78.32	85.15
0.60	46.89	62.86	66.24	89.40
Average	68.89	67.56	81.90	91.37

increasing trend with the age. After mixing the coal gangue and cement with water, $Ca(OH)_2$ produced by the hydration of the calcareous material had a secondary reaction with the active SiO_2 and Al_2O_3 in the coal gangue, forming the stable water-insoluble hydrated calcium silicate gel and hydrated calcium aluminate gel. The cross reaction, joint production, and mutual filling of the secondary hydration products reduce the porosity of the hydration products, increasing the late stage strength.

The field test results show that the working faces of GER with fully mechanized gangue backfilling mining stabilize at the 28-day age after backfilling, and the working load of the RSB is fundamentally stable. Hence, in studying the influence rules of the age on the gangue concrete's strength in gangue backfilling GER engineering, it is necessary to determine whether the 28-day strength of the gangue concrete can meet the working load requirements in the stable period and to analyze whether the development degree of the gangue concrete's early stage strength can satisfy the mining-induced pressure requirements in the GER process. In this paper, the development degree of the gangue concrete's strength is expressed as a percentage of the early age strength of the samples to the strength at the 28-day age:

$$f_{7d} = \frac{s_{7d}}{s_{28d}} \times 100\%,$$

$$f_{14d} = \frac{s_{14d}}{s_{28d}} \times 100\%, \tag{2}$$

where f_{7d} and f_{14d} represent the development degrees of the gangue concrete's strength at the 7-day and 14-day ages, respectively. s_{7d}, s_{14d}, and s_{28d} depict the concrete's compressive strengths at the 7-day, 14-day, and 28-day ages, respectively.

Table 3 shows the strength development conditions of the concrete with different water-cement ratios and a gangue content of 50%. Within the range of 0.40–0.60, the average value of f_{14d}, 91.37%, for the concrete samples with a gangue content of 50% is notably higher than that of the pure cement samples which is 81.90%. Their average values of f_{7d} are

essentially the same, which are 67.56% and 68.89%, respectively. From Figures 4 and 5, it can be seen that the gangue concrete has basically reached its late stage strength at the 14-day age, after which the strength has a very small increase, indicating that with the addition of the gangue aggregate, the concrete's hydration and hardening processes are significantly accelerated.

In a general GER condition, because backfilling measures are not taken for the gob, the roof shearing needs to be done in the initial stage of the roadside support. The field test results show that the roof shearing resistance is commonly required at 3–6 MPa, so the early stage strength of the RSB material is particularly critical. During the roof shearing process, a low strength of the RSB material will lead to insufficient roadside support, excessive roof deformation, and in a serious scenario roof breaking along the solid coal side, resulting in an accident. In the condition of GER with fully mechanized gangue backfilling mining, due to the support of the backfilling area, the gob-side roof subsidence is smaller than in the general GER condition. Moreover, with continuous support, the roof will only bend and subside to a certain degree on the whole without obvious weighting. Therefore, the RSB does not serve as the main carrier of the roof load at the beginning of entry retaining, and so the early stage RSB strength required is smaller than that of GER with roof shearing. It can be seen from Figure 4 that the early average strength of the gangue concrete material with various mixing ratios reaches 3.1–8.1 MPa, which is greater than the roof shearing resistance. Therefore, the compressive strength of the material fundamentally meets the early stage vertical load bearing requirements.

Moreover, for solid backfilling GER, it is necessary to guarantee the structural stability of the RSB under the lateral pressure of the granular backfilling material in the gob. Therefore, to ensure that the RSB does not slip toward the inner side of the roadway and loose stability due to rotation, it is required to carry a design load within a reasonable range. The results of the existing practices show that the vertical stress is gradually increasing from the working face to the backfilling area, which essentially stabilizes 30 m–50 m back from the working face (Figure 9). The lateral pressure exerted by the granular material on the RSB is also fundamentally consistent with this trend, mainly because the

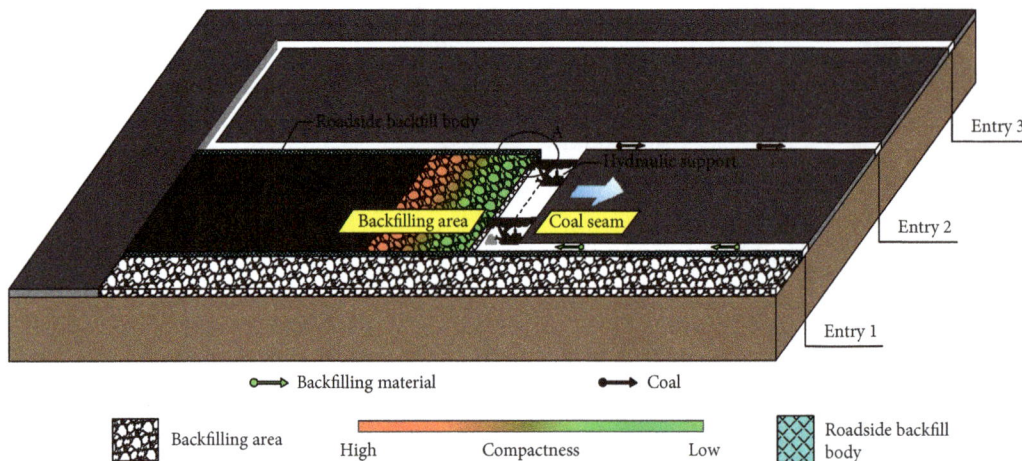

FIGURE 9: Schematic diagram of the vertical stress distribution in the backfilling area.

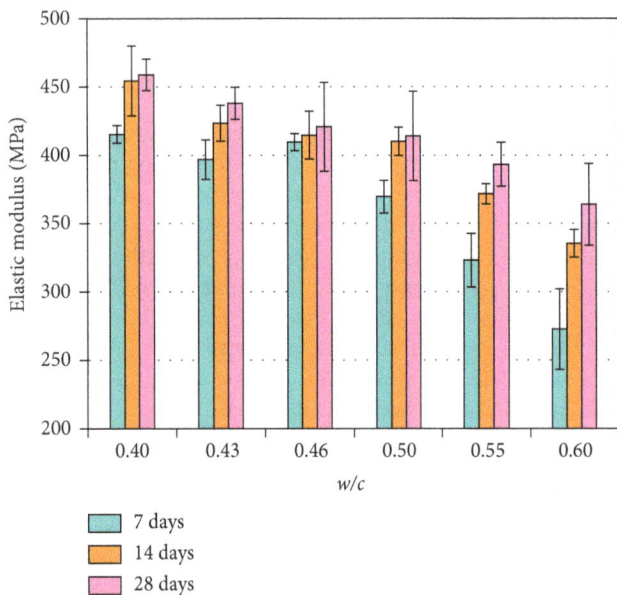

FIGURE 10: Average elastic modulus and error of the gangue concrete at various ages.

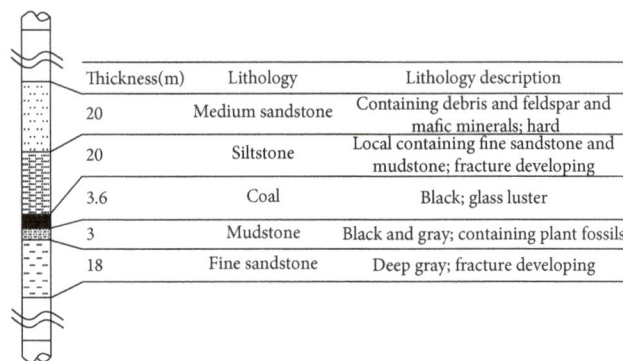

FIGURE 11: Strata geological log.

reason that the lateral pressure coefficient of the granular material with a certain property does not change with the vertical stress. Therefore, as a RSB material, gangue concrete is required to have a sufficient compressive strength to withstand a certain vertical load and prevent slippage and side turn under the lateral pressure of the granular backfilling material in the gob, which will affect the safe use of the retained entry. Alternatively, it is required to have a certain degree of contractibility to adapt to the "given deformation" of the roof in the A area when the vertical support is insufficient in the backfilling area in the early stages. In engineering practices, the strength proportion should be chosen in an economical and rational manner according to the above principles while considering parameters, such as the RSB width and aspect ratio and the surface slope angle on the gob side.

3.3.2. Effect of Age on the Contractibility of the Gangue Concrete. From the stiffness development with the gangue concrete age as shown in Figure 10, it can be seen that the stiffness of the 7-day gangue concrete is obviously lower than the stiffness of the 14-day gangue concrete. Since cement hydration is related to the age, the increasing trends of the elastic modulus with time are apparent before the 14-day age, and the elastic modulus stabilizes when the age is greater than 14 days. This property can adapt to the early stage deformation of the roof in GER with fully mechanized gangue backfilling mining. In a general GER condition, since the roadside support is a passive support and the roof will produce a certain amount of given deformation, the roof-floor displacement of the roadway accounts for 80% of the total surrounding rock deformation for the area 20 m in front of the working face and the area from the back of the backfilling roadside support to 80 m in back of the working face. In contrast, with the gob backfilling effect, the roof subsidence mainly concentrates 30 m in back of the working face. So, at age 0–14 days, a relatively high contractibility is required for the roadside support, while in the late stable stage and the second working, a relatively high strength is required. In the initial stage of entry retaining, the roadway support can be adaptable to the given deformation of the roof in

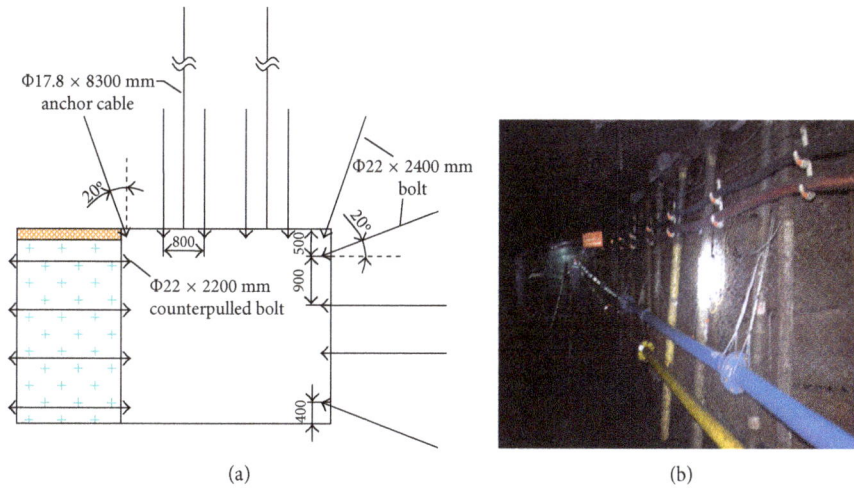

FIGURE 12: Roadway support design and effect. (a) Roadside support design scheme. (b) Roadside support effect.

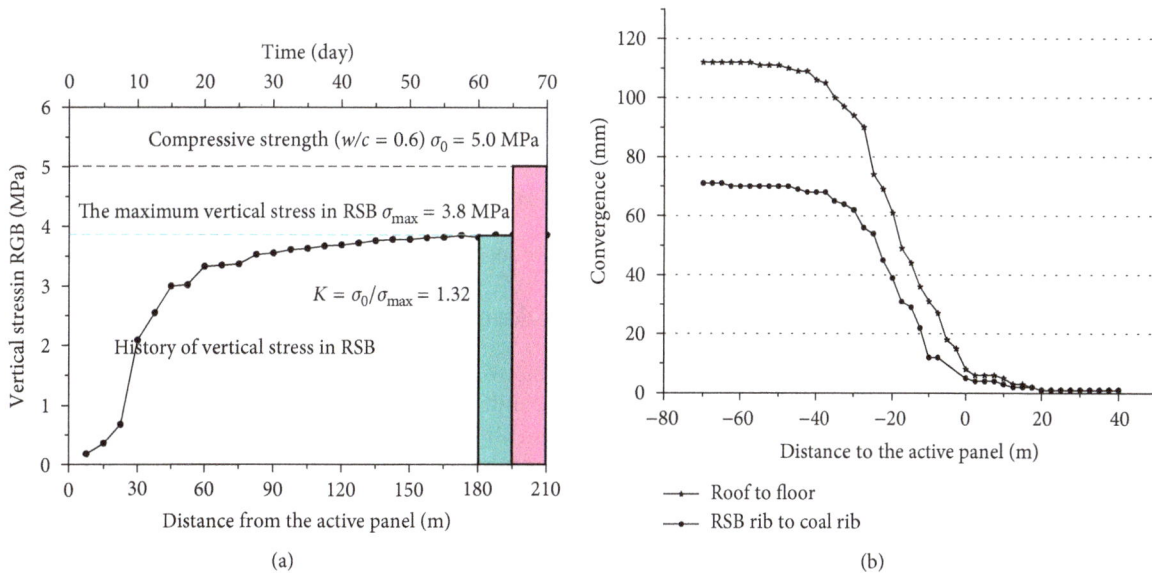

FIGURE 13: Monitoring result during advancement of the working face. (a) Vertical stress in RSB. (b) Roadway deformation curve.

GER with fully mechanized gangue backfilling mining through appropriate downward contracting and yielding, fully utilizing the carrying capacities of the backfilling gob and the surrounding rock. The characteristics of the high strength and large elastic modulus of the gangue concrete in the late stage can satisfy the roadside support strength required by the GER in the second mining process.

4. Field Test and Monitoring

4.1. Mining Geological Conditions. The test mine is in the Shandong Province of China. The ground elevation of the test zone is +32.93 to +33.08 m; the underground elevation is −642 to −636 m; the coal seam strike approximates East–West, trending South; and true dip is at 0°–3°. The main mining coal seam is seam 3_{lower}, coefficient of

hardness $f = 1–2$, and thickness is at 3.08–4.10 m. Seam 3_{lower} of the coal floor consists of mudstone and siltstone, and Seam 3_{lower} of the coal roof consists of mudstone and siltstone. The hardness coefficient of the mudstone and siltstone $f = 4–6$. The advancing speed of the working face was 3 m per day, and the gob backfilling ratio was designed as 100%. The strata geological log is as shown in Figure 11.

4.2. Roadside Support Design. The width of the RSB is 2 m, and the height is 3.6 m. Crushed continuously graded native gangues at $n = 0.6$ were used as aggregate to prepare concrete of 50% aggregate content. Cement grade is p.c32.5, and water-cement ratio is 0.60. Flexible cushion at height of 200 mm is arranged at the upper part of the gangue concrete for roof contact (Figure 12).

4.3. Field Monitoring. A safe RSB material should meet the following three requirements according to different mining geological conditions.

First of all, the strength of RSB should be greater than roof load at its location. Evaluation criterion of RSB stability is whether the RSB material breaks down in the process of loading and unloading caused by roof movement.

Secondly, shrinkage should be larger than the field-measured deformation of basic roof. Ability of RSB to adapt to the deformation of basic roof can make full use of the bearing capacity of surrounding rock, reducing the roof load on RSB. This is an effective way to ensure that its strength is greater than the roof load.

Thirdly, there is no mutation in the postpeak strength. This means that the material can ensure that the global impact damage will not occur to RSB even if there is local high stress at the gob side. The local slow destruction can be repaired by means of reinforcement, which ensures the safety of RSB.

To analyze the effect of the roadside support, monitoring stations are arranged in the roadway to record the vertical stress and converging deformation during advancement of the working face, as shown in Figure 13.

During backfill GER, the compressive strengths of the roadside support material of various ages satisfy the onsite construction requirement, the max vertical stress in RSB is 3.8 MPa, and the strength safety coefficient is

$$K = \frac{\sigma_0}{\sigma_{max}} = 1.32, \tag{3}$$

where K is the strength safety coefficient of the gangue concrete material, σ_0 is the average compressive strength of the concrete used in the construction, and σ_{max} is the max vertical stress in RSB, that is, the max workload.

During advancement of the working face, the roof-floor displacement of the roadway was larger than the two-side displacement. The maximum roof-floor displacement was 112 mm, and the maximum two-side displacement was 71 mm. The roadway using gangue cement as RSB material satisfies the design and application requirements.

5. Conclusions

Based on the current experimental research and field measurements, the following can be concluded:

(1) The most effective way to adjust the strength of gangue concrete is to change the water-cement ratio of the cementing material. For a fixed gangue content, with the raising of the water-cement ratio, the strength of gangue concrete demonstrates a non-linear decreasing trend. The water-cement ratio within the range of 0.46–0.60 has the most significant regulation effect on the strength of the gangue concrete.

(2) Mixing concrete with a certain amount of coal gangue as the aggregate has an important effect on the destruction mode and the postpeak carrying characteristics of the concrete. In the postpeak

stage, the stress demonstrates a decreasing trend with an increase of the strain, the process of the crack development is relatively slow, and the gangue concrete has a certain postpeak carrying capacity.

(3) Notwithstanding the water-cement ratio, the compressive strength of the gangue concrete demonstrates an increasing trend with age. With the addition of the gangue aggregate, the concrete's hydration and hardening processes are significantly accelerated. At the age of 14 days, the stiffness and strength of the gangue concrete essentially stabilize.

(4) Based on the laboratory test results and the law of underground pressure, industrial tests were carried out at the working face of gob-backfilled GER. The monitoring results demonstrate that with a rational proportion design, the use of gangue concrete as a RSB material can meet the design and application requirements of the GER. The feasibility and rationality of gangue concrete as a RSB material for the GER were proved. A RSB material featuring high safety, high waste utilization rate, fast construction speed, and low costs was provided, guaranteeing the RSB stability for the GER.

Conflicts of Interest

The authors declare that they have no conflicts of interest.

Acknowledgments

This work is supported by the National Natural Science Foundation of China (nos. 51323004, 51674250, 51574228, and 51074163), Major Program of National Natural Science Foundation of China (no. 50834005, 51734009), the Graduate Innovation Fund Project of Jiangsu Province (no. CXZZ13_0924), and Open Fund of State Key Laboratory for Geomechanics and Deep Underground Engineering (SKLGDUEK1409). The authors sincerely acknowledge the former researchers for their excellent works, which greatly assisted our academic study.

References

[1] P. Gong, Z. Ma, R. R. Zhang, X. Ni, F. Liu, and Z. Huang, "Surrounding rock deformation mechanism and control technology for gob-side entry retaining with fully mechanized gangue backfilling mining: a case study," *Shock and Vibration*, vol. 2017, Article ID 6085941, 15 pages, 2017.

[2] H. Yang, S. Cao, Y. Li, C. Sun, and P. Guo, "Soft roof failure mechanism and supporting method for gob-side entry retaining," *Minerals*, vol. 5, no. 4, pp. 707–722, 2015.

[3] H. Li, D. Jiang, and D. Li, "Analysis of ground pressure and roof movement in fully-mechanized top coal caving with large mining height in ultra-thick seam," *Journal of China Coal Society*, vol. 39, no. 10, pp. 1956–1960, 2014.

[4] H. Yang, S. Cao, S. Wang, Y. Fan, S. Wang, and X. Chen, "Adaptation assessment of gob-side entry retaining based on geological factors," *Engineering Geology*, vol. 209, pp. 143–151, 2016.

[5] S. Xie, G. Zhang, S. He et al., "Surrounding rock control mechanism and its application of gob-side retaining entry in deep backfilling with large mining height," *Journal of China Coal Society*, vol. 39, no. 12, pp. 2362–2368, 2014.

[6] D. Zhang, X. Miao, and X. Mao, "Simulation on roof activities of gob-side entry retaining in fully-mechanized top-coal caving faces," *Journal of China University of Mining & Technology*, vol. 30, no. 3, pp. 47–50, 2001.

[7] C. Han, N. Zhang, G. Li, B. Li, and H. Wu, "Stability analysis of compound bearing structure of gob-side entry retaining with large mining height," *Chinese Journal of Geotechnical Engineering*, vol. 36, no. 5, pp. 969–976, 2014.

[8] Y. Li and X. Hua, "Mechanical analysis of stability of key blocks of overlying strata for gob-side entry retaining and calculating width of roadside backfill," *Rock and Soil Mechanics*, vol. 33, no. 4, pp. 1134–1140, 2012.

[9] Z. Ma, R. Gu, Z. Huang, G. Peng, L. Zhang, and D. Ma, "Experimental study on creep behavior of saturated disaggregated sandstone," *International Journal of Rock Mechanics & Mining Sciences*, vol. 66, no. 1, pp. 76–83, 2014.

[10] X. Zhu, G. Guo, and Q. Fang, "Coupled discrete element-finite difference method for analyzing subsidence control in fully mechanized solid backfilling mining," *Environmental Earth Sciences*, vol. 75, no. 8, pp. 1–12, 2016.

[11] G. Guo, W. Feng, J. Zha, Y. Liu, and Q. Wang, "Subsidence control and farmland conservation by solid backfilling mining technology," *Transactions of Nonferrous Metals Society of China*, vol. 21, no. S3, pp. 665–669, 2011.

[12] G. Li, S. Cao, Y. Li, and Z. Zhang, "Load bearing and deformation characteristics of granular spoils under unconfined compressive loading for coal mine backfill," *Advances in Materials Science & Engineering*, vol. 2016, Article ID 8530574, 11 pages, 2016.

[13] F. W. Solesbury, "Coal waste in civil engineering works: 2 case histories from South Africa," in *Advances in Mining Science & Technology*, Elsevier, Vol. 1987pp. 207–218, Elsevier, Amsterdam, Netherlands, 1987.

[14] W. Yin, Z. Chen, K. Quan, and X. Mei, "Strata behavior at fully-mechanized coal mining and solid backfilling face," *SpringerPlus*, vol. 5, no. 1, pp. 1–12, 2016.

[15] X. Zhu, G. Guo, J. Zha, T. Chen, Q. Fang, and X. Yang, "Surface dynamic subsidence prediction model of solid backfill mining," *Environmental Earth Sciences*, vol. 75, no. 12, pp. 1–9, 2016.

[16] D. Wu, B. Yang, and Y. Liu, "Transportability and pressure drop of fresh cemented coal gangue-fly ash backfill slurry in pipe loop," *Powder Technology*, vol. 284, pp. 218–224, 2015.

[17] M. Rezaei, M. F. Hossaini, and A. Majdi, "Determination of longwall mining-induced stress using the strain energy method," *Rock Mechanics and Rock Engineering*, vol. 48, no. 6, pp. 2421–2433, 2015.

[18] J. Bai, H. Zhou, C. Hou, X. Tu, and D. Yue, "Development of support technology beside roadway in goal-side entry retaining for next sublevel," *Journal of China University of Mining & Technology*, vol. 33, no. 2, pp. 59–62, 2004.

[19] X. He and L. Song, "Status and future tasks of coal mining safety in China," *Safety Science*, vol. 50, no. 4, pp. 894–898, 2012.

[20] J. Ning, J. Wang, T. Bu, S. Hu, and X. Liu, "An innovative support structure for gob-side entry retention in steep coal seam mining," *Minerals*, vol. 7, no. 5, p. 75, 2017.

[21] H. Wang, D. Zhang, and L. Liu et al., Stabilization of gob-side entry with an artificial side for sustaining mining work," *Sustainability*, vol. 8, no. 7, p. 627, 2016.

[22] X. Querol, M. Izquierdo, E. Monfort et al., "Environmental characterization of burnt coal gangue banks at Yangquan, Shanxi Province, China," *International Journal of Coal Geology*, vol. 75, no. 2, pp. 93–104, 2008.

[23] C. Zhou, G. Liu, Z. Yan, T. Fang, and R. Wang, "Transformation behavior of mineral composition and trace elements during coal gangue combustion," *Fuel*, vol. 97, no. 2, pp. 644–650, 2012.

[24] C. Zhou, G. Liu, S. Wu, and P. K. Lam, "The environmental characteristics of usage of coal gangue in bricking-making: a case study at Huainan, China," *Chemosphere*, vol. 95, no. 1, pp. 274–280, 2014.

[25] J. Wang, Q. Qin, S. Hu, and K. Wu, "A concrete material with waste coal gangue and fly ash used for farmland drainage in high groundwater level areas," *Journal of Cleaner Production*, vol. 112, pp. 631–638, 2015.

[26] G. Li, C. Fang, X. Zhao, Y. An, and Y. Liu, "Cyclic behavior of rebar-penetrated connection between gangue concrete filled steel tubular column and reinforced gangue concrete beam," *Advanced Steel Construction*, vol. 11, no. 1, pp. 54–72, 2015.

[27] D. Wu, S. Cai, and Y. Liu, "Effects of binder on suction in cemented gangue backfill," *Magazine of Concrete Research*, vol. 68, no. 12, pp. 1–11, 2015.

[28] S. Walker, D. L. Bloem, R. D. Gaynor, and J. E. Gray, "Relationships of concrete strength to maximum size aggregate," *Highway Research Board, Proceedings of the Annual Meeting*, vol. 38, pp. 367–385, 1959.

Mechanical Characterization of Lightweight Foamed Concrete

Marcin Kozłowski (ID)[1] **and Marta Kadela** (ID)[2]

[1]*Department of Structural Engineering, Faculty of Civil Engineering, Silesian University of Technology, 5 Akademicka St., 44-100 Gliwice, Poland*
[2]*Building Research Institute, 1 Filtrowa St., 00-611 Warszawa, Poland*

Correspondence should be addressed to Marta Kadela; m.kadela@itb.pl

Academic Editor: Kai Wei

Foamed concrete shows excellent physical characteristics such as low self weight, relatively high strength and superb thermal and acoustic insulation properties. It allows for minimal consumption of aggregate, and by replacement of a part of cement by fly ash, it contributes to the waste utilization principles. For many years, the application of foamed concrete has been limited to backfill of retaining walls, insulation of foundations and roof tiles sound insulation. However, during the last few years, foamed concrete has become a promising material for structural purposes. A series of tests was carried out to examine mechanical properties of foamed concrete mixes without fly ash and with fly ash content. In addition, the influence of 25 cycles of freezing and thawing on the compressive strength was investigated. The apparent density of hardened foamed concrete is strongly correlated with the foam content in the mix. An increase of the density of foamed concrete results in a decrease of flexural strength. For the same densities, the compressive strength obtained for mixes containing fly ash is approximately 20% lower in comparison to the specimens without fly ash. Specimens subjected to 25 freeze-thaw cycles show approximately 15% lower compressive strengths compared to the untreated specimens.

1. Introduction

Foamed concrete is known as light-weight or cellular concrete. It is commonly defined as a cementitious material with a minimum of 20% (by volume) mechanically entrained foam in the mortar mix where air-pores are entrapped in the matrix by means of a suitable foaming agent [1]. It shows excellent physical characteristics such as low self weight, relatively high strength, and superb thermal and acoustic insulation properties. It allows for minimal consumption of aggregate, and by replacement of a part of cement by fly ash, it contributes to the waste utilization principles [2]. By a proper selection and dosage of components and the foaming agent, a wide range of densities (300–1600 kg/m^3) can be achieved for various structural purposes, insulation, or filling applications [2].

Foamed concrete has been known for almost a century and was patented in 1923 [3]. The first comprehensive study of foamed concrete was carried out in the 1950s and 1960s by Valore [3, 4]. Following this research, more detailed evaluation regarding the composition, properties, and applications of cellular concrete was reported by Rudnai [5], as well as by Short and Kinniburgh [6] in 1963. New mixtures were developed in the late 1970s and early 1980s, which led to the increased commercial use of foamed concrete in building constructions [7, 8].

For many years, the application of foamed concrete has been limited to backfill of retaining walls, insulation of foundations, and sound insulation [8]. However, in the last few years, foamed concrete has become a promising material also for structural purposes [7, 9], for example, stabilization of weak soils [10, 11], a base layer of sandwich solutions for foundation slabs [12], industrial floors [13], and highway as well as subway engineering applications [14, 15].

With the increasing environmental challenges, it is paramount that sustainable materials are researched for a wider range of applications to offer feasible alternatives alongside conventional materials.

TABLE 1: Mix proportions.

Mix symbol	Foaming agent content (l/100 kg C)	Cement (kg)	Fly ash (kg)	Water (kg)	Foaming agent (kg)	w_{eff}/c (–)
FC1	2.00	25.00	0.00	10.50	0.50	0.44
FC2	4.00	25.00	0.00	10.00	1.00	0.44
FC3	6.00	25.00	0.00	9.50	1.50	0.44
FC4	8.00	25.00	0.00	9.00	2.00	0.44
FC5	10.00	25.00	0.00	8.50	2.50	0.44
FCA1	2.00	25.00	1.25	10.50	0.50	0.44
FCA2	4.00	25.00	1.25	10.00	1.00	0.44
FCA3	6.00	25.00	1.25	9.50	1.50	0.44
FCA4	8.00	25.00	1.25	9.00	2.00	0.44
FCA5	10.00	25.00	1.25	8.50	2.50	0.44

TABLE 2: Cement chemical composition (%).

SiO_2	Al_2O_3	Fe_2O_3	CaO	MgO	SO_3	Na_2O	K_2O	Cl
19.5	4.9	2.9	63.3	1.3	2.8	0.1	0.9	0.05

TABLE 4: Fly ash chemical composition (%).

SiO_2	Al_2O_3	Fe_2O_3	CaO	MgO	SO_3	Na_2O	K_2O
76.5	1.42	5.80	3.61	1.63	0.263	0.038	0.096

TABLE 3: Physical properties of cement.

Specific surface area (m²/kg)	Specific gravity (g/cm³)	Compressive strength (MPa) After days	
		2	28
3840	3.06	28.0	58.0

Foamed concrete, being an alternative to ordinary concrete, fulfills the criteria of the principles of sustainability in building constructions [16–18]. The general principles, based on the concept of sustainable development as it applies to the life cycle of buildings and other construction works, are identified in ISO 15392:2008. First, foamed concrete consumes relatively low amount of raw material in relation to the amount of hardened state. Second, during its production, recycled materials such as fly ash can be used. In this way, foamed concrete contributes to the disposal of waste products of thermal power plants. Third, foamed concrete can be recycled and used as replacement of sand in insulation materials. Moreover, the manufacturing of foamed concrete is nontoxic, and the product does not emit toxic gases when it is exposed to fire. At last, it is cost-effective not only during the construction stage but also throughout lifetime operation and maintenance of the structure.

Besides contribution to the disposal of the waste products of thermal power plants, the addition of fly ash improves the workability of the fresh foamed concrete mix and has positive effect on drying shrinkage [2, 19]. On one hand, the only drawback of this mineral additive is lower early strength of mortar in comparison to the mix without fly ash [20]. On the other, it has been proven that the long-term strength is improved [19, 21].

Despite its favourable and promising strength and physical properties, foamed concrete is still utilized in limited scale, particularly for structural applications. This is mainly due to the insufficient knowledge regarding its mechanical properties and small number of research on its fracture behaviour [22–28].

The main objective of this work is to investigate the mechanical characteristics of foamed concrete with varying density (400–1400 kg/m³). A series of tests was performed to examine compressive strength, elastic modulus, flexural strength, and material degradation characteristics after freeze-thaw cycles.

2. Experimental Program

2.1. Specimens Preparation and Concrete Mix Composition. The materials used in this study were Portland cement, fly ash, water, and foaming agent. The compositions of the mix are presented in Table 1. The industrial Portland cement was CEM I 42.5 R [29], according to PN-EN 197-1: 2011. Its chemical composition and physical properties, measured as per PN-EN 196-6:2011 and PN-EN 196-6:2011-4, are given in Tables 2 and 3. Tap water was used in all experiments. Compressive strength of cement was determined according to PN-EN 196-1:2016-07 (Table 3).

To improve the workability and reduce shrinkage, fly ash was used in some mixes. The ash used met the requirements of PN-EN 450-1:2012. Its chemical composition is given in Table 4.

A commercial foaming agent was used to produce foam. The liquid agent was pressurized with air at approximately 5 bars in order to make the stable foam with a density of approximately 50 kg/m³. Cement pastes with 2 ÷ 10 litres of liquid foaming agent for 100 kg of cement were prepared.

Two different types of concrete mixes (one without fly ash and the other with fly ash) were used. In total, 10 mixes were produced, five specimens for one concrete mix (Table 1). A constant $w_{eff}/c = 0.44$ ratio was used for all mixes (w_{eff} includes water and liquid foaming agent; c is the cement content). It was based on the results of Jones and McCarthy [7] and Xianjun et al. [30]. The target densities of hardened foamed concrete to be produced in this study were from 400 to 1400 kg/m³.

The entire manufacturing process of foamed concrete must carefully consider the densities of the mix, the foaming production rate, and other factors in order to prepare high-

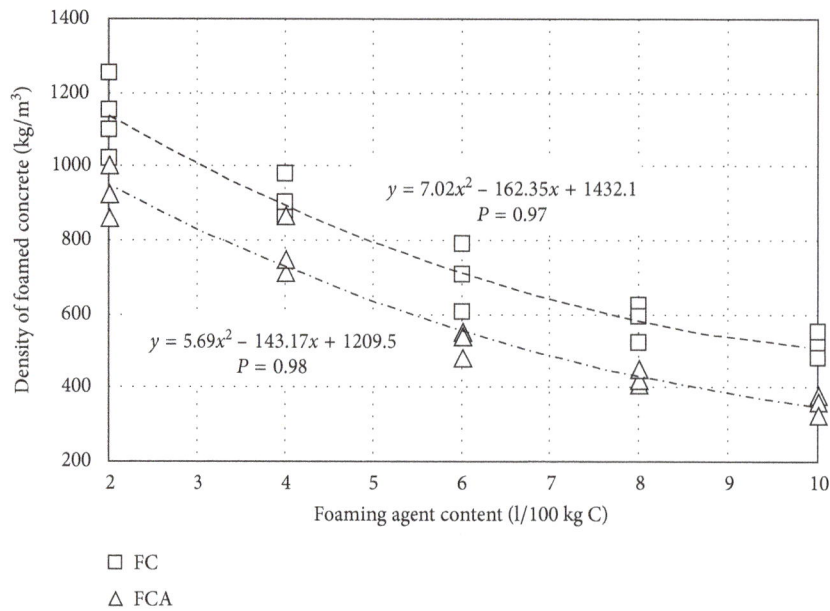

FIGURE 1: Apparent density of foamed concrete specimens FC and FCA as a function of foaming agent content.

quality foamed concrete. The key factors to produce stable foamed concrete were pressurizing of foaming agent at stable pressure and constant rotational speed of mixing the components.

All specimens, after casting in steel moulds, were covered and stored in a curing room at $20 \pm 1°C$ and 95% humidity for 24 hours. Subsequently, the samples were removed from the moulds and stored in ambient conditions (at $20 \pm 1°C$ and $60 \pm 10\%$ humidity) for 28 or 42 days before testing.

2.2. Tests. Foamed concrete is a relatively new material, and currently there are no standardized test methods to measure its physical and mechanical properties. Therefore, procedures for preparation of specimens and testing methods, usually used for ordinary concrete, were adapted in this research. The compressive strength, modulus of elasticity, and flexural strength were determined according to the recommendations: PN-EN 12390-3:2011 + AC:2012, Instruction of Research Building Institute No. 194/98, PN-EN 12390-13:2014, and PN-EN 12390-5:2011, respectively. The density was measured as per PN-EN 12390-7:2011.

Compressive strength was measured with $150 \times 150 \times 150$ mm standard cubes as stated in PN-EN 12390-3:2011 + AC:2012. The loading rate was assumed according to PN-EN 772-1: 2015 + A1:2015 as for cellular concrete masonry units.

Elasticity modulus was determined according to the Instruction of Research Building Institute No. 194/98 and PN-EN 12390-13:2014-02 with cylindrical specimens with the dimensions of 150×300 mm. The loading rate was 0.1 ± 0.05 MPa/s, according to PN-EN 679:2008 as for cellular concrete masonry units. Two electrical resistance strain gauges with 100 mm measurement length were bonded on two opposite sides of the specimens at mid-height. The stress-strain characteristic was recorded for the evaluation of modulus of elasticity.

FIGURE 2: Typical failure pattern observed during compression tests with cube specimens.

Flexural strength was tested in three-point bending setup with beams $100 \times 100 \times 500$ mm, according to PN-EN 12390-5:2011. The nominal distance between the supports was 300 mm. The rollers allowed for free horizontal movement. The specimens were loaded at constant displacement rate of 0.1 mm/min as an optimum value determined experimentally.

Degradation characteristics under freeze-thaw cycles were evaluated with $150 \times 150 \times 150$ mm standard cubes. The compressive strength was determined with the procedure as described before. The test campaign consisted of 25 cycles of freezing and thawing. Each cycle included cooling of the specimens to the temperature of $-18°C$ within 2 h. The samples were then kept frozen for 8 h at $-18 \pm 2°C$ and thawed in water at the temperature of $+19°C \pm 1°C$ for 4 h. Reference specimens were kept immerse in water as references.

3. Results and Discussion

3.1. Apparent Density. The dosage of foaming agent highly influences the density of mix and hardened foamed concrete.

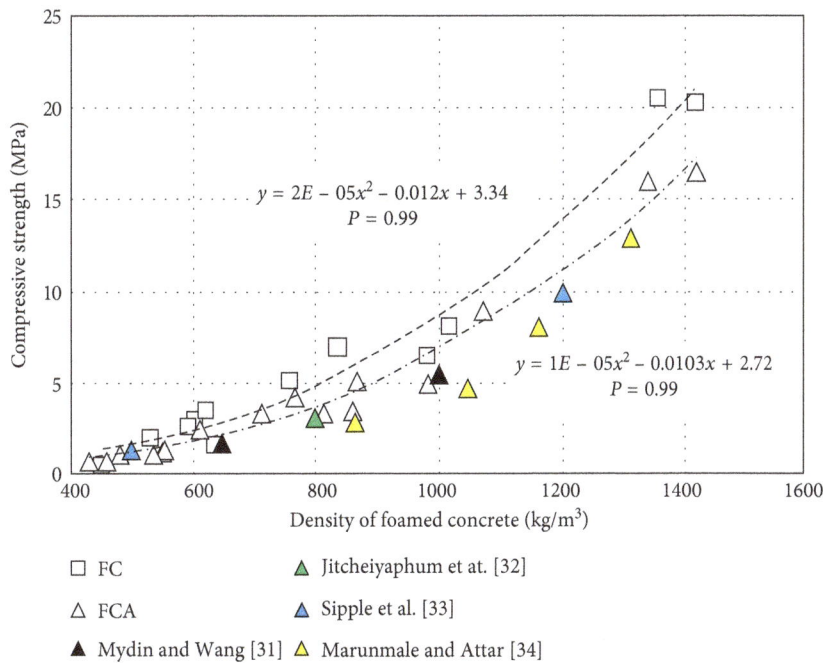

FIGURE 3: Compressive strengths of foamed concrete FC and FCA as a function of density of foamed concrete.

Figure 1 shows the relationship between the dosage of foaming agent and the apparent density of hardened foamed concrete for the specimens without fly ash (FC) and the other with fly ash (FCA). The apparent density of hardened foamed concrete is strongly correlated with the foam content and the composition of cement paste and air voids in fresh mix. The increase of foam content is accompanied by the increase of volume of fresh concrete, which results in a decrease of density of hardened foamed concrete. It can be observed that there are exponential relationships for FC and FCA specimens. Moreover, results obtained in FC show density level of approximately 20% higher than FCA. This can be explained by the fact that the process of hardening is slowed down in the specimens containing fly ash. The physical reaction between fly ash and air-pores results in larger number of air-pores entrapped in the mix. It was also found that the mixes with the foaming agent content above 10 litres per 100 kg of cement resulted in unstable mix. The results were approximated with polynomial functions as shown in Figure 1.

3.2. Compressive Strength.
Cube foamed concrete specimens tested in compression present the mechanism of failure similar to ordinary concrete. A typical conical postbreakage failure pattern was observed for all specimens (Figure 2).

The compressive strengths of foamed concrete without ash (FC) and foamed concrete with addition of fly ash (FCA) as a function of apparent density are presented in Figure 3. It can be noticed that there are exponential relationships for both FC and FCA; however, there seems to be a difference between the strengths obtained from FC and FCA samples. The specimens without ash seem to show higher strengths than the mixtures containing ash. This is due to the fact that the process of hardening is slowed down due to the presence

FIGURE 4: Typical failure pattern observed during compression tests with cylindrical specimens.

of fly ash [20]. In addition, this difference increases along with the density. The values of compressive strengths obtained correspond to the results of the works of others [31–34]. The results were approximated with polynomial functions as shown in Figure 3.

3.3. Modulus of Elasticity.
Cylindrical foamed concrete specimens tested in compression present the mechanism of failure similar to ordinary concrete. A typical conical postbreakage failure pattern was observed for all specimens

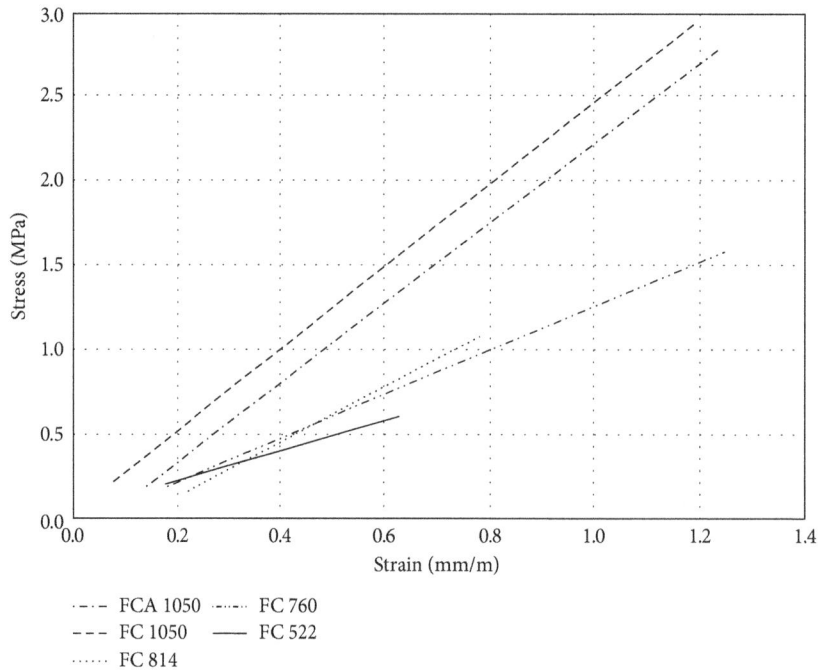

FIGURE 5: Stress-strain relationships of cylindrical specimens FC and FCA.

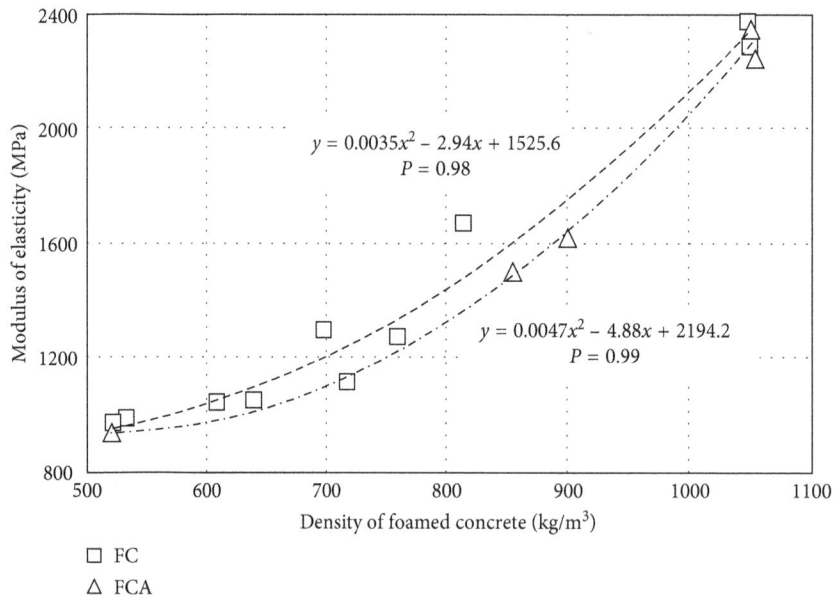

FIGURE 6: Modulus of elasticity of foamed concrete FC and FCA as a function of density of foamed concrete.

(Figure 4). Stress-strain relationships of cylindrical specimens are presented in Figure 5. The plots show the relations in the range of 0.2 MPa until failure, according to PN-EN 12390-13: 2014-02.

Figure 6 shows the relationships between the modulus of elasticity of foamed concrete and its density. It can be observed that there are exponential relationships for FC and FCA. The specimens without fly ash seem to have higher modulus of elasticity than the mixtures containing fly ash

[35]. The values of modulus of elasticity obtained correspond to the results of the works of Aldridge [8].

3.4. Flexural Strength. Figure 7 presents the relationship between the density of foamed concrete and the flexural strength. The tests were carried out on specimens without fly ash. Figure 7 includes also the results of experiments carried out by authors and reported in [23–28]. The decrease of

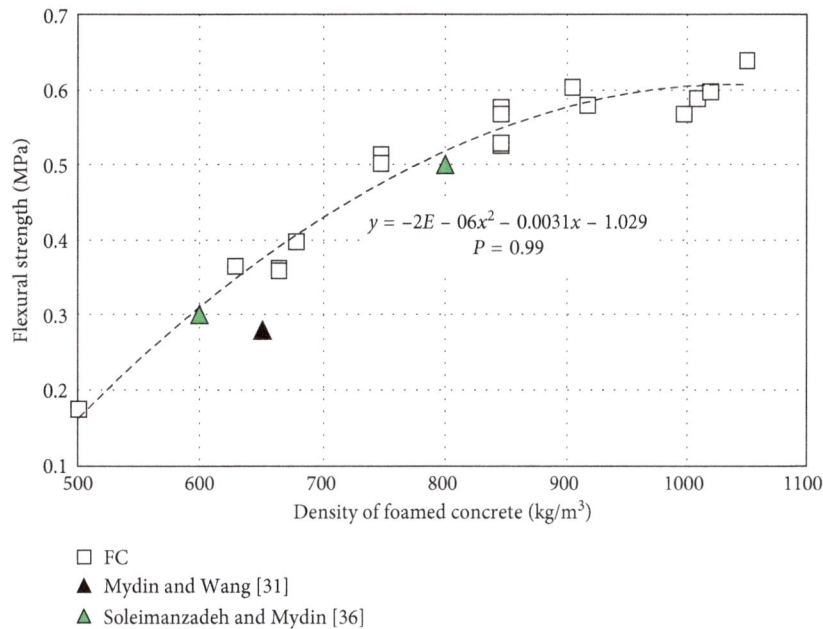

FIGURE 7: Flexural strength as a function of density of foamed concrete.

$$y = -2E - 06x^2 - 0.0031x - 1.029$$
$$P = 0.99$$

□ FC
▲ Mydin and Wang [31]
▲ Soleimanzadeh and Mydin [36]

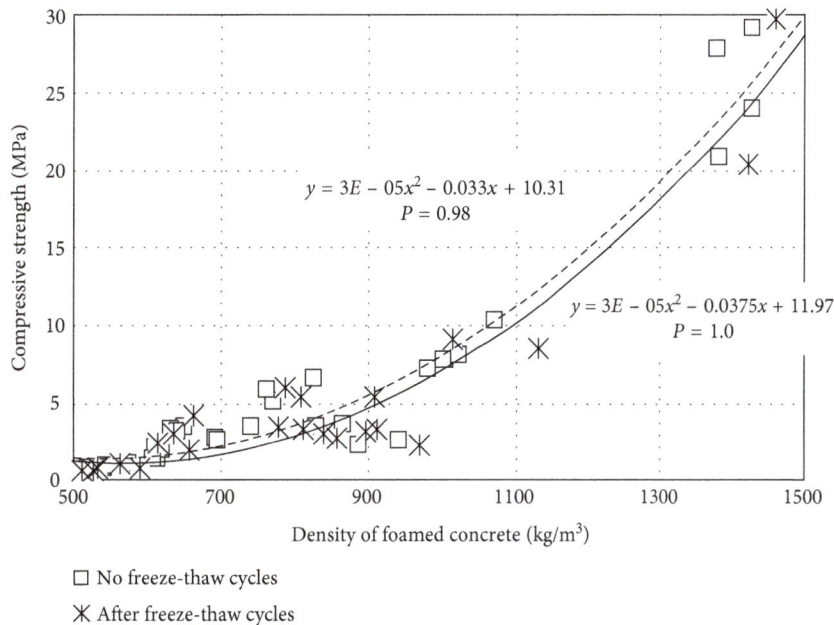

$$y = 3E - 05x^2 - 0.033x + 10.31$$
$$P = 0.98$$

$$y = 3E - 05x^2 - 0.0375x + 11.97$$
$$P = 1.0$$

□ No freeze-thaw cycles
✕ After freeze-thaw cycles

FIGURE 8: Compressive strength of foamed concrete after 25 freeze-thaw cycles as a function of density.

flexural tensile strength with the decrease of the density of the foamed concrete can be noted. The values of flexural strengths correspond to the results of works of Mydin and Wang [31] and Soleimanzadeh and Mydin [36].

3.5. Degradation Characteristics under Freeze-Thaw Cycles. Figure 8 shows the results of compressive strength of foamed concrete after 25 freeze-thaw cycles as a function of density. As a reference, results from untreated samples are shown in Figure 8. The freeze-thaw treatment of the specimens has only minor influence on the compressive strength of foamed concrete. The strengths obtained for the specimens subjected to freeze-thaw cycles showed approximately 15% lower values. The results were approximated with polynomial functions as shown in Figure 8.

4. Conclusions

Foamed concrete can achieve much lower densities (400 to 1400 kg/m³) in comparison to conventional concrete. A series of tests was carried out to examine the mechanical parameters of foamed concrete: compressive strength, flexural strength, and modulus of elasticity. Furthermore, the influence of 25 cycles of freezing and thawing on the compressive strength was examined.

The main conclusions that can be drawn from this study are the following:

(i) The dosage of foaming agent influences the density of mix and hardened foamed concrete. The density of foamed concrete is strongly correlated with the foam content in the mix.

(ii) The compressive strength, modulus of elasticity, and flexural strength decreased with the decrease of the density of the foamed concrete; the polynomial functions were suggested to describe these relationships.

(iii) The compressive strength and modulus of elasticity of foamed concrete were slightly decreased by the addition of 5% of fly ash.

(iv) The compressive strength of foamed concrete subjected to freeze-thaw tests shows the values only approximately 15% lower comparing to untreated specimens.

Conflicts of Interest

The authors declare that they have no conflicts of interest.

Acknowledgments

This work was supported by the ongoing research project "Stabilization of weak soil by application of layer of foamed concrete used in contact with subsoil" (LIDER/022/537/L-4/NCBR/2013) financed by the National Centre for Research and Development within the LIDER Programme. The authors gratefully acknowledge the skills and commitment of laboratory technician Alfred Kukiełka, without whom the present study could not have been successfully completed.

References

[1] S. Van Deijk, *Foamed Concrete: In A Dutch View*, British Cement Association, Blackwater, UK, 1992.

[2] R. Ramamurthy, E. K. Kunhanandan Nambiar, and G. Indu Siva Ranjani, "A classification of studies on properties of foam concrete," *Cement and Concrete Composites*, vol. 31, no. 6, pp. 388–396, 2009.

[3] R. C. Valore, "Cellular concrete part 1 composition and methods of production," *ACI Journal Proceedings*, vol. 50, no. 5, pp. 773–796, 1954.

[4] R. C. Valore, "Cellular concrete part 2 physical properties," *ACI Journal Proceedings*, vol. 50, no. 6, pp. 817–836, 1954.

[5] G. Rudnai, *Lightweight Concretes*, Akademikiado, Budapest, Hungary, 1963.

[6] A. Short and W. Kinniburgh, *Lightweight Concrete*, Asia Publishing House, Delhi, India, 1963.

[7] M. R. Jones and A. McCarthy, "Preliminary views on the potential of foamed concrete as a structural material," *Magazine of Concrete Research*, vol. 57, no. 1, pp. 21–31, 2005.

[8] D. Aldridge, "Introduction to foamed concrete: what, why, how?," in *Use of Foamed Concrete in Construction: Proceedings of the International Conference, Dundee, Scotland, UK*, K. Ravindra, D. Moray, and M. Aikaterini, Eds., vol. 5, pp. 1–14, July 2005.

[9] R. K. Dhir, M. D. Newlands, and A. McCarthy, *Use of Foamed Concrete in Construction*, Thomas Telford, London, UK, 2005.

[10] M. Drusa, L. Fedorowicz, M. Kadela, and W. Scherfel, "Application of geotechnical models in the description of composite foamed concrete used in contact layer with the subsoil," in *Proceedings of the 10th Slovak Geotechnical Conference on Geotechnical Problems of Engineering Constructions*, Bratislava, Slovakia, May 2011.

[11] L. Fedorowicz, M. Kadela, and Ł. Bednarski, "Modeling of the foamed concrete behavior for the layered structures cooperating with subsoil," in *Technical Notes of Katowice School of Technology*, vol. 6, pp. 73–81, Katowice School of Technology, Katowice, Poland, 2014.

[12] J. Hulimka, A. Knoppik-Wróbel, R. Krzywoń, and R. Rudišin, "Possibilities of the structural use of foamed concrete on the example of slab foundation," in *Proceedings of the 9th Central European Congress on Concrete Engineering*, pp. 67–74, Wroclaw, Poland, June 2013.

[13] M. Kadela and M. Kozłowski, "Foamed concrete layer as substructure of industrial concrete floor," *Procedia Engineering*, vol. 161, pp. 468–476, 2016.

[14] M. R. Jones and A. McCarthy, "Behaviour and assessment of foamed concrete for construction applications," in *Use of Foamed Concrete in Construction: Proceedings of the International Conference, Dundee, Scotland, UK*, K. Ravindra, D. Moray, and M. Aikaterini, Eds., vol. 5, pp. 61–88, July 2005.

[15] W. Tian, L. Li, X. Zhao, M. Zhou, and N. Wamg, "Application of foamed concrete in road engineering," in *Proceedings of International Conference on Transportation Engineering, ASCE*, pp. 2114–2120, Chengdu, China, July 2009.

[16] K. K. B. Siram and K. Arjun Raj, "Concrete + Green = Foam Concrete," *International Journal of Civil Engineering and Technology*, vol. 4, pp. 179–184, 2013.

[17] A. S. Moon and V. Varghese, "Sustainable construction with foam concrete as a green building material," *International Journal of Modern Trends in Engineering and Research*, vol. 2, pp. 13–16, 2014.

[18] A. S. Moon, V. Varghese, and S. S. Waghmare, "Foam concrete as a green building material," *International Journal of Research in Engineering and Technology*, vol. 2, pp. 25–32, 2015.

[19] P. Chindaprasirt, S. Homwuttiwong, and V. Sirivivatnanon, "Influence of fly ash fineness on strength, drying shrinkage and sulfate resistance of blended cement mortar," *Cement and Concrete Research*, vol. 34, no. 7, pp. 1087–1092, 2004.

[20] P. Chindaprasirt and S. Rukzon, "Strength, porosity and corrosion resistance of ternary blend Portland cement, rice husk ash and fly ash mortar," *Construction and Building Materials*, vol. 22, no. 8, pp. 1601–1606, 2008.

[21] E. P. Kearsley and P. J. Wainwright, "The effect of high fly ash content on the compressive strength of foamed concrete," *Cement and Concrete Research*, vol. 31, no. 1, pp. 106–112, 2001.

[22] N. A. Rahman, Z. M. Jaini, and N. N. Zahir, "Fracture energy of foamed concrete by means of the three-point bending tests on notched beam specimens," *Journal of Engineering and Applied Sciences*, vol. 10, pp. 6562–6570, 2015.

[23] M. Kozłowski, M. Kadela, and A. Kukiełka, "Fracture energy of foamed concrete based on three-point bending test on notched beams in Proceedings of the 7th Scientific-Technical Conference on Material Problems in Civil Engineering MATBUD'2015," *Procedia Engineering*, vol. 108, pp. 349–354, 2015.

[24] M. Kozłowski, M. Kadela, and M. Gwóźdź-Lasoń, "Numerical fracture analysis of foamed concrete beam using XFEM method," *Applied Mechanics and Materials*, vol. 837, pp. 183–186, 2016.

[25] M. Kadela, A. Cińcio, and M. Kozłowski, "Degradation analysis of notched foam concrete beam," *Applied Mechanics and Materials*, vol. 797, pp. 96–100, 2016.

[26] A. Cińcio, M. Kozłowski, M. Kadela, and D. Dudek, "Numerical degradation analysis of foamed concrete beam," in *Proceedings of the 13th International Conference on New Trends in Statics and Dynamics of Buildings, Slovak university of Technology*, Bratislava, Slovakia, October 2015.

[27] M. Kozłowski, M. Kadela, and M. Gwóźdź-Lasoń, "XFEM fracture analysis of notched foamed concrete beams," in *Proceedings of the 13th International Conference on New Trends in Statics and Dynamics of Buildings, Slovak university of Technology*, Bratislava, Slovakia, October 2015.

[28] M. Kozłowski and M. Kadela, "Experimental and numerical investigation of fracture behavior of foamed concrete based on three-point bending test of beams with initial notch," in *Proceedings of the International Conference on Mechanical, Civil and Material Engineering*, Barcelona, Spain, August 2015.

[29] Technical Data Sheet CEM I 42.5 R, http://www.gorazdze.pl.

[30] T. Xianjun, C. Weizhong, H. Yingge, and W. Xu, "Experimental Study of Ultralight (<300 kg/m3) Foamed Concrete," *Advances in Materials Science and Engineering*, vol. 2014, Article ID 514759, 7 pages, 2014.

[31] M. A. O. Mydin and Y. C. Wang, "Mechanical properties of foamed concrete exposed to high temperatures," *Construction and Building Materials*, vol. 26, no. 1, pp. 638–654, 2012.

[32] K. Jitchaiyaphum, T. Sinsiri, and P. Chindaprasirt, "Cellular lightweight concrete containing pozzolan materials," *Procedia Engineering*, vol. 14, pp. 1157–1164, 2011.

[33] M. A. Sipple, "High strength self-compacting foam concrete. initial thesis report," ACME, UNSW@ADFA, https://www.researchgate.net/publication/265483433_Structural_Strength_Self-Compacting_Foam_Concrete, 2009.

[34] A. K. Marunmale and A. C. Attar, "Designing, developing and testing of cellular lightweight concrete brick (CLC) wall built in rat-trap bond," *Current Trends in Technology and Sciences*, vol. 3, pp. 331–336, 2014.

[35] M. Kadela and A. Kukiełka, "Influence of foaming agent content in fresh concrete on elasticity modulus of hard foam concrete," in *Brittle Matrix Composites 11-Proceedings of the 11th International Symposium on Brittle Matrix Composites BMC 2015, Institute of Fundamental Technological Research PAS*, pp. 489–496, Warszawa, Poland, September 2015, ISBN: 978-838968796-8.

[36] S. Soleimanzadeh and M. A. O. Mydin, "Influence of high temperatures on flexural strength of foamed concrete containing fly ash and polypropylene fiber," *International Journal of Engineering*, vol. 26, no. 2, pp. 117–126, 2013.

3

Models for Strength Prediction of High-Porosity Cast-In-Situ Foamed Concrete

Wenhui Zhao [iD],[1,2] **Junjie Huang** [iD],[1,2] **Qian Su**,[1,2] **and Ting Liu**[1,2]

[1]*School of Civil Engineering, Southwest Jiaotong University, Chengdu, China*
[2]*MOE Key Laboratory of High-Speed Railway Engineering, Southwest Jiaotong University, Chengdu, China*

Correspondence should be addressed to Junjie Huang; jjhuang_swjtu@126.com

Academic Editor: João M. P. Q. Delgado

A study was undertaken to develop a prediction model of compressive strength for three types of high-porosity cast-in-situ foamed concrete (cement mix, cement-fly ash mix, and cement-sand mix) with dry densities of less than $700 \, \text{kg/m}^3$. The model is an extension of Balshin's model and takes into account the hydration ratio of the raw materials, in which the water/cement ratio was a constant for the entire construction period for a certain casting density. The results show that the measured porosity is slightly lower than the theoretical porosity due to few inaccessible pores. The compressive strength increases exponentially with the increase in the ratio of the dry density to the solid density and increases with the curing time following the composite function $A_2 (\ln t)^{B_2}$ for all three types of foamed concrete. Based on the results that the compressive strength changes with the porosity and the curing time, a prediction model taking into account the mix constitution, curing time, and porosity is developed. A simple prediction model is put forward when no experimental data are available.

1. Introduction

Foamed concrete is an important type of geotechnical material [1]. It is a light solidification material mainly composed of cement, a filler, and a percentage of stable tiny bubbles and possesses the advantages of being light weight, vertically stable, and convenient for construction [2–5]. The statistical results compiled by the China Concrete and Cement Products Association (CCPA) indicate that the annual production volume of foamed concrete was over 40 million·m³ in China in 2016, of which more than 80% was cast-in-situ foamed concrete. Due to the increase in large-scale construction and civil engineering projects, the applications of foamed concrete will increase in the future. Therefore, it is important to control the quality of foamed concrete. In the application of foamed concrete for engineering projects, the quality indicators are the casting density during the casting process and the compressive strength during the design stage. However, it is important to know the compressive strength of the foamed concrete at different times after the casting process and during the initial construction period. Therefore, it is

necessary to develop a compressive strength prediction model by considering the important parameters and the mix compositions. The strength of foamed concrete is influenced by a number of parameters that have been determined in previous studies [6–8]. The uniaxial compressive strength is given by

$$\sigma = f\left(\sigma_0, p, m, \frac{m_w}{m_c}, \dots\right), \qquad (1)$$

where σ_0 is the uniaxial compressive strength of concrete at a porosity of 0; p is the porosity; m is the degree of hydration, $0 \leq m \leq 1$, $m = 0$ means the start of the hydration and $m = 1$ means the completion of the hydration; and m_w/m_c is the water/cement ratio.

Based on (1), for the same curing conditions and mix composition, the strength of the foamed concrete is mainly influenced by the porosity, the degree of hydration, and the water/cement ratio. Several models have been proposed to express this ratio, and they are listed in Table 1. It is evident that several compressive strength prediction models for foamed concrete are based on the Powers model and the

TABLE 1: Review of compressive strength prediction models for foamed concrete.

Author	Material composition	Models
Balshin [8]	Cement	$\sigma = \sigma_0 (1 - p)^n$
Neville [9]	Cement, sand, fly ash	$\sigma = kg^n$
Tam et al. [10]	Cement	$\sigma = k (1/(1 + m_w/m_c + m_a/m_c))^n$
Durack and Weiqing [11] Nambiar and Ramamurthy [12]	Cement, sand, fly ash	$\sigma = k ((2.06\alpha V_c)/(1 - V_{fl} - V_c (1 - \alpha)))^n$
Hoff [13]	Cement	$\sigma = \sigma_0 ((d_c (1 + 0.2\rho_c))/((1 + k_s)\rho_c \gamma_w))^b$
Kearsley and Wainwright [14]	Cement, sand, fly ash	$\sigma = \sigma_0 ((d_c (1 + 0.2\rho_c + s_v))/((1 + k_s)(1 + s_w)\rho_c \gamma_w))^b$

σ = the uniaxial compressive strength of foamed concrete; p = porosity; g = gel-space ratio; m_a/m_c = the air/cement ratio; d_c = casting density; ρ_c = the specific gravity of cement; γ_w = unit weight of water; k_s = water-solid ratio by weight; σ_0 = uniaxial compressive strength of concrete at a porosity of 0; s_w = filler-cement ratio by weight; s_v = admixture-cement ratio by volume; k = gel strength; α = hydration water-cement ratio by weight; V_c = volume of cement; V_{fl} = filler volume of unit volume; d_c = fresh density; b, n = empirical constant.

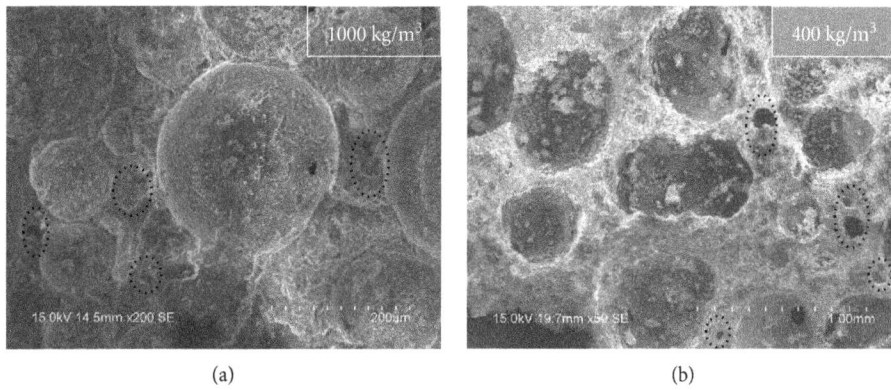

(a)

(b)

FIGURE 1: Comparison of voids in (a) high-density and (b) low-density casting of foamed concrete. Red dotted circles: small voids in the pore walls.

Balshin model; most prediction models have limitations in terms of the influencing factors of porosity, the degree of hydration, and the water/cement ratio.

As can be seen in Figure 1, foamed concrete is a typical noncompacting type of concrete; therefore, the porosity of foamed concrete is controlled by the volume of the voids in the concrete. The voids include gel pores, microcapillaries, macrocapillaries, and artificial air pores [15, 16]. Based on (1), its compressive strength is related to the proportion occupied by voids. Odler and Rößler [7] established a relationship between porosity and strength for a series of cement pastes with different water/cement ratios after periods of hydration. The research showed that the relationship between the compressive strength and the porosity is linear for porosity values between 5% and 28%. Fagerlund [17] stated that it is necessary to determine a limit for the porosity. When the porosity is below the limit, an equation fits the experimental data. For higher porosities, a different equation is required. As the foamed concrete density increases, the pore spaces become smaller and the pore walls become thicker. When the dry density of foamed concrete is more than 700 kg/m^3 (the relative density is about 0.3), there is a transitional change and the material changes from a porous structure to a solid structure containing isolated pores [18]. The dry density of foamed concrete used as roof insulation material ranges from 160 to 300 kg/m^3. For heat insulation material, the dry density ranges from 300 to 500 kg/m^3. When used as geotechnical fill material, its dry density ranges from 400 to 600 kg/m^3 [19, 20]. As we can see, all values all smaller than 700 kg/m^3.

For a given casting density of foamed concrete, when the water/cement ratio is low, the added bubbles will burst while the mixture is being stirred and the flow requirement of the foamed concrete cannot be reached. When the water/cement ratio is high, instability will occur for the fresh foamed concrete and the bubbles float on top of the foamed concrete slurry. This separation phenomenon between the foam and the cement slurry influences the casting results [21, 22]. These two phenomena related to bubble instability are shown in Figure 2. For the cast-in-situ foamed concrete during construction, the flow value is regulated between 160 and 180 mm. As for the cast-in-situ foamed concrete, a superplasticizer was not added during the production process due to limitations in the construction conditions and the construction equipment; therefore, the water/cement ratio was a constant value for a given casting density.

The degree of hydration of the foamed concrete is a function of time, curing temperature, and other parameters. As the curing time increases, the cement hydration in the foamed concrete may produce solid products that fill the pores of the sample. At the same time, the self-weight consolidation and evaporation of water may significantly increase the stiffness and density of the samples [23]. A study on the effect of the relationship between water permeability and pore connectivity under different curing times indicated that the sample had a coarse structure during the early stage,

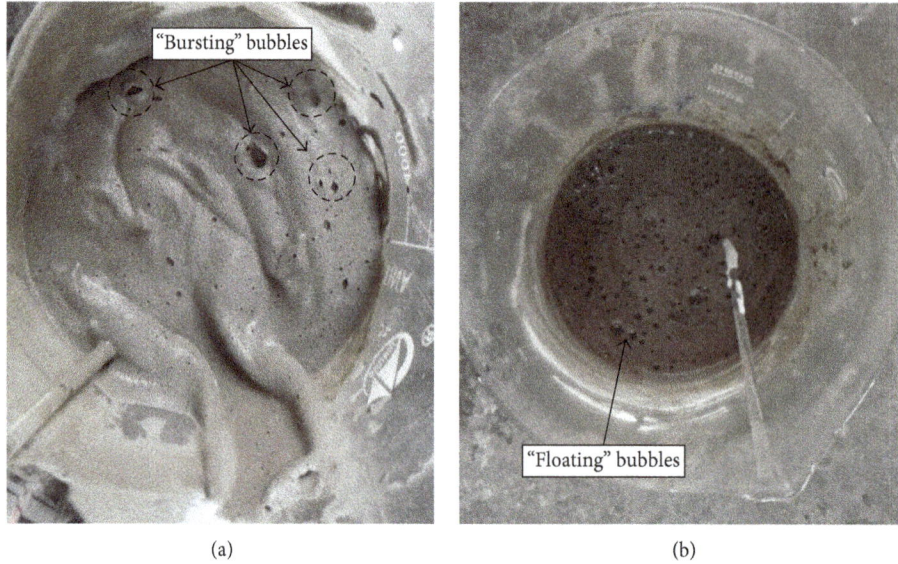

FIGURE 2: Bubble instability phenomena: (a) bubble bursting; (b) bubble floating.

TABLE 2: The constituent materials of the foamed concrete.

Materials	Remarks
Cement	Type I Portland cement conforming to GB 175-2007
Sand	Finer than 300 microns, specific gravity = 2.5
Fly ash	Class F Type I conforming to GB/T 1596-2005
Foaming agent	Synthetic type, specific gravity = 1.06
Water	Tap water

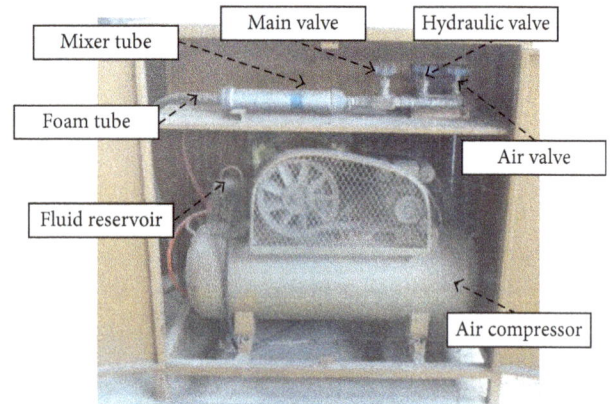

FIGURE 3: Air bubble generator.

whereas the pore structure in the hardened cement paste became denser as the curing time increased [24, 25]. For the same casting density and curing conditions, the compressive strength of foamed concrete varies with the curing time; therefore, it is necessary to know the compressive strength for different curing times.

In view of these observations, it is necessary to develop a model for compressive strength prediction of high-porosity cast-in-situ foamed concrete that should consider the porosity and the degree of hydration. The model can help determine the mix composition of the foamed concrete, the casting density, and the compressive strength. At the same time, it will provide a reference for the initial construction time of the engineering project.

2. Materials and Methods

2.1. Materials. The foamed concrete used in this study was made from ordinary Portland cement, fine sand, fly ash, and bubbles. The constituent materials used in the experiments are shown in Table 2. In this research, a synthetic type of foaming agent was used because it was highly eco-friendly and its air bubbles were strong. The bubbles were entrained or entrapped within the slurry to promote lightness [26, 27].

2.2. Mix Design Procedure. To produce the bubbles for the production of the samples, the foaming agent was diluted with water at a ratio of 1 : 60 (namely, the multiple of dilution equals 60). A prefoaming method was used to produce the foamed concrete. In this method, the air bubbles were first foamed by a bubble generator, which is shown in Figure 3. The density of air bubbles was set at $35 \pm 5 \, \text{kg/m}^3$.

A flowchart of the foamed concrete mix method is shown in Figure 4. The cement, fly ash, and sand were mixed with water at a certain ratio to produce a cement slurry as shown in Table 3. After that, the air bubbles were mixed well with the cement slurry by stirring with an electric blender to produce the foamed concrete.

2.3. Samples and Maintenance Procedure. Twenty-one densities of foamed concrete were cast; the corresponding mix proportions and major parameters are listed in Table 3. For each density, the number of groups was decided by the testing time. Six identical samples (100 mm long × 100 mm wide × 100 mm high) were prepared for each group and

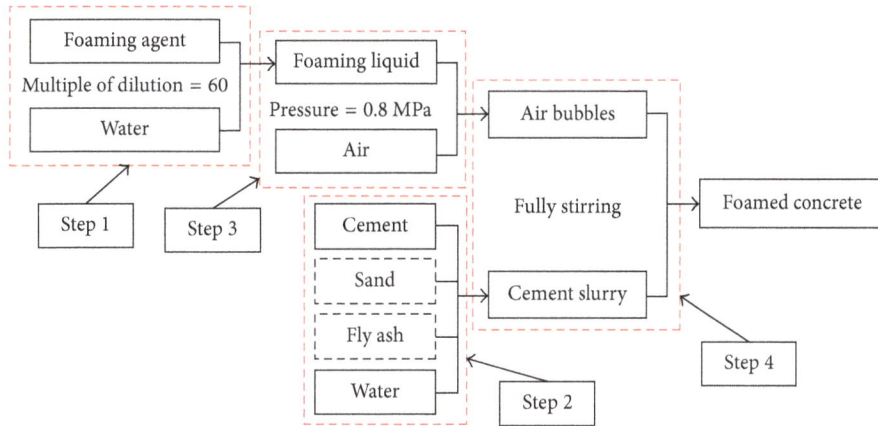

FIGURE 4: Flowchart of the foamed concrete mix method [28].

testing time. Three samples were used to test the compressive strength, and the others were used to measure the dry densities. The samples were demoulded after 24 h to ensure that they were sufficiently hard for further handling. Then, all the test samples were subjected to standard curing.

2.4. Testing Method

2.4.1. Compressive Strength.
The cubes were tested by using a compressive machine at the pace rates of 2.00 kN/s according to the "Test method of autoclaved aerated concrete" (GBT 11969-2008). The stress data were recorded to determine the unconfined compressive strength. The data used for the analysis consisted of the average of three sample test results.

2.4.2. Porosity.
The measured porosity of the foamed concrete was determined using a vacuum saturation apparatus. The samples were placed in an electric, constant-temperature drying oven at 60°C for 24 h. After that, the temperature was increased to 80°C and maintained for 24 h. The samples were dried at 100°C until the weight of the samples was constant. After the samples were placed into the vacuum saturation apparatus (Figure 5), the pressure was maintained at −1 MPa for 2 h. After turning on the air valve, water flowed into the device slowly until the samples were immersed in the liquid. Prior to the weighing tests, the samples were saturated for another 22 h. The measured porosity was calculated using the following formula [29]:

$$P = \frac{\left(W_{\text{sat}} - W_{\text{dry}}\right)}{\left(W_{\text{sat}} - W_{\text{wat}}\right)} \times 100, \tag{2}$$

where P is the vacuum saturation porosity (%), W_{sat} is the weight in air of the saturated sample, W_{wat} is the weight in water of the saturated sample, and W_{dry} is the weight of oven-dried sample.

3. Results and Discussion

Nambiar and Ramamurthy [12, 30] expressed the theoretical porosity of foamed concrete using the variables of casting density, water-solid ratio, filler-cement ratio of the freshly foamed concrete mixture, specific gravity of the cement, and unit weight of water, which can be seen in (3) and (4). According to previously determined data, the porosity values vary from 0.18 to 0.23 for different kinds of raw materials and different proportions of ingredients [10, 31, 32]. The ratio of hydration water to cement by weight is assumed to be 0.2.

$$n = 1 - \frac{d_{\text{c}}\left(1 + 0.2\rho_{\text{c}}\right)}{\left(1 + k_{\text{s}}\right)\rho_{\text{c}}\gamma_{\text{w}}}, \tag{3}$$

$$n = 1 - \frac{d_{\text{c}}\left(1 + 0.2\rho_{\text{c}} + s_{\text{v}}\right)}{\left(1 + k_{\text{s}}\right)\left(1 + s_{\text{w}}\right)\rho_{\text{c}}\gamma_{\text{w}}}. \tag{4}$$

These equations are inconvenient and flawed. First of all, there are many variables that need to be calculated. Secondly, it is assumed that the hydration of the cement is complete, which ignores the degree of cement hydration. Lastly, these equations do not consider the difference in the hydration water between the cement and the admixtures. In order to avoid these problems, the following equation can be used to determine the theoretical porosity [18]:

$$n = 1 - \frac{\rho^{*}}{\rho_{\text{s}}}, \tag{5}$$

where ρ^{*} is the dry density of foamed concrete and ρ_{s} is the dry density of the solid material.

In (5), the dry density can be considered a certain value for the same mix constitution for the same brand of material when it is completely hydrated. When the constitution proportions of the materials change, ρ_{s} can be obtained based on the proportions of the constitutions. The hydrations of the cement and admixtures are functions of the curing time, and the degree of hydration will affect the dry density. When the hydration is complete and the testing time is long enough, (3) and (4) are applicable. However, (5) applies in all cases. In addition, this equation is simple and intuitive.

The relationship between the measured and theoretical porosities (after 1 year) of foamed concrete is shown in Figure 6. It is evident that deviations exist between the measured and theoretical porosities. For all mixes, the

TABLE 3: Mix proportions and major parameters of foamed concrete.

Mix number	$m_c + m_m + m_s$	f/c	s/c	w/binder	Water (kg)	Cement (kg)	Fly ash (kg)	Sand (kg)	Air bubbles (l)	Dry density (kg/m³)	Casting density (kg/m³)	Flow value (cm)	Testing time (d)
1	A	0	0	0.75	173.0	230.7	0.0	0.0	752.6	291.3	430.1	16.1	7, 28, 90, 180, 270, 365
2		0	1:1	0.63	144.2	115.4	0.0	115.4	772.5	255.8	401.9	17.7	7, 28, 90, 180, 270, 365
3		1:1	0	0.75	173.0	115.4	115.4	0.0	745.4	274.5	429.8	16.8	7, 28, 90, 180, 270, 365
4	B	0	0	0.70	204.5	292.2	0.0	0.0	701.2	367.3	521.3	16.2	28, 365
5		0	1:1	0.59	172.4	146.1	0.0	146.1	722.0	324.8	489.9	16.5	28, 365
6		1:1	0	0.70	204.5	146.1	146.1	0.0	692.1	345.4	521.0	16.9	28, 365
7	C	0	0	0.65	229.9	353.7	0.0	0.0	656.0	427.6	606.6	16.5	28, 365
8		0	1:1	0.56	196.4	176.9	0.0	176.9	675.8	391.8	573.8	16.6	28, 365
9		1:1	0	0.65	229.9	176.9	176.9	0.0	645.0	421.3	606.2	17.2	28, 365
10	D	0	0	0.60	249.1	415.2	0.0	0.0	616.9	514.9	685.9	16.8	7, 28, 90, 180, 270, 365
11		0	1:1	0.52	215.9	207.6	0.0	207.6	634.1	461.5	653.3	17.5	7, 28, 90, 180, 270, 365
12		1:1	0	0.60	249.1	207.6	207.6	0.0	604.1	496.2	685.5	17.4	7, 28, 90, 180, 270, 365
13	E	0	0	0.55	262.2	476.7	0.0	0.0	584.0	585.4	759.3	16.8	28, 365
14		0	1:1	0.49	233.6	238.4	0.0	238.4	594.1	526.4	731.2	17.1	28, 365
15		1:1	0	0.55	262.2	238.4	238.4	0.0	569.3	561.2	758.8	17.7	28, 365
16	F	0	0	0.54	290.6	538.2	0.0	0.0	535.8	645.6	847.6	17.1	28, 365
17		0	1:1	0.48	258.3	269.1	0.0	269.1	547.2	597.2	815.7	16.8	28, 365
18		1:1	0	0.54	290.6	269.1	269.1	0.0	519.1	638.1	847.0	17.6	28, 365
19	G	0	0	0.53	317.8	599.7	0.0	0.0	488.7	705.4	934.6	16.7	7, 28, 90, 180, 270, 365
20		0	1:1	0.47	281.9	299.9	0.0	299.9	501.5	663.8	899.1	17.5	7, 28, 90, 180, 270, 365
21		1:1	0	0.53	317.8	299.9	299.9	0.0	470.1	687	934.0	17.4	7, 28, 90, 180, 270, 365

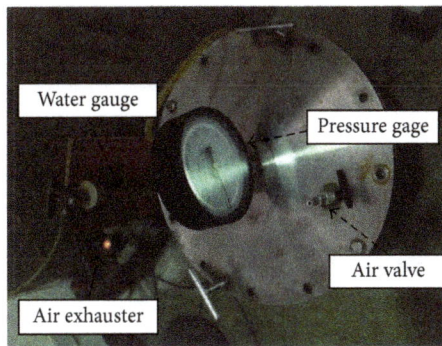

FIGURE 5: Vacuum saturation device.

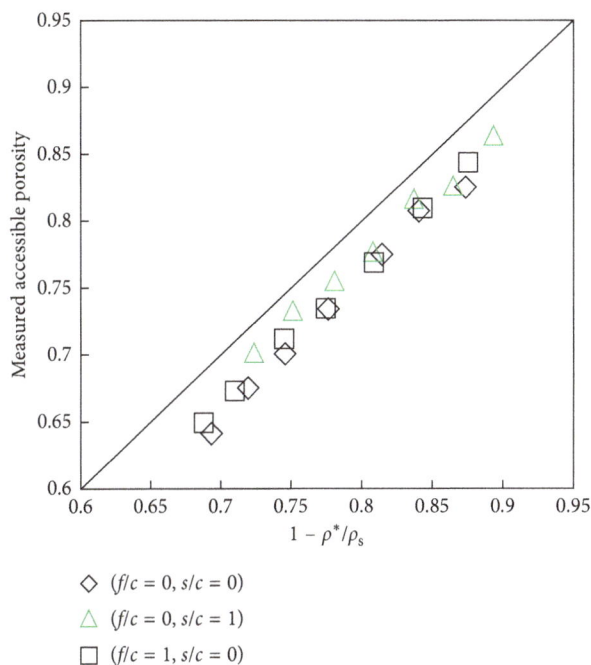

◇ $(f/c = 0, s/c = 0)$
△ $(f/c = 0, s/c = 1)$
□ $(f/c = 1, s/c = 0)$

FIGURE 6: Relationship between measured porosity and $1 - \rho^*/\rho_s$.

measured values are lower than the theoretical values. The reason is that foamed concrete has closed pores, which are difficult to fill entirely with water. The closed pores differ for different casting densities. In high-density casting, the closed pores occur in the pore structures and pore walls, while in low-density casting, the closed pores mainly occur in the pore walls. A comparison of the differences between the measured and theoretical porosities for the three types of foamed concrete of the same casting density indicates that the difference is largest for the cement mix, followed by the cement-fly ash mix and the cement-sand. After the addition of fly ash, the initial setting time of the foamed concrete increases resulting in instability of the foamed concrete slurry and an increase in the number of connected holes [14]. For the foamed concrete containing sand, the pore walls become thinner and the number of the connected holes increases due to the reduction in the cement proportion and the increase in the density of the solid material [33, 34]. When the number of connected holes increases, there is

a higher likelihood that the voids are filled with water, which decreases the deviations between the measured and the theoretical porosities.

3.1. Effect of Porosity on Compressive Strength. The relationship between the theoretical porosity and the compressive strength (28 days and 1 year) of the foamed concrete is shown in Figure 7. The data show that the compressive strength increases exponentially with the increase in ρ^*/ρ_s as described in (6); this occurs for all types of foamed concrete for the 1 yr duration and the 28 d duration.

$$\sigma = A_1 \left(\frac{\rho^*}{\rho_s} \right)^{B_1}. \tag{6}$$

A comparison between Figures 7(a) and 7(b) shows that the dry density is slightly higher for 1 yr than for 28 d for the same density of foamed concrete because the hydration reaction is not fully complete at 28 d, which is consistent with previously reported results [2]. For the three conditions $(f/c = 0, s/c = 0; f/c = 0, s/c = 1; f/c = 1, s/c = 0)$, the A_1 value is higher for 1 yr than for 28 d. The B_1 value is lowest for the cement-sand mix, followed by the cement mix and the cement-fly ash mix.

Table 4 lists the constants of the strength-porosity models of the foamed concrete with various mix constituents. It shows that the values of the constants obtained in this study are close to the values obtained by Hoff [13] using cement paste. However, the constants obtained in this study are lower than those obtained by Nambiar and Ramamurthy [12] for cement-sand. Nambiar and Ramamurthy [21] stated that foamed concrete with a fly ash additive was less dependent on the pore parameters than foamed concrete with sand; these results differ from our results. Figure 8 shows the microscopic photographs of the pore wall structures of the foamed concrete with different casting densities. The pore wall structure is dense for the high-density casting and less dense for the low-density casting. In the low-porosity foamed concrete $(f/c = 0, s/c = 1)$, the sand has an interlocking effect under pressure. In the high-porosity foamed concrete $(f/c = 0, s/c = 1)$, the hole walls are thinner and less dense, which reduces the interlocking effect. Therefore, in this study, the σ_0 parameter of high-porosity foamed concrete only reflects the uniaxial compressive strength of the mix constitution.

3.2. Effect of Curing Time on Compressive Strength. Most of the existing compressive strength prediction models have been used to predict the strength of the foamed concrete after casting at particular times, while the compressive strength of foamed concrete increases with the curing time. Because the use of the foamed concrete begins at 28 d after casting or earlier, it is important to predict the strength of foamed concrete immediately after casting and thereafter.

The effects of the curing time on the compressive strengths of three casting densities are shown in Figure 9. It can be seen that the compressive strength increases with the

(a)

(b)

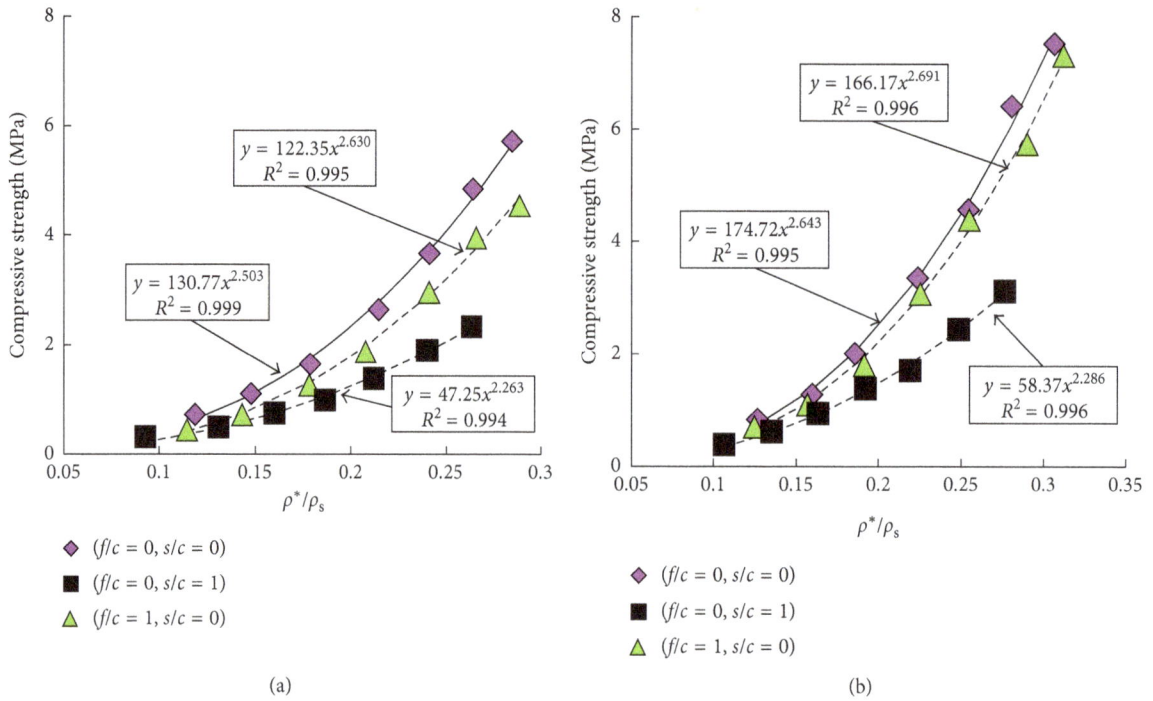

FIGURE 7: Effect of porosity on compressive strength at (a) 28 days and (b) 1 year.

TABLE 4: Comparison of the constants of various strength-porosity models.

Model by	Mix constituents	Constants	
		σ_0	b
Hoff [13]	Cement	115–290	2.7–3
Kearsley and Wainwright [14]	Cement with and without fly ash	188	3.1
Nambiar and Ramamurthy [12]	Cement-sand	155.66	4.3
	Cement-sand-fly ash	105.14	2.58
	Cement	174.72	2.643
Present study	Cement-sand	58.37	2.286
	Cement-fly ash	166.17	2.691

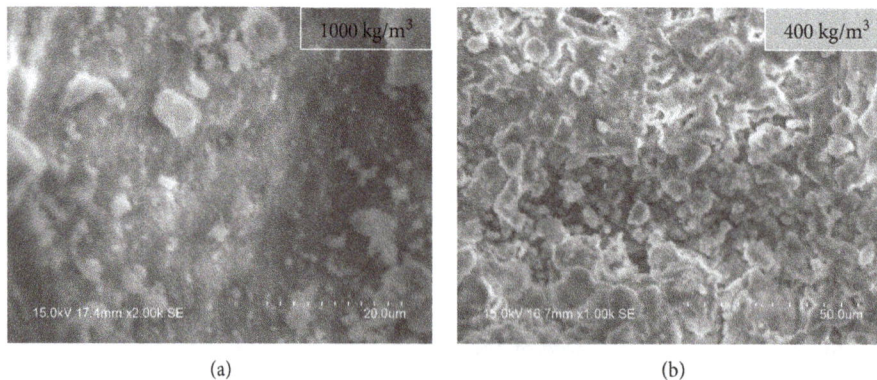

(a)

(b)

FIGURE 8: Microscopic photographs of the pore wall structures of (a) high-density and (b) low-density casting foamed concrete.

curing time; initially, the rate of increase in the compressive strength is large and it slows as the curing time increases. The relationship between the compressive strength and the curing time can be expressed by the following equation for a given density and mix constitution:

$$\sigma = A_2 \left(\ln t \right)^{B_2}. \tag{7}$$

A comparison of the results shown in Figures 9(a)–9(c) indicates that for the same casting density and curing time, the compressive strength is highest for the cement mix, followed by

(a)

(b)

(c)

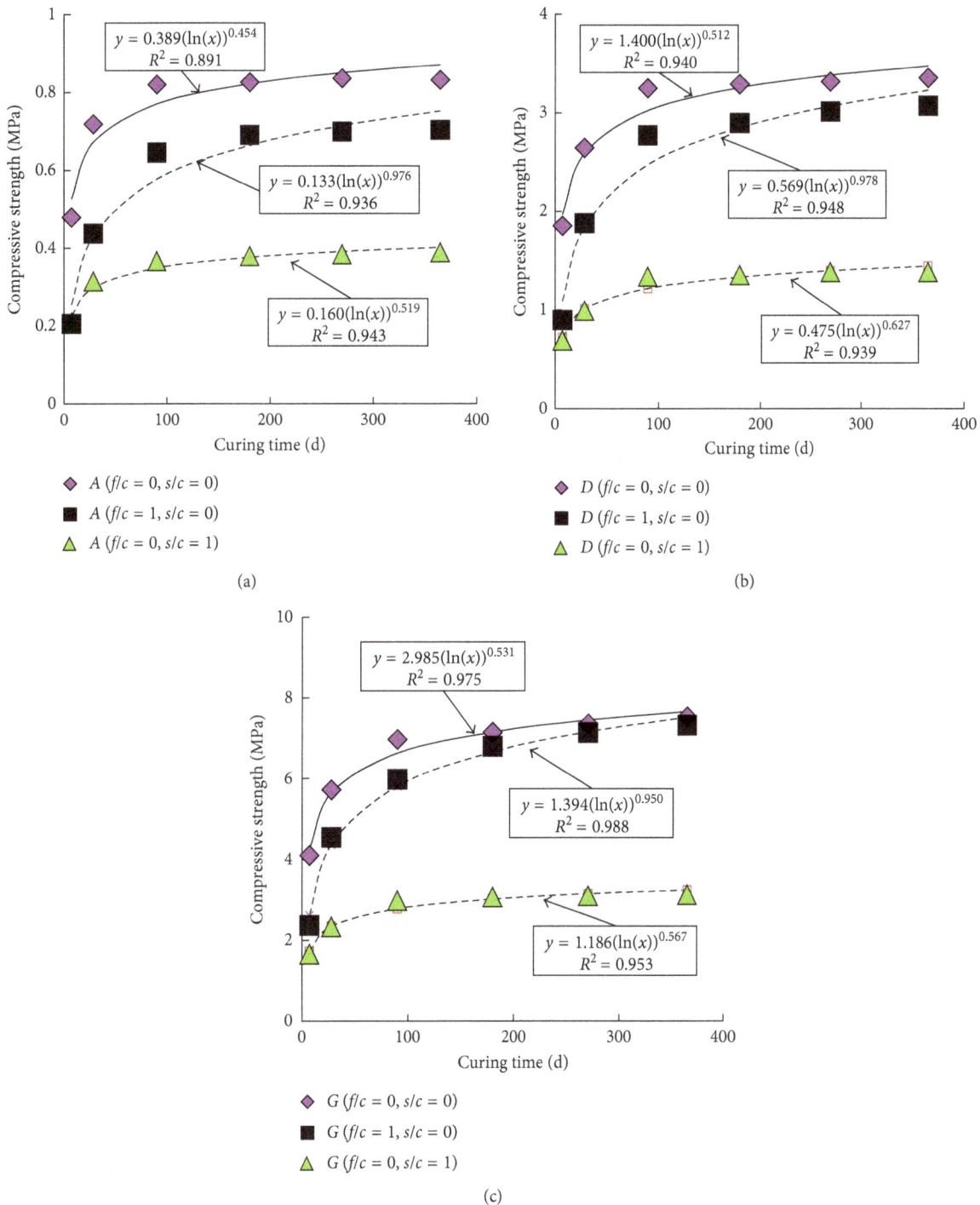

FIGURE 9: Effect of curing time on compressive strength for the three casting densities: (a) A, (b) D, and (c) G.

the cement-fly ash mix and the cement-sand mix. The compressive strength is almost stable at 90 d for the cement mix and the cement-sand mix; the compressive strength does not increase after 180 d for the cement-fly ash mix. It is evident that the values provided by the prediction model are lower during the early stage and higher during the late stage compared with the test values. The compressive strength decreases when fly ash is added, but the attenuation ratio decreases with an increase in the casting density. The B_2 values are similar for the cement

mix and the cement-sand mix, but the B_2 value is lower for the cement-fly ash mix. Because the hydration reaction material in the cement mix and the cement-sand mix is cement, the hydration rates are very similar under the same curing conditions [35], while the hydration rate of the fly ash is lower. The B_2 value represents the hydration rate of the raw material.

3.3. *Proposed Model.* Based on the above results, the compressive strength of foamed concrete is a function of the mix

TABLE 5: Comparison of the constants of various mix constituents.

Mix constituents	A	B	C	$A(\ln 365)^B$	R^2
Cement	110.255	0.204	2.568	158.362	0.997
Cement-sand	34.989	0.305	2.292	60.123	0.995
Cement-fly ash	59.551	0.531	2.603	152.830	0.993

TABLE 6: Comparison of the constants of various mix constituents ($\rho^* = \rho_{365}^*$).

Mix constituents	A	B	C	$A(\ln 365)^B$	R^2
Cement	60.640	0.470	2.459	139.655	0.988
Cement-sand	22.771	0.518	2.244	57.106	0.991
Cement-fly ash	33.315	0.821	2.548	143.055	0.982

constitution, curing time, and porosity. The following equation is derived from (6) and (7):

$$\sigma = A(\ln t)^B \left(\frac{\rho^*}{\rho_s}\right)^C, \quad (8)$$

where t is the curing time, $1 < t \leq 365$; A is the parameter associated with the compressive strength of concrete; B is the parameter reflecting the hydration rate of the mix constituents; and C is the parameter reflecting the porosity and pore structure, which is associated with the pore quality including pore size, pore shape, and so on.

When (8) is used to fit the test data, ρ^* is measured when the compressive strength is tested. Table 5 shows the constants of various mix constituents. It can be inferred that there is no obvious difference in the B value for cement and cement-sand, whereas the B value of the cement-fly ash is high, which is in agreement with the foregoing conclusions. Compared with the B values shown in Figure 9, the values are lower in Table 5 because the hydration rate of the mix constituent is reflected by the parameters B and C. When parameter C is added to the equation, the value of B decreases.

The parameter ρ^* is rarely measured outside the laboratory when a strength prediction model is used. It is important to reduce the number of measurements. The dry densities are similar for different curing times for the same casting densities. When the differences are ignored, the dry densities are only measured after 1 yr. The results are shown in Table 6.

A comparison of the data shown in Tables 5 and 6 shows that the A and C values are lower in Table 6 for the same mix constitutions, while the B values are higher. The degree of fit decreases for all mix constitution, although a high degree of fit is still ensured. Therefore, the assumption of $\rho^* = \rho_{365}^*$ is reasonable in order to simplify the model. At the same time, ρ^* can be simply calculated using (9) according to the "Technical specification for application of foamed concrete" (JGJT341-2014):

$$\rho^* = S_a(m_c + m_m) + m_s, \quad (9)$$

where S_a is an empirical constant for certain cement and admixture factories, m_c is the cement dosage of foamed concrete per cubic meter, m_m is the admixture dosage of foamed concrete per cubic meter, and m_s is the fine aggregate dosage of foamed concrete per cubic meter.

In this study, S_a equals to 1.211, 1.216, and 1.175 for the cement mix, cement-sand mix, and cement-fly ash mix, respectively. The values are very similar for the cement mix and the cement-sand mix. The value is slightly lower for the cement-fly ash mix because cement is a hydraulic binding material and fly ash is the active mineral admixture; the addition of fly ash reduces the value of S_a. Cement is the hydration reaction material in the cement-sand mix, and sand is the nonactive admixture; therefore, the S_a values are similar for the cement mix and the cement-sand mix. Combined with the other results, (9) can be used to calculate the dry density, in which S_a is an empirical constant for a given mix constitution and a given cement factory. When there are no measurements of dry density, the following equation can be used to predict the compressive strength of foamed concrete:

$$\sigma = A(\ln t)^B \left(\frac{S_a(m_c + m_m) + m_s}{\rho_s}\right)^C. \quad (10)$$

4. Conclusion

The following conclusions can be drawn based on the experimental and comparative results:

(1) The measured porosity is slightly lower than the theoretical porosity due to few inaccessible pores.

(2) The compressive strength increases with the increase in the ratio of dry density to solid density following the equation $A_1(\rho^*/\rho_s)^{B_1}$ for all three types of foamed concrete. The B_1 value is similar for the cement mix and the cement-fly ash mix. The value of B_1 is lowest for the cement-sand mix, followed by the cement mix and the cement-fly ash mix.

(3) The compressive strength increases with the curing time following the composite function $A_2(\ln t)^{B_2}$ for all three types of foamed concrete. The B_2 values are similar for the cement mix and cement-sand mix, but the B_2 value is lower for the cement-fly ash mix.

(4) Based on the results that the compressive strength changes with the porosity and the curing time, a prediction model taking into account the mix constitution, curing time, and porosity is proposed. A simple prediction model is put forward when no experimental data are available.

Conflicts of Interest

The authors declare that there are no conflicts of interest regarding the publication of this paper.

Acknowledgments

This work was supported by the Sichuan Science and Technology Support Project (2016JY0005) and the China Railway Science and Technology Research Plan Project (2015G002-K).

References

[1] C. Hu, H. Li, Z. Liu, and Q. Wang, "Research on properties of foamed concrete reinforced with small sized glazed hollow beads," *Advances in Materials Science and Engineering*, vol. 2016, Article ID 5820870, 8 pages, 2016.

[2] K. Ramamurthy, E. K. K. Nambiar, and G. I. S. Ranjani, "A classification of studies on properties of foam concrete," *Cement and Concrete Composites*, vol. 31, no. 6, pp. 388–396, 2009.

[3] Z. Liu, K. Zhao, C. Hu, and Y.-F. Tang, "Effect of water-cement ratio on pore structure and strength of foam concrete," *Advances in Materials Science and Engineering*, vol. 2016, Article ID 9520294, 9 pages, 2016.

[4] P. Onprom, K. Chaimoon, and R. Cheerarot, "Influence of bottom ash replacements as fine aggregate on the property of cellular concrete with various foam contents," *Advances in Materials Science and Engineering*, vol. 2015, Article ID 381704, 11 pages, 2015.

[5] Y. S. Jeong and H. K. Jung, "Thermal performance analysis of reinforced concrete floor structure with radiant floor heating system in apartment housing," *Advances in Materials Science and Engineering*, vol. 2015, Article ID 367632, 7 pages, 2015.

[6] M. Rößler and I. Odler, "Investigations on the relationship between porosity, structure and strength of hydrated Portland cement pastes I. Effect of porosity," *Cement and Concrete Research*, vol. 15, no. 2, pp. 320–330, 1985.

[7] I. Odler and M. Rößler, "Investigations on the relationship between porosity, structure and strength of hydrated Portland cement pastes. II. Effect of pore structure and of degree of hydration," *Cement and Concrete Research*, vol. 15, no. 3, pp. 401–410, 1985.

[8] M. Y. Balshin, "Relation of mechanical properties of powder metals and their porosity and the ultimate properties of porous metal-ceramic materials," *Doklady Akademii Nauk SSSR*, vol. 67, no. 5, pp. 831–834, 1949.

[9] A. Neville, *Properties of Concrete*, Wiley, New York, NY, USA, 1996.

[10] C. T. Tam, T. Y. Lim, R. Sri Ravindrarajah, and S. L. Lee, "Relationship between strength and volumetric composition of moist-cured cellular concrete," *Magazine of Concrete Research*, vol. 39, no. 138, pp. 12–18, 1987.

[11] J. M. Durack and L. Weiqing, "The properties of foamed air cured fly ash based concrete for masonry production," in *Proceedings of the Fifth Australasian Masonry Conference*, Gladstone, QLD, Australia, July 1998.

[12] E. K. K Nambiar and K. Ramamurthy, "Models for strength prediction of foam concrete," *Materials and Structures*, vol. 41, no. 2, pp. 247–254, 2008.

[13] G. C. Hoff, "Porosity-strength considerations for cellular concrete," *Cement and Concrete Research*, vol. 2, no. 1, pp. 91–100, 1972.

[14] E. P. Kearsley and P. J. Wainwright, "The effect of high fly ash content on the compressive strength of foamed concrete," *Cement and Concrete Research*, vol. 31, no. 1, pp. 105–112, 2001.

[15] A. A. Hilal, N. H. Thom, and A. R. Dawson, "On entrained pore size distribution of foamed concrete," *Construction and Building Materials*, vol. 75, pp. 227–233, 2015.

[16] Z. Pan, H. Li, and W. Liu, "Preparation and characterization of super low density foamed concrete from Portland cement and admixtures," *Construction and Building Materials*, vol. 72, pp. 256–261, 2014.

[17] G. Fagerlund, "Strength and porosity of concrete," in *Proceedings of the International Symposium on Pore Structure and Properties of Materials (RILEM/IUPAC)*, pp. 51–141, Prague, Czech Republic, September 1973.

[18] L. J. Gibson and M. F. Ashby, *Cellular Solids: Structure and Properties*, Cambridge University Press, Cambridge, UK, 1999.

[19] X. Tan, W. Chen, Y. Hao, and X. Wang, "Experimental study of ultralight (<300 kg/m³) foamed concrete," *Advances in Materials Science and Engineering*, vol. 2014, Article ID 514759, 7 pages, 2014.

[20] B. Dolton and C. Hannah, "Cellular concrete: Engineering and technological advancement for construction in cold climates," in *Proceedings of the 2006 Annual General Conference of the Canadian Society for Civil Engineering*, Calgary, AB, Canada, May 2006.

[21] E. K. K. Nambiar and K. Ramamurthy, "Fresh state characteristics of foam concrete," *Journal of Materials in Civil Engineering*, vol. 20, no. 2, pp. 111–117, 2008.

[22] M. R. Jones, K. Ozlutas, and L. Zheng, "Stability and instability of foamed concrete," *Magazine of Concrete Research*, vol. 68, no. 11, pp. 542–549, 2016.

[23] E. K. K. Nambiar and K. Ramamurthy, "Sorption characteristics of foam concrete," *Cement and Concrete Research*, vol. 37, no. 9, pp. 1341–1347, 2007.

[24] K. Kurumisawa and K. Tanaka, "Three-dimensional visualization of pore structure in hardened cement paste by the gallium intrusion technique," *Cement and Concrete Research*, vol. 36, no. 2, pp. 330–336, 2006.

[25] T. H. Wee, S. B. Daneti, and T. Tamilselvan, "Effect of w/c ratio on air-void system of foamed concrete and their influence on mechanical properties," *Magazine of Concrete Research*, vol. 63, no. 8, pp. 583–595, 2011.

[26] E. Kuzielová, L. Pach, and M. Palou, "Effect of activated foaming agent on the foam concrete properties," *Construction and Building Materials*, vol. 125, pp. 998–1004, 2016.

[27] D. K. Panesar, "Cellular concrete properties and the effect of synthetic and protein foaming agents," *Construction and Building Materials*, vol. 44, pp. 575–584, 2013.

[28] W. H. Zhao, Q. Su, T. Liu, and J. J. Huang, "Experimental study on the frost resistance of cast-in-situ foamed concrete," *Electronic Journal of Geotechnical Engineering*, vol. 22, pp. 5509–5523, 2017.

[29] J. G. Cabrera and C. J. Lynsdale, "A new gas permeameter for measuring the permeability of mortar and concrete," *Magazine of Concrete Research*, vol. 40, no. 144, pp. 177–182, 1988.

[30] E. P. Kearsley and P. J. Wainwright, "The effect of porosity on the strength of foamed concrete," *Cement and Concrete Composites*, vol. 32, no. 2, pp. 233–239, 2002.

[31] H. F. W. Taylor, *Cement Chemistry*, Thomas Telford Ltd., London, UK, 1997.

[32] P. Hewlett, *Lea's Chemistry of Cement and Concrete*, Butterworth-Heinemann, Oxford, UK, 2003.

[33] E. K. K. Nambiar and K. Ramamurthy, "Models relating mixture composition to the density and strength of foam concrete using response surface methodology," *Cement and Concrete Composites*, vol. 28, no. 9, pp. 752–760, 2006.

[34] F. Batool and V. Bindiganavile, "Air-void size distribution of cement based foam and its effect on thermal conductivity," *Construction and Building Materials*, vol. 149, pp. 17–28, 2017.

[35] E. K. K. Nambiar and K. Ramamurthy, "Influence of filler type on the properties of foam concrete," *Cement and Concrete Composites*, vol. 28, no. 5, pp. 475–480, 2006.

4

Improvement for Construction of Concrete-Wall with Resistance to Gas-Explosion

Daegeon Kim (iD)

Architecture Engineering, Dongseo University, Busan, Republic of Korea

Correspondence should be addressed to Daegeon Kim; gun43@gdsu.dongseo.ac.kr

Academic Editor: Lijing Wang

The research was initiated to investigate the performance of fiber-reinforced concrete for protecting people or assets in the building against the explosion or debris missiles. The fiber-reinforced concrete has the difficulty with being applied in the actual construction conditions with the normal ready-mixed concrete system. The fibers for the protection performance require high toughness to endure the huge energy from an explosion, but the large amount of the fiber is required. The required amount of fibers can result in decreased workability and insufficient dispersion of fibers. It has been difficult to apply fiber-reinforced concrete on field placing with the ready-mixed concrete system of plant mixing, delivering, and placing. This research carried out the investigation of properties of combined fiber of steel and polymeric fiber to improve workability and agitating in the mixer. Based on the preliminary experimental test results in a laboratory, combined fiber-reinforced concrete was applied on the actual field construction of chemical plant. According to the results from the laboratory tests and application in the real construction project, it is expected to introduce the combined fiber for desirable mechanical performance with less adverse effect on workability of the mixture.

1. Introduction

Against terrorism or warfare, structures or facilities with special purpose should have sufficient protecting performance for shock wave or debris missiles from bombing or explosion. In addition to the special purposed facilities, many people who work at the plants where handling explosive substances such as explosive gas or massive structures should be protected against explosion. Generally, to secure the enough protecting performance against these kinds of forces, the wall should be constructed thicken enough with normal strength-ranged reinforced concrete. For the lateral stress caused by earthquake or explosion, fiber-reinforced concrete (FRC) is known as a solution with its high energy absorption capacity and high tensile strength [1]. Comparing to the normal concrete without fiber reinforcement, FRC has high tensile strength and toughness. Generally, for FRC, the fiber content is a key of improving mechanical properties of the material [2].

On the contrary, addition of fiber in concrete mixture causes reduction of workability with increasing both viscosity and yield stress. Because of poor yield stress, FRC with increased fiber content has been reported a fiber ball effect during the mixing process and unfavorable consolidating performance. Therefore, as a method of achieving the maximum mechanical performance without workability issue, slurry infiltrated fiber concrete was introduced [3, 4]. Likewise, for FRC, the fiber content should be balanced between mechanical properties and workability.

The reinforcing fibers for improving the performance of cementitious materials have different roles or performances depending on their aspect ratios (length-to-diameter ratio), materials, or shapes (straight, bent, or hooked). Especially, regarding the materials, the reinforcing fiber can be categorized into metallic and polymeric fibers. First, metallic fiber, mainly steel fiber, increases the toughness of the mixture. The metallic fiber itself has a high tensile strength and elastic modulus; thus it provides increasing tensile strength and elastic modulus while it is pulled out from the cement matrix. Since the metallic fiber has a higher tensile strength than cement matrix, the fail behavior of metallic fiber is pulling out of the fibers, so there are various geometries of the metallic fibers such as hooked, bent, or various cross sections. Otherwise, polymeric fibers, such as polypropylene, polyethylene, or nylon fibers, have a relatively lower tensile strength and elastic modulus than metallic fiber. Hence, the polymeric fiber cannot improve the mechanical properties of the mixture as the metallic fiber does; however, because of the advantage of good dispersion inside of the fresh state cementitious materials, it contributes on improving mechanical properties of the mixture. Especially, the polymeric fiber can be produced with the high aspect ratio with thin diameter, and because of flexibility of the shape, it does not decrease the workability of the mixture rather than metallic fiber. The hybrid fiber or cocktailed fiber means the combined fibers of different types to achieve synergetic effect. For instance, Banthia et al. and Markovic et al. reported improved mechanical properties of FRC with two fibers with different materials [5, 6], and Peng et al. reported two different polymeric fibers with different aspect ratios and melting points for improved performance of preventing spalling damage of high-performance concrete mixture [7]. These studies were showing improvement in desired properties of FRC with combined fibers or hybrid fiber with decreased fiber content for the achievement of improved workability. Therefore, combining different types of fibers has been used as a solution of decreased workability by decreasing fiber content with equivalent performances [8].

Although various researches were reported using combined fiber for improving mechanical properties, the study based on sufficient workability by field placement is not reported enough. Especially, because of the issue of securing the quality of fiber dispersion and relatively low workability, it was difficult to apply the FRC on field placement using the ready-mixed concrete system including plant mixing, agitator truck delivering, and placing with pump. In this research, with the goal of providing protectable concrete against flying debris or missiles, combined fiber was applied for sufficient performance with favorable workability. Therefore, both the protecting performance against flying debris and the field applicability of fresh concrete placing were evaluated. Especially, for field applicability, the experiment was conducted for an actual plant construction. From the result of this research, it is expected to provide a technique of manufacturing high-performance fiber-reinforced cementitious composites (HPFRCCs) with favorable protecting performance and workability.

2. Experiment on Protecting Performance

2.1. Experimental Plan.
To evaluate the protection performance of combined FRC, mixture conditions with the combination of three different fibers were prepared for the experimental test as shown in Table 1. The water-to-binder ratio (w/b) was fixed as 0.50 and the target concrete compressive strength was 24 MPa for the concrete wall. The water content was 220 kg/m^3 to satisfy 150 ± 20 mm of the target slump. According to the preliminary test, the sand-to-aggregate ratio (S/a) was designed as 0.55 for stable viscosity and the target air content was $4.5 \pm 1.5\%$. For cementitious

TABLE 1: Experimental plan.

Mixing properties		Test items	
w/b	0.50	Fresh concrete	Slump
Water content (kg/m^3)	220		Air content
S/a	0.55		
Target slump (mm)	150 ± 20		Compressive strength at 7 and 28 days
Target air content (%)	4.5 ± 1.5		
Binder composition (by weight)	OPC : BS : FA = 7 : 2 : 1	Hardened concrete	Flexural strength at 7 and 28 days
Fiber content (%)	1		Tensile strength at 7 and 28 days
Fiber combination	SF only PF only SF + PF		Impact of high-velocity projectile

binder, ternary mix design was used with ASTM type I cement (OPC) including 20% of blast-furnace slag (BS) and 10% of fly ash (FA) by weight to expect the better workability [9]. Using steel fiber (SF) and polyaramid fiber (PF), three mix combinations were set up, and the fiber content was fixed to 1 percent of entire mixture volume which was determined by the preliminary test and the study of Yusof et al. [10]. The tests of the slump and air content were conducted to confirm the fresh concrete performance. For hardened concrete performance, fundamental mechanical properties were evaluated with compressive, flexural, and tensile strengths at 7 and 28 days of age, and protecting performance was evaluated with impacting high-velocity projectile.

2.2. Materials and Sample Preparation.
According to the information provided from the cement manufacturer, the specific gravity was 3.15 and Blaine was 3,650 cm^2/g. Blast-furnace slag was 2.88 of specific gravity and 4,469 cm^2/g of fineness. Fly ash was similar to the class F fly ash designated by ASTM C618; the specific gravity was 2.27, and fineness was 3,381 cm^2/g. The properties of admixture are shown in Table 2. For concrete mixture, coarse and fine aggregates were used with the use of manufactured aggregate and river sand, respectively, and the physical properties of those aggregates are shown in Table 3. As fiber reinforcement, two different types of fiber were used: steel and polyaramid fibers. The SF was bent, and the PF was twisted. The properties and shapes of fibers are provided in Table 4 and Figure 1, respectively.

Three different mixtures with different fiber conditions were prepared. Based on the single concrete mixture, SF, PF, and the binary fiber of SF and PF were added 1% of entire mixture volume. The mix proportions of three mixtures are shown in Table 5. For FRC mixing, a 60-liter pan-type mixer was used. The mixing protocol is shown in Figure 2. For the first step, cementitious binder and aggregate were introduced into the mixer and mixed with low speed (20 rpm) for 30 seconds. Next, mixing water was added and mixed with medium speed (30 rpm) for 60 seconds, and as the last

TABLE 2: Physical properties of admixture.

Classification	Form	Property	Color	pH	Density (g/cm^3)
SP agent	Liquid	Polycarbonate	Ivory white	6.5	1.06

SP: superplasticizer.

TABLE 3: Physical properties of aggregates.

Aggregate	Density (g/cm^3)	Fineness modulus	Unit volume weight (kg/m^3)
Fine aggregate	2.57	2.57	2.84
Coarse aggregate	2.71	6.78	1.57

TABLE 4: Physical properties of fibers.

Fiber	Aspect ratio	Length (mm)	Diameter (mm)	Tensile strength (MPa)
SF	66	35	0.53	1,108
PF	61	30	0.49	623

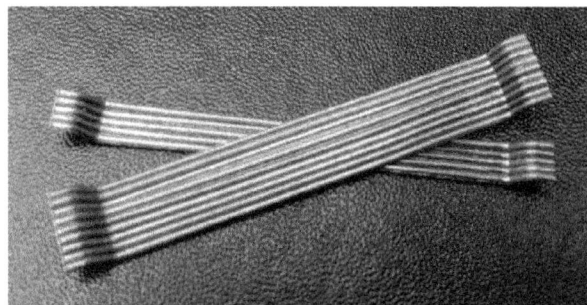

(a)

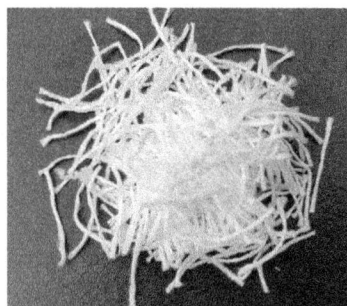

(b)

FIGURE 1: Shape of the fibers. (a) Steel fiber (SF). (b) Polyaramid fiber (PF).

step, superplasticizer was added and the entire mixture was mixed with high speed (40 rpm) for 90 seconds during the fiber was spread into the mixer to prevent fiber ball production, which can be found in FRC [11].

2.3. Test Methods. For fresh state properties, slump and air content of the mixtures were measured as ASTM C143 and C138 standards, respectively. The mechanical properties of mixtures were inclusive of compressive, flexural, and tensile strengths at 7 and 28 days of ages. Each test was conducted following ASTM C39 and C78 for the compressive and flexural strength tests, and JSCE-E-531 [12] standard for the direct tensile test from Japan Society of Civil Engineering (JSCE) was used for tensile strength measurement. Each test was conducted with three specimens for a single averaged value. To evaluate the protecting performance against missiles resulted from explosion in building, the direct impact test was conducted by shooting 25 mm diameter of spherical iron projectiles to the concrete panel of height 200 mm, width 200 mm, and depth 50 mm. The projectile was shot by compressed gas, and the velocity of projectile at impact was 170 m/s. The evaluation of protecting performance was executed by observing both the front and back surfaces of the concrete panel after impact. The impact test was conducted with two samples for a single case's result.

3. Results and Discussion

3.1. Fresh Concrete Properties. To investigate the influence of different fiber combinations, slump of each mixture was measured, and the result is shown in Figure 2. Figure 2 shows that slump decreased for the mixture of only PF in concrete mixture. As the well-known theory, slump of concrete is related with yield stress of the mixture, and as the report of Tattersall et al. [13], adding fiber causes increasing yield stress with viscosity, especially, the yield stress is increased significantly [14]. Between the SF and PF, the biggest difference is rigidity of fiber. Therefore, it is considered that SF can be oriented and has relatively lower resistance of collapse of the mixture than PF because of its rigidity. Comparing with two mixtures with solely SF and PF, the concrete mixture with combined SF and PF showed a relatively favorable result of slump. Although the slump value of the mixture was slightly lower than the mixture with SF, still the slump value of the mixture with combined SF and PF showed a similar value to the slump value of the mixture with SF, and the value (130 mm) was higher than the average slump value (100 mm) of the mixture with single fiber.

The air contents of concrete mixtures were measured and are shown in Figure 3. Generally, all three cases satisfied the target air content range. As shown in the figure, the concrete mixture with SF showed the highest air content while the concrete mixture with PF showed the lowest air content. According to the research of Balaguru and Shah [15], the steel fiber contributes to the increase in air content while the polymeric fiber does not influence the air content. It is also considered as a result of different rigidity of fibers, and the test results agreed with the reference. For the concrete mixture with combined fibers of SF and PF, the air content was relatively close to the mixture with SF. From this result, it can be stated that SF influences more than PF on air content of the concrete mixture. Hence, the air content of the mixture with both SF and PF showed 9 % higher air content than the average air content of the mixtures with each SF and PF. Therefore, summarizing the fresh-state test results, combining SF and PF is having the properties of the dominated type of the fiber and it is mainly influenced by the rigidity of fiber. Furthermore, eventually, for slump and air

TABLE 5: Mix proportions.

Name	w/b	Fiber content (%)	S/a	SP (%/b)	Unit weight (kg/m³)							
					W	C	FA	BS	S	G	SF	PF
SF					220	293	44	88	829	693	79	0
PF	0.50	1.0	0.55	0.7	220	293	44	88	829	693	0	11
SF + PF					220	293	44	88	829	693	39	5

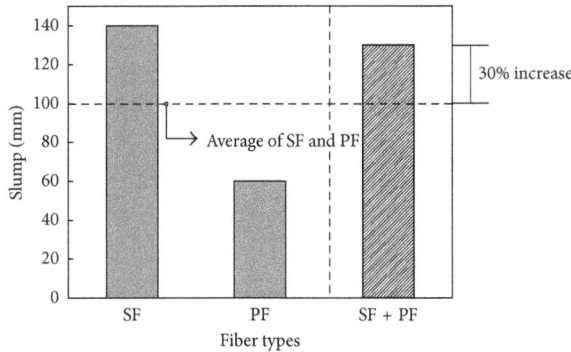

FIGURE 2: Influence of fiber combination on slump of concrete mixtures (the dotted line expresses the average value of SF and PF).

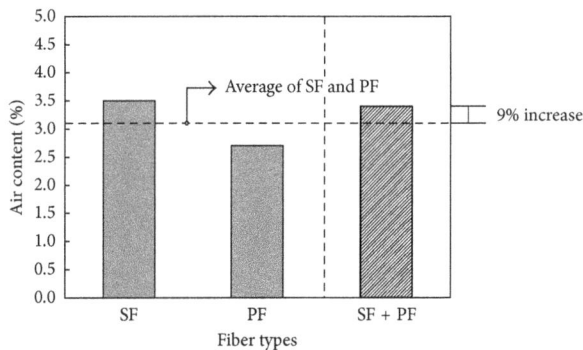

FIGURE 3: Influence of fiber combination on air content of concrete mixtures (the dotted line expresses the average value of SF and PF).

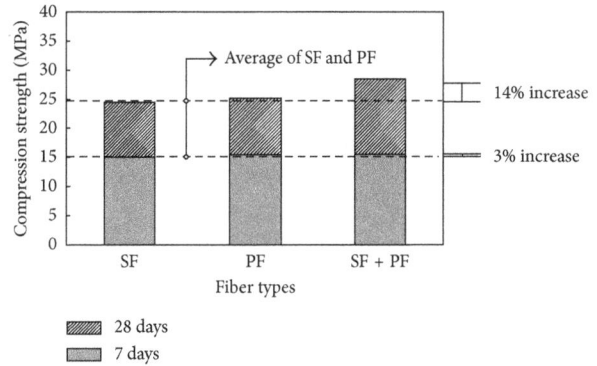

FIGURE 4: Influence of fiber combination on compressive strength of concrete mixtures (the dotted line expresses the average value of SF and PF).

content of concrete mixture, combining SF and PF shows favorable results on fresh-state concrete performances.

3.2. Mechanical Properties. To evaluate the mechanical properties of the concrete mixtures with various fiber conditions, compressive, flexural, and tensile strengths were measured. The compressive strength measurement results are shown in Figure 4. From the compressive strength results, it can be state that the different types of fiber do not influence on the compressive strength of the concrete mixture, in this research scope. However, when two types of fiber are combined, improved compressive strength can be obtained. At seven-day age, the average value of compressive strengths of the mixtures including each SF and PF was 15.3 MPa, and the compressive strength of the mixture with combined fiber of SF and PF was 15.8 MPa. Although there was no significant improvement at seven-day age, at 28-day age, the compressive strength of the mixture with combined fiber of SF and PF showed 28.7 MPa, approximately 14 % higher than the average value of each mixture. This improved

compressive strength with combined fiber is considered as a result of synergetic effect of different types of fiber for controlling cracks and confining the sample, and at 28-day ages, the increasing compressive strength due to the fiber reinforcement was maximized with hardened cementitious matrix.

From the flexural strength test result, the synergetic effect of combined fiber was also shown. As shown in Figure 5, the concrete mixture with SF showed relatively higher flexural strength than the concrete mixture with PF at both seven-day and 28-day ages. In spite of this different performance of the mixtures with different fibers, the concrete mixture with combined fibers showed improved flexural strength values at both seven-day and 28-day ages. Also, similar to the compressive strength result, the improvement of flexural strength at 28-day age was higher than the improvement of it at seven-day age with approximately 29 and 19 %, respectively.

As shown in Figure 6, unlike compressive and flexural strength results, from the tensile strength test result, at seven-day age, the concrete mixture with PF was slightly higher than the concrete mixture with SF. Generally, steel fiber is difficult to be broken by tensile forces but pulled out, while polymeric fiber is easily broken by tensile forces. Therefore, at seven-day age, the concrete with SF experienced pulling out of the fiber rather than the breaking of fiber, and the improvement of tensile strength of the mixture with SF between seven days and 28 days was higher than that of the mixture with PF. Since the main factor of resistance against tensile forces was pulling rather than breaking of the fiber for concrete mixture with PF, there was less improvement between seven days and 28 days. In spite of this different trend of the concrete mixtures with single-type fiber, the concrete mixture with combined fiber showed similar trend with the compressive and flexural strengths'

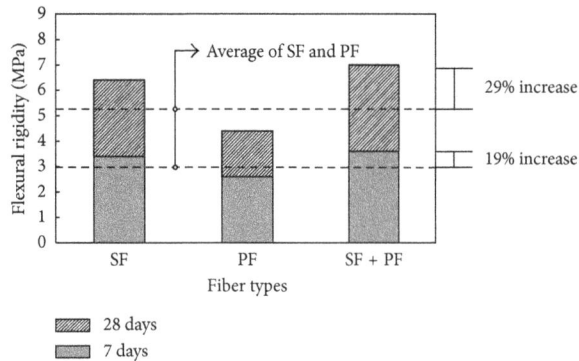

FIGURE 5: Influence of fiber combination on flexural strength of concrete mixtures (the dotted line expresses the average value of SF and PF).

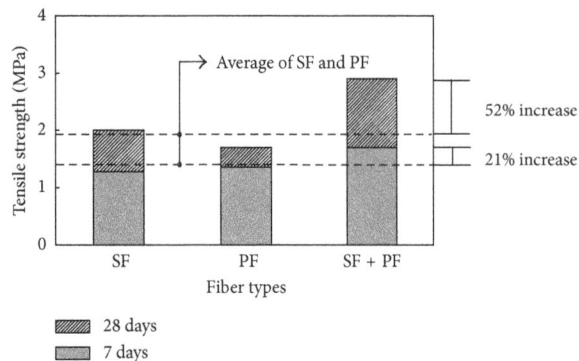

FIGURE 6: Influence of fiber combination on tensile strength of concrete mixtures (the dotted line expresses the average value of SF and PF).

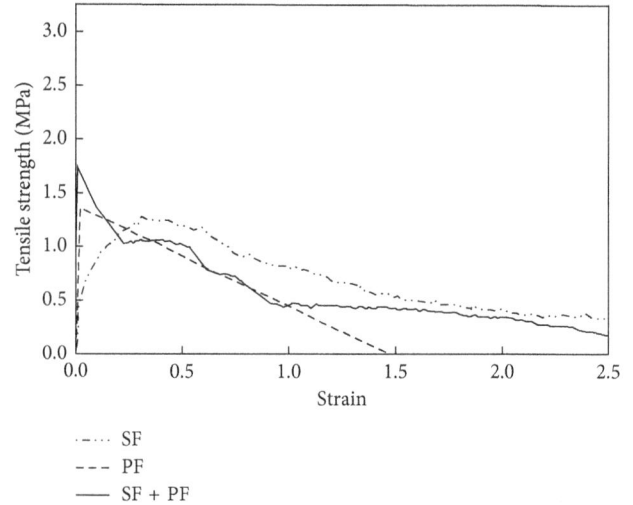

FIGURE 7: Influence of fiber combination on toughness of concrete mixture at seven-day age.

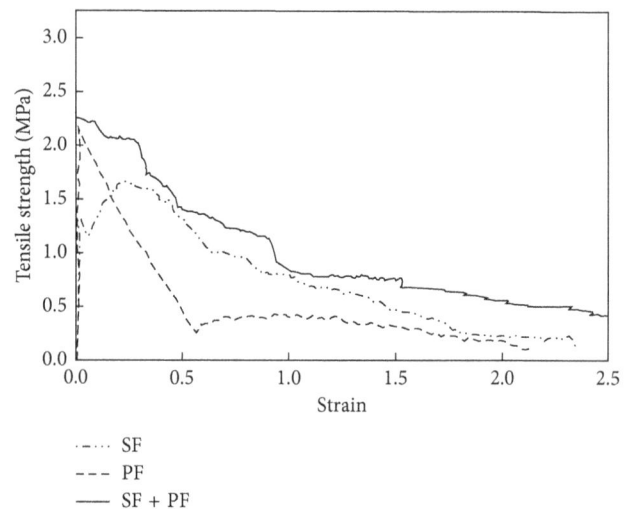

FIGURE 8: Influence of fiber combination on toughness of concrete mixture at 28-day age.

results of improving performance. At seven-day age, the tensile strength of the concrete mixture with combined fiber showed approximately 21 % higher than the average value of the mixtures with each SF and PF, and approximately 52 % higher than the average value of the mixtures with each SF and PF. Based on the tensile forces, strain-tensile strength relations were obtained and are shown in Figures 7 and 8 for seven-day and 28-day ages, respectively. From the results, generally, the concrete mixture with SF showed dropped tensile strength at very early stage, but after this, strain-hardening behavior was shown and continuously absorbed energy. On the contrary, in the case of the concrete mixture with PF, initially relatively higher strength was shown rather than the mixture with SF, but after the yield point, steep decrease of strength was observed. Hence, based on these results, it can be stated that some of SF could be pulled out at initial tensile forces, while the remaining fibers could be resisted and could absorb the energy while PF was hard to be pulled out by tensile forces, but it could be broken when the tensile forces were applied. Furthermore, at 28-day, the cementitious matrix became hardened and it grabbed the fibers hard to prevent pulling of fibers (by comparing tensile strength of all the cases in Figures 7 and 8). For the mixture with combined fiber, especially, synergetic effect was shown at both ages: high tensile strength at early age with PF and

high toughness and energy absorption at later period with SF. Therefore, based on the mechanical properties of fiber-reinforced concrete mixtures, it is considered that the reinforcement with combined fibers has better performance than the reinforcement with single fiber. Furthermore, considering with fresh-state test results, it can be suggested that using combined fiber of different types can achieve improved mechanical properties with favorable workability of the concrete rather than using single type of fiber.

3.3. Protecting Performance. In this paper, steel fiber was applied into a concrete protection wall to ensure the safety of workers in the explosion of Freon gas used in cosmetic raw material companies. The gas tank used is located away from the wall as shown in the magnified area in Figure 9. It does not directly affect the workers by the gas explosion when the gas explosion occurs, but a lot of piping and equipment

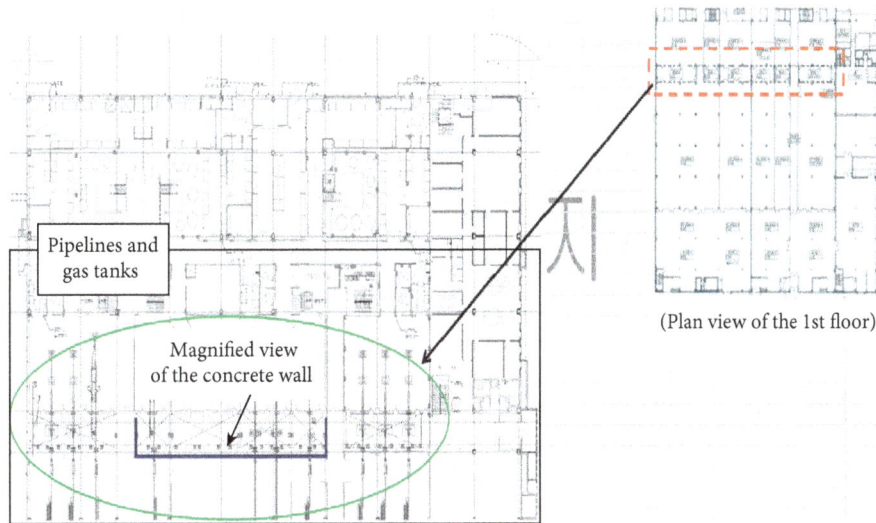

FIGURE 9: Plan view of the protection wall against flying debris due to explosion.

between the gas tank and the concrete wall are able to be affected at explosion due to gas explosion. Numerous debris from the explosion can impact the concrete, and scabbing or perforation on the other side of wall can damage workers on the other side of the concrete wall by an impact. As a result, the projectile test was performed to study the effect of debris damage for workers on the other side of wall.

Using the high-velocity projectile, the influence of different fiber combination on the protection performance of concrete was evaluated. Since there was no standardized measurement method, the concrete panels after the impact test were observed on the impacted face and the back side. As shown in Figure 10, the concrete panels including a single fiber type showed penetrated damage. Comparing the impacted side, for both concrete panels with single-type fiber, the back side suffered spalling out of the concrete while the impacted surface has a hole with the size of projectile. However, in the case of the concrete panel with combined fibers, clearly improved protecting performance was shown rather than two concrete panels with SF and PF. In the case of the concrete panel with combined fiber, on the impacted side, relatively wide damage was shown instead of penetration. Since the penetration of projectile was prevented, cracking from the center of the panel was occurred without loss of concrete. Hence, it is considered that because of high pulling resistance and toughness by PF and SF, respectively, in the concrete panel with combined fiber, the impact resistance is improved and efficiently absorbed the impact energy. Summarizing the mechanical properties and protecting performance of the concrete samples with various fiber conditions, the combined fiber of SF and PF showed relatively improved mechanical properties and protecting performance with the synergetic effect from combined fiber.

4. Field Applications

4.1. Field Conditions for Site and Mixture. The target building was the chemical plant for manufacturing cosmetics. Because of the process of manufacturing the cosmetics, high-

pressured gas should be injected, and the special area for this process should be protected against unexpected explosion or flying debris caused by the explosion. The main structure was reinforced concrete structure. In this research, the suggested HPFRCC was applied on the outer wall of the protected area as shown in Figure 11. The applied amount of HPFRCC was approximately $50\,m^3$ for 3 m depth of the protecting wall. The concrete mixture delivered by trucks with an agitator was placed using the ready-mixed concrete system from mixing at plant to placing by pump. The target concrete mixture was 25 MPa of target compressive strength (at 28 days), and 150 mm of target slump. Unlike the laboratory test, the field applied concrete mixture contained a maximum size 25 mm of coarse aggregate. For improving workability, combined fiber of SF to PF of 1 : 1 was replaced 1% of entire volume of the mixture. These levels of experimental tests were shown in Table 6.

The concrete mixture was mixed in the ready-mixed concrete plant with the central mix method. However, since the ready-mixed concrete plant should be used, the plant did not have the setting for fiber introduction, and thus, the fiber was introduced manually through the materials feeding entrance for the premeasured amount. The mixing time for HPFRCC was approximately one to two minutes instead of 30 to 40 seconds of normal concrete to provide sufficient workability and dispersion of fibers. Other processes of delivering and placing of concrete were conducted with agitator trucks and pump truck by following South Korean ready-mixed concrete standard of KSF 4009 [16]. For the ready-mixed concrete, cement with a density of $3.15\,g/cm^3$ was used. River and crushed sand were mixed in a proportion of 50 : 50 for fine aggregates, whereas coarse aggregates were a mixture of 5 to 10 mm and 10 to 25 mm aggregates in a proportion of 35 : 65.

4.2. Test Methods. To evaluate the properties of the mixtures for the actual field condition, slump and slump flow tests for workability, air content, and compressive strength for

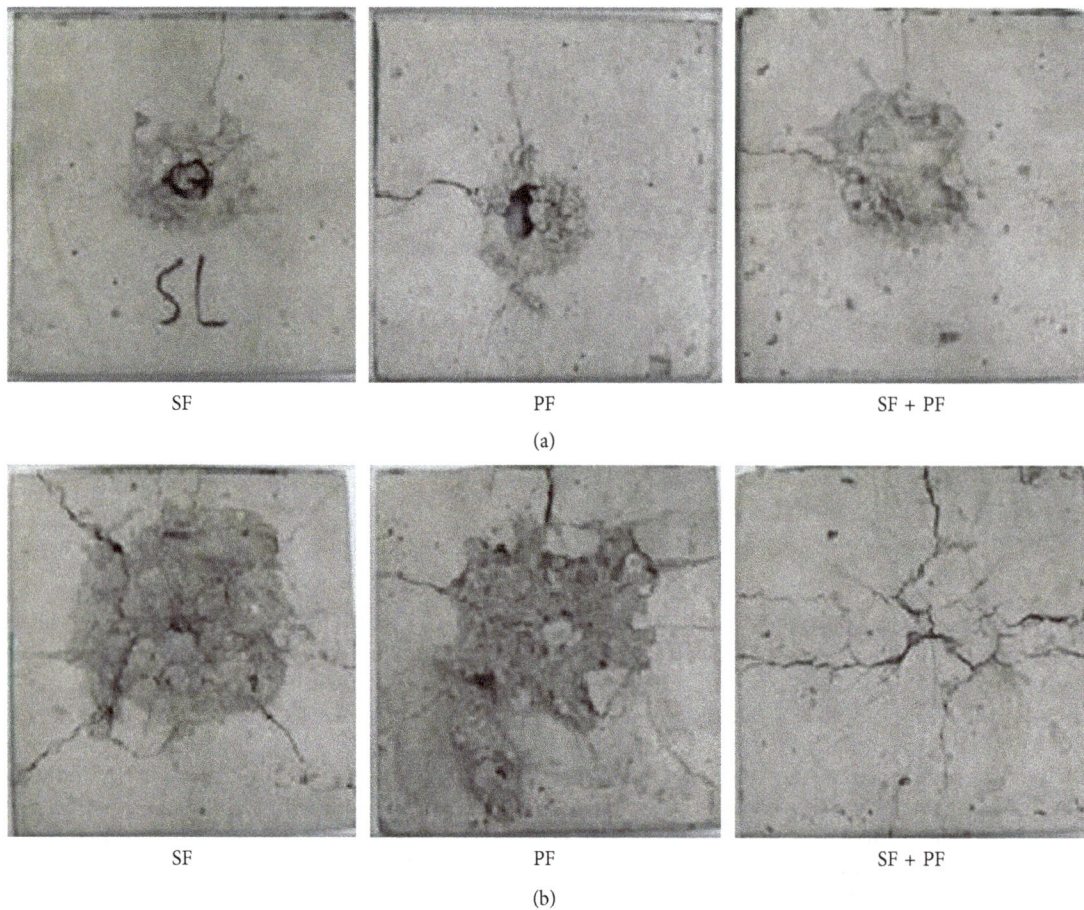

FIGURE 10: Influence of fiber combination on protecting performance of the concrete panel against the impact test. (a) Impacted side (front). (b) Back side.

FIGURE 11: (a) Field application architecture perspective and (b) cross-sectional view.

TABLE 6: Field test plan.

Experimental items		Level of experiment
Mixture	Ready-mixed concrete specification	25-24-150
	Fiber mixing ratio (%)	1.0
	Combination of fibers	SF + PF

mechanical properties were measured. The testing samples were obtained from the first and third agitator trucks arrived in sites such as in Table 6 and tested samples were obtained before and after the pumping. Each test was conducted following ASTM C143, C1611, C231, and C39 methods for slump, slump flow, air content, and compressive strength,

respectively. The compressive strength was tested at seven- and 28-day ages.

4.3. Result of Field Applications. First, from the slump and slump flow test results as shown in Table 7, respectively, the fluidity of the concrete mixture was increased after the pumping. Generally workability of concrete is decreased in slump or flow after the pumping. However, in this research with fiber-reinforced concrete, it is considered that the fibers in concrete mixture were oriented by the pressure of the pumping and it contributed to the improved fluidity of the fiber-reinforced concrete mixture. In spite of improved fluidity of the mixture, air content of concrete was decreased. It is similar trend of already reported results of studies.

TABLE 7: Experiment result of concrete.

Division		Slump (mm)	Slump flow (mm)	Air contents (%)	Compression strength (MPa)	
					7 days	28 days
First agitator truck arrived in site	Before pumping	130	225/220	4.0	20.9	30.9
	After pumping	170	240/300	3.6	23.8	32.7
Third agitator truck arrived in site	Before pumping	135	230/240	3.8	21.8	31.0
	After pumping	160	310/280	3.5	24.2	33.1

However, in general, the properties of fresh-state fiber-reinforced concrete mixture were acceptable to apply field construction, and there was no problem on placing process of the wall.

The field-processed HPFRCC's mechanical properties were evaluated with compressive strength. As shown in Table 7, all concrete samples showed over 30 MPa and it was higher than the target compressive strength of 25 MPa. For the concrete mixture obtained after the pumping, slightly increased compressive strength was observed. It can be stated that decreased air content and well-oriented fiber can contribute to the improved compressive strength. For more detail, although it is necessary to study the relation between pumping and performance of HPFRCC, in this research, the goal of the experiment was evaluating field applicability of HPFRCC, thus it is not discussed in this paper.

5. Conclusion

In this research with a goal of applying HPFRCC on field conditions, the workability, mechanical properties, and protecting performance of combined fiber-reinforced concrete mixtures were evaluated, and field application was conducted with a ready-mixed concrete system. According to a series of experiment, some conclusions can be obtained as follows:

(1) By using combined fiber of SF and PF, fresh-state properties of HPFRCC were improved rather than the case with the unfavorable result with a single fiber and showed better performances than the averaged value of each single-type fiber-reinforced mixture.

(2) For mechanical properties of compressive, flexural, and tensile strengths, the mixture with combined fiber showed improved values rather than any single-type fiber-reinforced mixtures.

(3) Regarding the protection performance against flying debris, the HPFRCC panel reinforced by combined fiber showed the most desirable performance of protecting the high-velocity projectile.

(4) The combined HPFRCC showed improved mechanical and protecting performances with favorable workability. Based on these improved features of combined fiber reinforcement, field application of combined HPFRCC was successful under the ready-mixed concrete system including agitators, delivering, and placing.

Conflicts of Interest

The author declares no conflicts of interest.

Acknowledgments

This research was supported by the Basic Science Research Program through the NRF funded under the grant NRF-2018R1C1B5045860.

References

[1] P. R. Tadepalli, Y. L. Mo, T. T. Hsu, and J. Vogel, "Mechanical properties of steel fiber reinforced concrete beams," in *Proceedings of Structures Congress 2009: Don't Mess with Structural Engineers: Expanding Our Role*, pp. 1–10, 2009.

[2] Z. Bayasi, "Development and mechanical characterization of carbon fiber reinforced cement composites and mechanical properties and structural applications of steel fiber reinforced concrete," Ph.D. dissertation, vol. 1, pp. 1–199, Michigan State University, 1989.

[3] D. R. Lankard, "Slurry infiltrated fiber concrete (SIFCON): properties and applications," *MRS Online Proceedings Library Archive*, vol. 42, 1984.

[4] A. E. Naaman and J. R. Homrich, "Tensile stress-strain properties of SIFCON," *ACI Materials Journal*, vol. 86, no. 3, pp. 244–251, 1989.

[5] N. Banthia, F. Majdzadeh, J. Wu, and V. Bindiganavile, "Fiber synergy in hybrid fiber reinforced concrete (HYFRC) in flexure and direct shear," *Cement and Concrete Composites*, vol. 48, pp. 91–97, 2014.

[6] I. Markovic, J. C. Walraven, and J. G. M. Van Mier, "Tensile behaviour of high performance hybrid fibre concrete," in *Proceedings of 5th International Symposium on Fracture Mechanics of Concrete and Concrete Structures*, pp. 1113–1121, 2004.

[7] G. F. Peng, W. W. Yang, J. Zhao, Y. F. Liu, S. H. Bian, and L. H. Zhao, "Explosive spalling and residual mechanical properties of fiber-toughened high-performance concrete subjected to high temperatures," *Cement and Concrete Research*, vol. 36, no. 4, pp. 723–727, 2006.

[8] B. S. Mohammed, M. F. Nuruddin, M. Aswin, N. Mahamood, and H. Al-Mattarneh, "Structural behavior of reinforced self-compacted engineered cementitious composite beams," *Advances in Materials Science and Engineering*, vol. 2016, Article ID 5615124, 12 pages, 2016.

[9] N. Bouzoubaa and M. Lachemi, "Self-compacting concrete incorporating high volumes of class F fly ash: preliminary results," *Cement and Concrete Research*, vol. 31, no. 3, pp. 413–420, 2001.

[10] M. A. Yusof, N. Norazman, A. Ariffin, F. M. Zain, R. Risby, and C. P. Ng, "Normal strength steel fiber reinforced concrete

subjected to explosive loading," *International Journal of Sustainable Construction Engineering and Technology*, vol. 1, no. 2, pp. 127–136, 2011.

[11] Z. Li, L. Wang, and X. Wang, "Compressive and flexural properties of hemp fiber reinforced concrete," *Fibers and Polymers*, vol. 5, no. 3, pp. 187–197, 2004.

[12] JSCE-E 531, *Test Method for Tensile Properties of Continuous Fiber Reinforcing Materials*, JSCE, Japan, 1995.

[13] G. H. Tattersall and P. F. Banfill, *The Rheology of Fresh Concrete*, Vol. 759, Pitman, London, 1983.

[14] A. W. Saak, H. M. Jennings, and S. P. Shah, "A generalized approach for the determination of yield stress by slump and slump flow," *Cement and Concrete Research*, vol. 34, no. 3, pp. 363–371, 2004.

[15] P. N. Balaguru and S. P. Shah, *Fiber-Reinforced Cement Composites*, Mc Graw Hill, New York, NY, USA, 1992.

[16] KS F 4009, *Ready-Mixed Concrete*, KS, Seoul, South Korea, 2016.

[17] ASTM C143, *Standard Specification for Portland Cement Standard Test Method for Slump of Hydraulic-Cement Concrete*, ASTM International, ASTM International, West Conshohocken, PA, USA.

[18] ASTM C1611, *Standard Test Method for Slump Flow of Self-Consolidating Concrete*, ASTM International, West Conshohocken, PA, USA.

[19] ASTM C231, *Standard Test Method for Air Content of Freshly Mixed Concrete by the Pressure Method*, ASTM International, West Conshohocken, PA, USA.

[20] ASTM C39, *Standard Test Method for Compressive Strength of Cylindrical Concrete Specimens*, ASTM International, West Conshohocken, PA, USA.

5

Prediction of Compressive Strength of Concrete in Wet-Dry Environment by BP Artificial Neural Networks

Chengyao Liang [ID],[1,2] **Chunxiang Qian** [ID],[1,2] **Huaicheng Chen** [ID],[1,2] **and Wence Kang**[1,2]

[1]*School of Material Science and Engineering, Southeast University, Nanjing 211189, China*
[2]*Research Institute of Green Construction Materials, Nanjing 211189, China*

Correspondence should be addressed to Chunxiang Qian; cxqian@seu.edu.cn

Academic Editor: Patrice Berthod

Engineering structure degradation in the marine environment, especially the tidal zone and splash zone, is serious. The compressive strength of concrete exposed to the wet-dry cycle is investigated in this study. Several significant influencing factors of compressive strength of concrete in the wet-dry environment are selected. Then, the database of compressive strength influencing factors is established from vast literature after a statistical analysis of those data. Backpropagation artificial neural networks (BP-ANNs) are applied to establish a multifactorial model to predict the compressive strength of concrete in the wet-dry exposure environment. Furthermore, experiments are done to verify the generalization of the BP-ANN model. This model turns out to give a high accuracy and statistical analysis to confirm some rules in marine concrete mix and exposure. In general, this model is practical to predict the concrete mechanical performance.

1. Introduction

Marine environment tends to have a negative effect on concrete structures' performance, which has been investigated in many researches. Compressive strength is used to describe the mechanical performance of concrete. In the marine environment, especially tidal and splash zones, concrete structure degradation is quite serious and the compressive strength of that is descending with exposure age. It is investigated in either laboratory or actual marine condition that compressive strength degrades with time.

In the early 1980s, British academic Mangat and Gurusamy [1] did a research on mechanical properties of steel fiber-reinforced concrete exposed to the marine splash zone and tidal zone in the Aberdeen beach, and the exposure age is up to three years (2000 wet-dry cycles). Results indicated that melt extract fibers are suitable for marine applications. In addition, another actual marine exposure experiment was done by Kuhail and Shihada [2] in the Gaza beach for a period of seven months. And it was found that the compressive strength of concrete shows a trend of rising early but declining later. Toutanji et al. [3] focused on studying the effect of different supplementary

cementitious materials on strength and durability of concrete. They found that proper mineral additives could improve the performance of concrete in the marine environment. Aye and Oguchi [4] investigated the effect of physical sulfate attack on the performance of plain and blended cement mortars. Specimens were exposed to 10% Na_2SO_4 and $MgSO_4$ solutions for 24 months. $MgSO_4$ was found to be more damaging than Na_2SO_4 considering chemical attack. However, Na_2SO_4 was more harmful than $MgSO_4$ as far as the physical attack was considered. Jiang and Niu [5] also did a research on the effect of different types of sulfate solutions on concrete performance under wet-dry cycles. Results show that the deterioration degree of concrete in magnesium sulfate solution is more severe than that in the other sulfate solutions. Chloride ions in the composite solution help decrease the deterioration rate of concrete effectively. Chen [6] established a constitutive model for concrete in wet-dry cyclic sulfate attack.

In this research, compressive strength of concrete served in the marine tidal zone and splash zone or exposed to the wet-dry cycle environment was focused on. Database is established from existing researches. An artificial neural network (ANN) is applied in this study. The ANN has been widely adopted in construction material property prediction by many researches [8, 9]. For example, Tavakoli et al. [10] have predicted the energy absorption capability of fiber-reinforced self-compacting concrete which contains nanosilica particles via an MLP- (multilayer perceptron-) type artificial neural network. Tavakoli et al. [11] simultaneously researched the mechanical properties of self-compacting concrete with nanosilica particles and various fibers via the MLP artificial neural network. In addition to that, many scholars utilized the ANN for concrete compressive strength prediction. Ni and Wang [12] utilized multilayer feed-forward neural networks (MFNNs) to predict the 28-day compressive strength of concrete, and the results conformed to some rules on mix of concrete. Lee [13] has developed the I-PreConS (Intelligent PREdiction system of CONcrete Strength) that provides in-place strength information of the concrete to facilitate concrete form removal and scheduling for construction based on the ANN. Alshihri et al. [14] have done a comparison on feed-forward backpropagation (BP) and cascade correlation (CC); CC is slightly better than BP, while both of them have good performance on light-concrete strength prediction. Öztaş et al. [15] have done a research on predicting compressive strength and slump of high-strength concrete. And the 187 sets of data used to establish a model come from literature. Except for that, Gaussian process regression (GPR) was applied by Hoang et al. [16], and it can well estimate the HPC strength. Nevertheless, those researches are mostly based on experimental data. Besides, these literature researches usually do not have a large database.

The backpropagation (BP) artificial neural network is selected in this research, for it has a mature application in various fields. The BP-ANN model is firstly put forward by Rumelhart and McClelland in 1986 [17]. Based on the gradient descent algorithm, the BP neural network is an error backpropagation multiple-layer feed-forward network, focused on calculating the minimum of the mean square error of actual outputs of the network and the target outputs. Compared with the former multilayer perceptron, the BP neural model is capable of dealing with a complex nonlinear problem. Besides, compared with multilayer feed-forward neural networks (MFNNs) [18], it is equipped with a better ability of classifying random complex models [19]. Moreover, good multidimensional function nonlinear versatile mapping capability and the flexible multilayer network structure are other two advantages of the BP neural model. After model development, experiments were done to validate the generalization of the prediction model.

2. Data Collection and Analysis

In order to establish a prediction model that can be widely used, the data were excerpted from large numbers of experiments carried out by researchers. 2167 sets of data were collected in all, which cover over 80 articles from China, United States, Europe, and other regions, and they are presented in Table 1 in detail.

2.1. Database Establishment. In this research, compressive strength of concrete in the wet-dry cycle exposure environment, as the target of prediction, is related to many

TABLE 1: Source of data collected.

Regions	China	USA	Europe	Others
Articles	17	19	23	27

factors, including material factors and environmental factors. Material factors include w/c, specimen sizes, initial strength, and fly ash dosage and slag dosage. Specific surface area is chosen to describe the specimen sizes. Environmental factors are various ion concentrations, including sodium, magnesium, chloride, and sulfate ions, exposure condition, and exposure age. Exposure condition contains five variables, including the wetting time, drying time, wetting temperature, drying temperature, and cycle period.

The detailed data of the factors mentioned above are excerpted from previous researches and are functioned as the database of the artificial neural network model.

2.2. Database Analysis

2.2.1. Statistical Analysis. Before establishing the model, a statistical analysis has been done on descriptive statistics of the database [7]. Statistical analysis is a multiplex method that can be used to describe the regularity and distribution of the sample database, and the fluctuation of database can be seen according to some statistics such as variance, standard deviation, skewness, and kurtosis [20]. Table 2 demonstrates some statistics of 1078 sets of data of factors influencing the compressive strength of concrete in the wet-dry environment. And those data were selected randomly from 2169 sets of data.

Table 2 mainly demonstrates six statistics of influencing factors. They are, respectively, standard deviation, variance, skewness, kurtosis, minimum, and maximum. Standard deviation is used to reflect the fluctuation of a series of data, and the instability of a statistic is measured by its variance [21]. In environmental factors, data of four types of solution ion concentrations share the similar discrete degree, so do the dosage of fly ash and slag in material factors. And compared to other factors, especially exposure age, variance and standard deviation of those six factors' data are much lower because almost the choice of ion concentration refers to seawater ion concentrations and the dosage of fly ash and slag is mainly concentrated at 0.3, which is considered as the optimum content. Nevertheless, the exposure age ranges from 2 days to several years, contributing to the high variance and standard deviation of exposure age. On top of this, the degree of dispersion and variation of initial strength is close to that of final strength.

Skewness and kurtosis are used to characterize the distribution of data [22, 23]. Normal distribution is the most common distribution. Skewness refers to the frequency distribution of asymmetric degree of skew direction. Deviation between normal distributions is often reflected by the coefficient of skewness. When the coefficient of skewness is higher than 0, the oblique direction of the distribution is right (positive); on the contrary, the oblique distribution is left (negative). Right oblique direction distribution has a thin

TABLE 2: Statistical analysis of 1078 sets of data.

1078 in total	Specific surface area	Initial strength	FA dosage	Slag dosage	Sodium ions	Magnesium ions	Chloride ions	Sulfate ions	Exposure age	Finial strength
Standard deviation	12.52	17.44	0.13	0.16	0.68	0.15	0.49	0.35	613.39	18.95
Variance	156.80	304.14	0.02	0.03	0.46	0.02	0.24	0.12	376251.72	359.12
Skewness	2.66	1.87	1.57	1.76	1.36	4.62	2.97	1.37	5.83	1.10
Kurtosis	9.17	8.32	2.28	1.75	1.91	20.49	13.85	0.77	37.65	4.55
Minimum	33.33	21.6	0.00	0.00	0.00	0.00	0.00	0.00	2.00	13.46
Maximum	120.00	163	0.60	0.80	4.42	0.83	4.42	1.37	5475.00	168.60

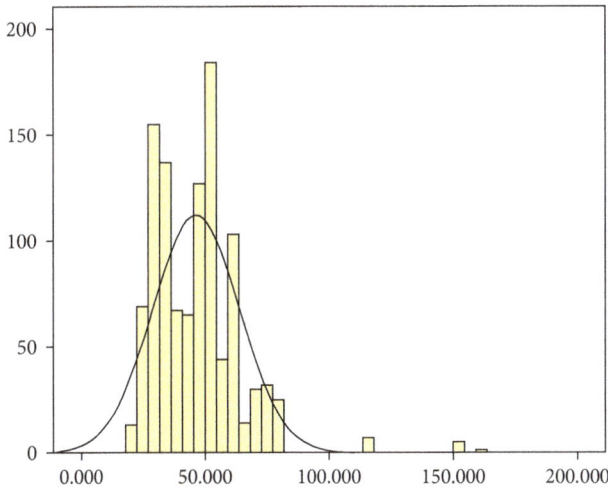

FIGURE 1: Distribution of initial strength.

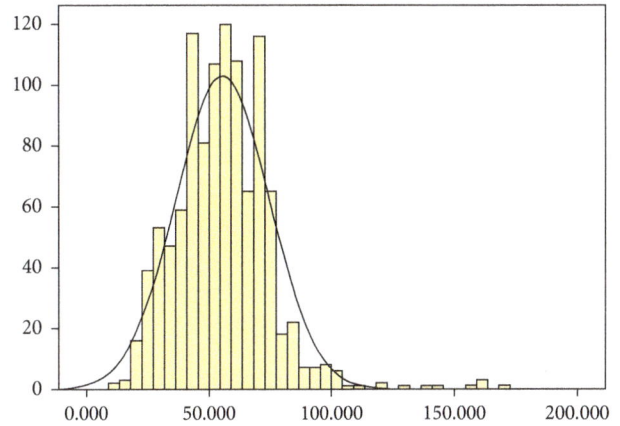

FIGURE 2: Distribution of final strength.

and long tail on the left, which means that the data mainly focused on the small figures range. And taking a general observation of the data distribution of the factors, all of them are right oblique direction distributed. The coefficient of kurtosis indicates the shape of data distribution. It can be divided into spire distribution, standard distribution, and flat distribution. When the coefficient of kurtosis is higher than 3, it belongs to spire distribution; conversely, it belongs to flat distribution. Throughout the statistics given above, it can be concluded that except for exposure age, the other factors' data are within normal limit. The coefficient of skewness of variables is all above 0, and it means that all the variables' distribution directions show a positive oblique trend. Nevertheless, according to the results of coefficient of kurtosis, the four variables, respectively, fly ash dosage, slag dosage, sodium ions, and sulfate ions, show the spire distribution and the five variables, respectively, specific surface area, initial strength, magnesium ions, chloride ions, and final strength, show the flat distribution. And due to the particularity of the distribution of exposure age, it belongs to neither spire nor flat distribution. And initial strength shares a similar distribution with the final strength. Figures 1 and 2 describe the distribution of initial strength and final strength. Both of them are close to normal distribution.

In Table 3, the strength range is divided into 7 intervals and the distribution of values of initial strength and final strength is demonstrated. It can be clearly seen that the samples whose values of initial strength are between 25 and

50 MPa account for the largest proportion. So are the final strength of samples. Moreover, there is a decrease in the numbers of final strength samples of high-strength concrete compared with that of initial strength samples of high-strength concrete.

2.2.2. Correlation Analysis. On the basis of statistical analysis, correlation between final strength and various influencing factors is made in this research. The Pearson correlation coefficient, Kendall coefficient, and Spearman coefficient are commonly used. Nevertheless, the Pearson correlation coefficient is used in linear relationships [24]. The Kendall coefficient and Spearman coefficient belong to nonparametric statistics [25]. In this study, Kendall and Spearman coefficients are used for a correlation analysis. The analysis results are demonstrated in Table 4.

The correlation coefficients calculated above are based on two different formulas, both of which can be adopted more widely in nonlinear relationships. The results are in absolute value. From the analysis results, it can be clearly seen that w/c and initial strength have a significant effect on the final strength. However, fly ash dosage is more effective than slag dosage. It is quite a complex relationship between mineral admixtures and compressive strength in the marine environment, and mineral admixtures are considered to have a positive influence on concrete properties. The difference between those two correlations is attributed to the fact that data collected focused more on fly ash dosage. Besides, seawater also plays a significant role in the evolution

TABLE 3: Sample numbers of initial and final strength distribution intervals.

Strength intervals (MPa)	0~25	25~50	50~75	75~100	100~125	125~150	>150
Sample numbers of initial strength	43	621	361	40	7	0	6
Sample numbers of final strength	83	588	405	2	0	0	0

TABLE 4: Kendall and Spearman correlation coefficients of influencing factors.

Correlation coefficient	Specific surface area	Initial strength	FA dosage	Slag dosage	Sodium ions	Magnesium ions	Chloride ions	Sulfate ions	Exposure age	w/c
Kendall	0.01	0.406**	0.172**	0.04	0.00	0.180**	0.01	0.00	0.078**	0.483**
Spearman	0.00	0.577**	0.226**	0.05	0.00	0.243**	0.02	0.00	0.126**	0.627**

**This factors has a quite high relevance with the final strength.

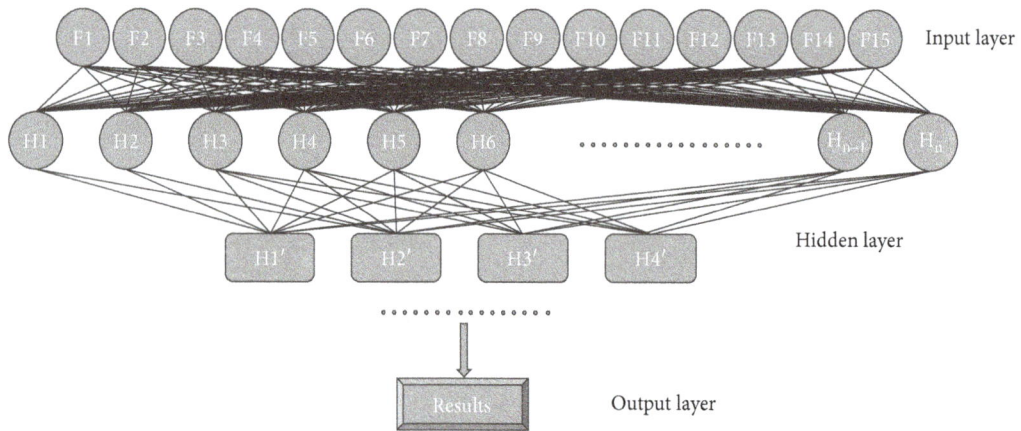

FIGURE 3: Backpropagation neural network abstract.

of compressive strength. Nevertheless, the correlation coefficients calculated seem hard to reflect that. Taking an observation of data collected, ion data are based on marine hydrographic data and almost have no change. Hence, correlation coefficient calculation formulas are built on data-changing trends [26]. Consequently, ion concentration seems to have a subtle effect.

In general, the correlation analysis results can be used as a simple reference for construction of artificial neural networks.

3. Construction of Artificial Neural Networks

Artificial neural network (ANN) model is a prediction model that has been widely applied in many fields. Similar to human brains, which respond to external stimulus through connections and exchanging information among ten billions neurons, the artificial neural network model is able to deal with the message inputted and realize result prediction [27]. One important training rule of the ANN is the delta rule, which is based on the idea of gradient descent. The delta rule is applied to determine the fraction of difference between the target and output [28].

The backpropagation (BP) neural network model is adopted in compressive strength in the research. The BP algorithm underpins the delta rule to neural nets with hidden nodes. The whole operation is demonstrated in Figure 3.

The operation can be divided into six steps [29]. Step 1: input training factors. Fifteen influencing factors are

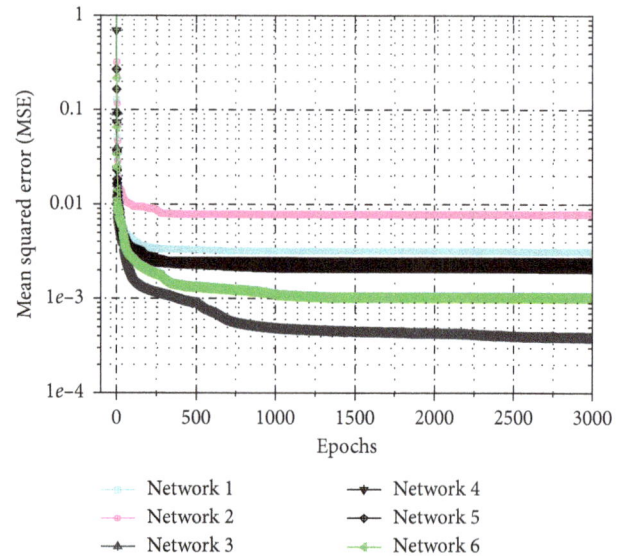

FIGURE 4: Comparison between six networks.

inputted into the model; Step 2: hidden nodes calculate the output. This is a quite complex process that detailed the calculated algorithm is invisible; Step 3: output nodes calculate outputs. Step 4: compare the outputs with targets and figure out the difference; Step 5: adjust the model parameters on the basis of the training rule using the results of Step 4

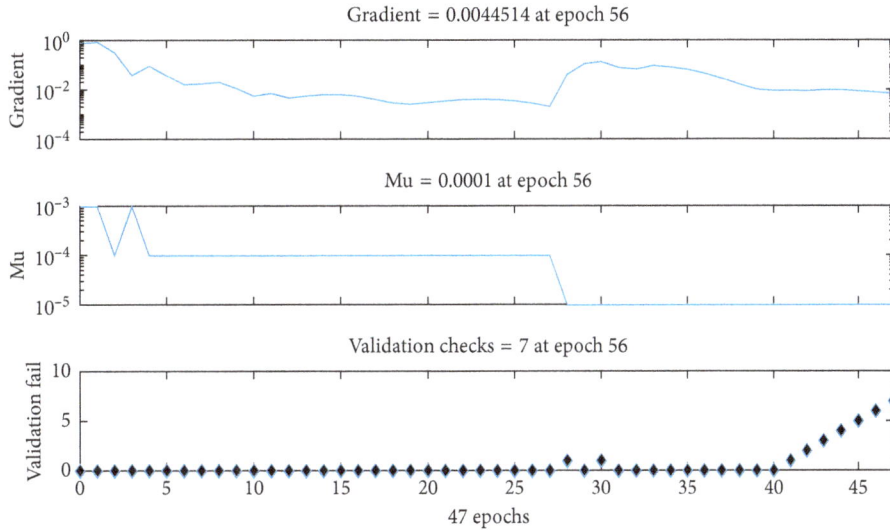

FIGURE 5: Significant controlling parameters changing with epochs.

and calculate every $d(n)$ in hidden nodes in this step; Step 6: the results of Step 5 are used to carry on a second training until the error is suitably small.

3.1. Determination of Input Neurons and Output Neurons and Pattern Classification.

As shown in Figure 3, fifteen inputs are given and they represent fifteen influencing parameters. 2169 sets of data are collected for the establishment of the BP neural network model. The datasets are categorized into three parts: training set, validation set, and testing set, respectively. There are various pattern classification methods. For example, Chen et al. [30] used 86% dataset for training and 14% dataset for testing to establish an ANN model to predict the strength of concrete. Hence, the conventional and widely recognized division ratio is 50% training set, 25% validation set, and 25% testing set [31]. Prechelt has done a research on some benchmark rules and problems of the neural network model establishment and come up with a basic pattern classification. The division ruler is following the convention in the literature of using half of the images for training and half for testing. And he proved that this pattern classification method can well avoid the overfitting of the model [32]. Hence, in this research, 50% of the data are used as the training set, 25% are used as the testing set, and the rest 25% are used as the validation set to predict the compressive strength of concrete that is exposed to the wet-dry environment.

3.2. Model Establishment.

The training and testing program is written and put into MATLAB. The inner network structure is discussed. Ash used the DNC method to select the proper node parameters [33]. DNC means dynamic node creation. DNC sequentially adds nodes one at a time to the hidden layer(s) of the network until the desired approximation accuracy is achieved. There are six networks and they are discussed in Figure 4. Detailed structure information is given below, and Figure 4 presents the training results.

Network 1: 15-25-10-1

Network 2: 15-15-5-1

Network 3:15-30-15-1

Network 4:15-30-1

Network 5:15-25-1

Network 6:15-20-1

From Figure 4, Network 3, 15-30-15-1, turns out to have the lowest mean square error and is applied into the BP neural network model. Because the mean squared error tends to be a plateau after 3000 epochs, the epochs are set as 3000 epochs in the parameter setting.

In this model, trainlm is a network training function that updates weight and bias values according to the Levenberg–Marquardt optimization. trainlm is often the fastest backpropagation algorithm in the toolbox and is highly recommended as a first-choice supervised algorithm, although it does require more memory than other algorithms. And the detailed trainlm training parameters are given as follows:

net.trainParam.show = 10

net.trainParam.lr = 0.05

net.trainParam.goal = $1e - 4$

net.trainParam.epochs = 3000

net.trainParam.max_fail = 7

In this model, performance goal $1e - 4$ is set as the threshold value. And the best validation performance is reached at the 49th epoch. The maximum validation failure is set as 7; thus, the model is calculated to the 56th epoch before ending. And the validation error reached the setting value. Significant parameters are demonstrated in Figure 5. After the 56th epoch run, the gradient reaches 0.0044514. And the decrease factor Mu is 0.0001 in this research. Validation vectors are used to stop training early if the network performance on the validation vectors fails to

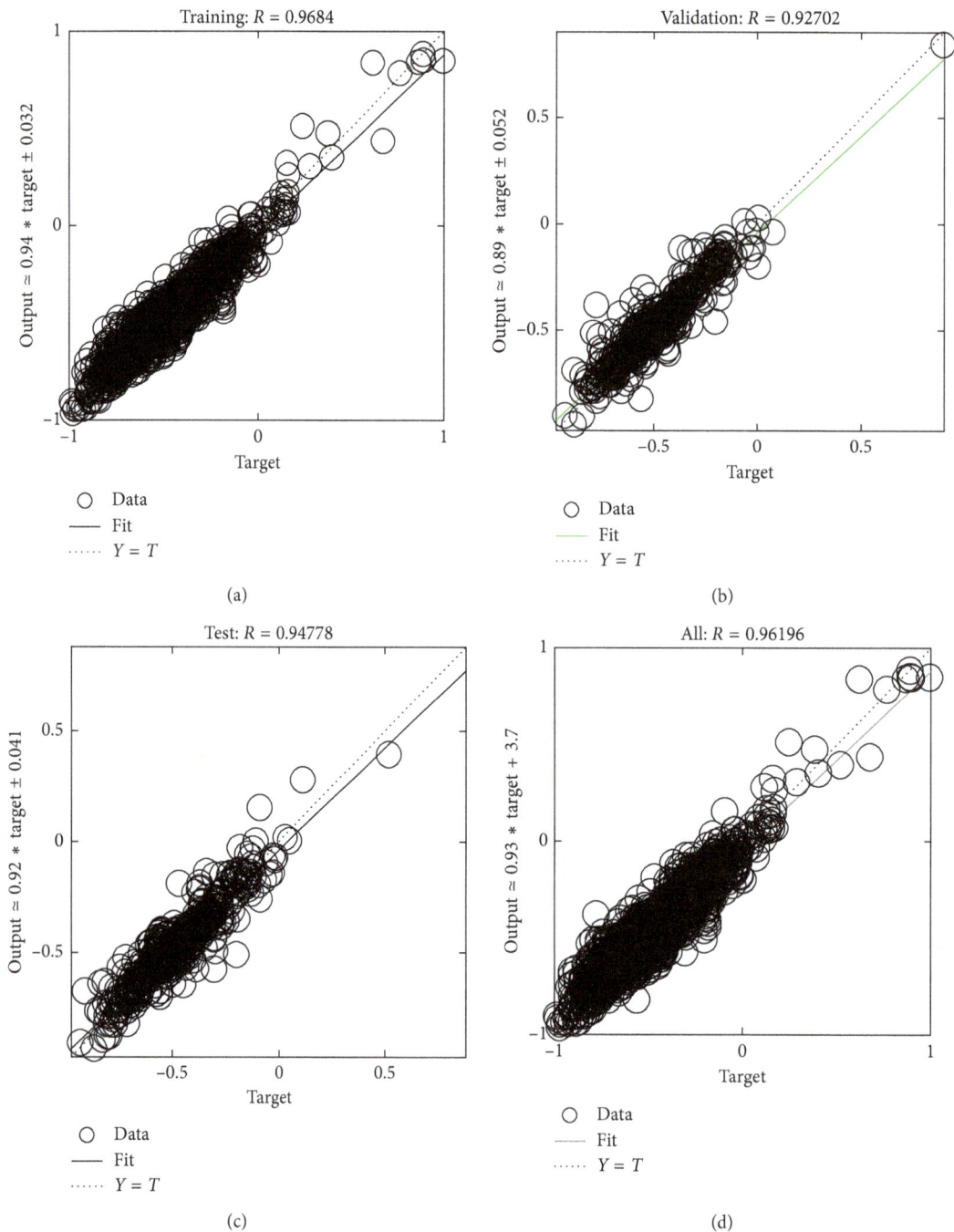

Figure 6: Results of training, validation, and testing.

improve or remains the same for max_fail epochs in a row. Test vectors are used as a further check that the network is generalizing well but does not have any effect on training.

3.3. Prediction Results.

After the establishment of this prediction model, the prediction results are demonstrated in Figure 6.

In the training set, the correlation coefficient R is 0.9684 (Figure 6), and the training model is quite appropriate. Then, the validation set is used to optimize the training set until it reaches the setting error value. And the final correlation coefficient of the validation set is 0.92702. After the prediction model establishment, testing set data is to test the performance of the neural network model. The correlation coefficient of that is 0.94778. The general prediction is listed in Figure 7 ($R = 0.962$). And this neural network model achieves a good prediction of 2169 sets of data.

In addition, the three error indexes, RMSE, MAPE, and MSE, of this model, respectively, are calculated in Table 5.

The MAE of training data, validation data, and testing data was all less than 4.05 MPa, the MAPE of training data, validation data, and testing data was all less than 5.5%, and

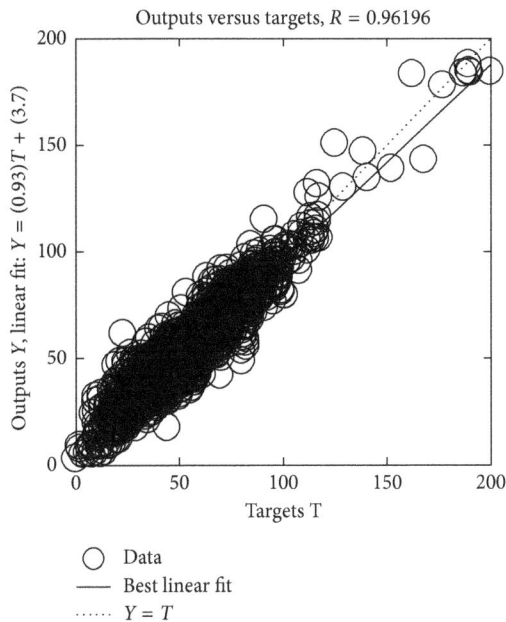

FIGURE 7: Fitting curve of outputs and targets.

TABLE 5: Prediction performance of different patterns of this model.

	R^2	MAE (MPa)	MAPE (%)	RMSE (MPa)
Training database	0.9684	2.47	3.52	2.51
Validation database	0.9270	4.02	5.13	4.13
Testing database	0.9478	3.87	4.97	4.04
All databases	0.9620	2.65	3.42	3.15

the RMSE of training data, validation data, and testing data was all less than 4.5 MPa. These results indicate that this model has an excellent testing performance in predicting compressive strength of concrete exposed to the wet-dry environment.

3.4. Comparison with the Linear Regression Model. Relationship between multiple input variables (explanatory variables) and one output variable (a response variable) also can be expressed by the linear regression model [34–36]. Linear regression model is a statistical model determined as follows:

$$Y = \alpha_0 + \alpha_1 X_1 + \alpha_2 X_2 + \cdots + \alpha_n X_n + \beta, \tag{1}$$

where Y is the output value and responds to compressive strength of concrete in this research; X_i represents different influencing factors, such as w/c and initial strength; α_i are the weighing coefficients of different influencing factors; and β is the error-modifying coefficient. Enter, stepwise, forward, and backward are four regression methods for ordinary least squares estimation.

A linear model of compressive strength responding to various influencing factors is established through SPSS 22.0.

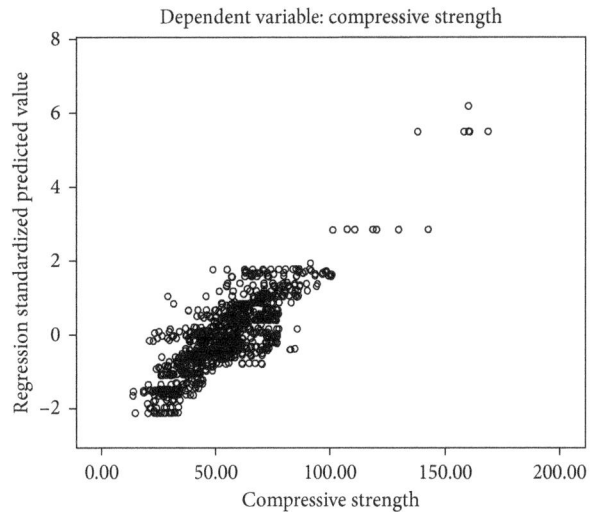

FIGURE 8: Fitting curve of outputs and targets of the linear model.

TABLE 6: Prediction performance comparison between LR and BP.

Model	R^2	MAE (MPa)	MAPE (%)	RMSE (MPa)
Linear regression	0.5842	8.63	20.8	10.99
BP neural network	0.9620	2.65	3.42	3.15

A stepwise method is chosen, and the fitting curve of outputs and targets of the linear model is shown in Figure 8.

The fitting result is obviously worse than that of the BP neural networks. And the MAE, RMSE, and MAPE are calculated in Table 6. From the comparison with the prediction results of the BP neural model, it is concluded that the BP neural model has a much better prediction performance.

4. Experiment Validation

In order to verify the generalization ability of this prediction model, a set of experiments were done.

4.1. Experiment Work and Data Collection

4.1.1. Raw Materials. Concrete specimens with three different strength grades were cast in the experiments. Three strength grades are, respectively, C30, C50, and C80. P52.5 Portland cement coming from Huaxin cement factory was chosen. Fine aggregates are fluvial sands and have a modulus of fineness of 2.7. The sediment content of the fluvial sands is 1%. Coarse aggregates are 5~25 continuously graded limestones and the sediment content is 0. A water-reducing agent with a solid content of 30% is polycarboxylate high-performance water-reducing admixture, manufactured by Subote New Materials Co., Ltd.

4.1.2. Mix Proportions and Specimen Preparation. Nine 100 mm × 100 mm × 100 mm cubes were cast to test the

TABLE 7: Mix design of raw materials.

Grades	Raw materials (kg/m^3)							w/c
	Cement	Silica fume	Water	Sand	Coarse aggregate	Gravel	Water reducer	
C30	375	—	200	750	675	450	—	0.53
C50	450	—	156.1	675	675	450	2.0	0.35
C80	514	57	141.65	684	615.6	410.4	9.5	0.26

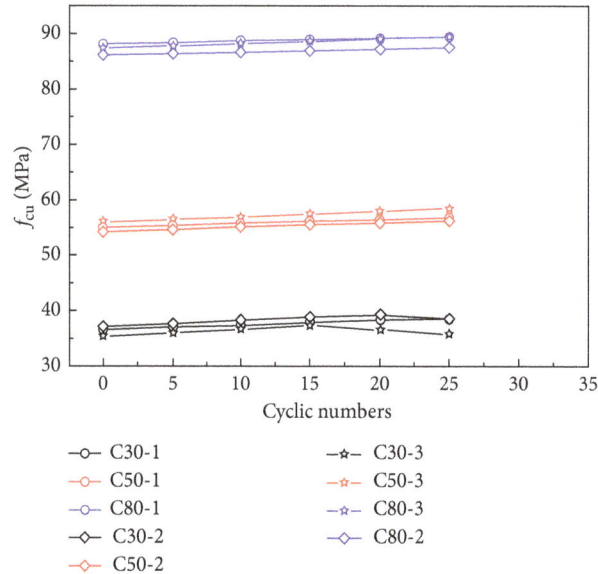

FIGURE 9: Compressive strength of concrete exposed to cyclic numbers.

mechanical properties, and three specimens were prepared at every strength grade level. Table 7 shows the mix design of concrete specimens.

Water-reducing admixture dosage is adjusted according to the rheological properties of fresh concrete in trial mix, and the final mix design was determined. The specimens are curing according to GB/T50081-2002. Curing temperature is $20 \pm 2°C$, and relative humidity is 95%. After curing for 90 days, cement hydration is almost completed, and the strength development reaches a plateau.

4.1.3. Details of Wet-Dry Exposure. Cube specimens are exposed to a wet-dry environment after 90-day standard curing. The wet-dry cycle system is in accordance with GB/T50082-2009 but a little different from that. In this system, the specimens were immersed in solution for 16 hours, where the solution temperature is 20°C. Then, the specimens were put into the oven where the baking temperature is 50°C for 8 hours. The whole process is treated as a complete cycle, and one wet-dry cycle takes one day.

4.1.4. Testing Procedure. After mixing and standard curing, all hardened specimens from each group were tested to estimate the compressive strength. In order to achieve nondestructive testing of concrete, the ultrasonic method was chosen to measure compressive strength of concrete. Compressive strength at 90 days is regarded as initial strength. After 5,

10, 15, 20, and 25 wet-dry cycles, compressive strength was tested through the ultrasonic method.

4.1.5. Experimental Results and Discussion. Figure 9 shows the compressive strength of cube specimens after wet-dry exposure. An assemble list including experiment results and influencing factors is shown in Table 8, and lists A~O are 15 influencing parameters of final compressive strength.

4.2. Consistency between Experiment Results and Prediction Results. The experimental parameters were brought into the input neurons of this model to test the difference between prediction values and actual values. Figure 10 shows the prediction value and actual value. These white dots are prediction values and red asterisks are actual values. It turns out to be a good compatibility. In order to assess the performance of this BP-ANN model, three statistical indictors are calculated.

Table 9 shows the detailed values and error percentages. A maximum error percentage is 3.93%, and all the errors are within acceptable range. The data fitting is ideal. Hence, this model could be used for the wet-dry environment concrete compressive strength prediction.

The MAE of the prediction is 0.6356 MPa, the RMSE is 0.8144 MPa, and the average error percentage is 1.09%. From the calculation results of MAE and RMSE, it can be concluded that this model is quite accurate for predicting the strength.

TABLE 8: Assemble list of influencing factors and results.

Number	A	B	C	D	E	F	G	H	I	J	K	L	M	N	O	Results
1	0.53	60	36.51	0	0	0.2816	0	0	0.1408	16	20	8	50	1	5	36.93
2	0.35	60	55.14	0	0	0.2816	0	0	0.1408	16	20	8	50	1	5	55.53
3	0.26	60	88.25	0	0	0.2816	0	0	0.1408	16	20	8	50	1	5	88.43
4	·0.53	60	37.14	0	0	1.408	0	0	0.7042	16	20	8	50	1	5	37.72
5	0.35	60	54.35	0	0	1.408	0	0	0.7042	16	20	8	50	1	5	54.76
6	0.26	60	86.14	0	0	1.408	0	0	0.7042	16	20	8	50	1	5	86.45
7	0.53	60	35.36	0	0	2.296	0	0	1.148	16	20	8	50	1	5	36.04
8	0.35	60	56.13	0	0	2.296	0	0	1.148	16	20	8	50	1	5	56.54
9	0.26	60	87.45	0	0	2.296	0	0	1.148	16	20	8	50	1	5	87.86
10	0.53	60	36.51	0	0	0.2816	0	0	0.1408	16	20	8	50	1	10	37.26
11	0.35	60	55.14	0	0	0.2816	0	0	0.1408	16	20	8	50	1	10	55.89
12	0.26	60	88.25	0	0	0.2816	0	0	0.1408	16	20	8	50	1	10	88.79
13	0.53	60	37.14	0	0	1.408	0	0	0.7042	16	20	8	50	1	10	38.29
14	0.35	60	54.35	0	0	1.408	0	0	0.7042	16	20	8	50	1	10	55.18
15	0.26	60	86.14	0	0	1.408	0	0	0.7042	16	20	8	50	1	10	86.77
16	0.53	60	35.36	0	0	2.296	0	0	1.148	16	20	8	50	1	10	36.72
17	0.35	60	56.13	0	0	2.296	0	0	1.148	16	20	8	50	1	10	57.02
18	0.26	60	87.45	0	0	2.296	0	0	1.148	16	20	8	50	1	10	88.28
19	0.53	60	36.51	0	0	0.2816	0	0	0.1408	16	20	8	50	1	15	37.88
20	0.35	60	55.14	0	0	0.2816	0	0	0.1408	16	20	8	50	1	15	56.12
21	0.26	60	88.25	0	0	0.2816	0	0	0.1408	16	20	8	50	1	15	89.01
22	0.53	60	37.14	0	0	1.408	0	0	0.7042	16	20	8	50	1	15	38.85
23	0.35	60	54.35	0	0	1.408	0	0	0.7042	16	20	8	50	1	15	55.54
24	0.26	60	86.14	0	0	1.408	0	0	0.7042	16	20	8	50	1	15	87.02
25	0.53	60	35.36	0	0	2.296	0	0	1.148	16	20	8	50	1	15	37.39
26	0.35	60	56.13	0	0	2.296	0	0	1.148	16	20	8	50	1	15	57.53
27	0.26	60	87.45	0	0	2.296	0	0	1.148	16	20	8	50	1	15	88.69
28	0.53	60	36.51	0	0	0.2816	0	0	0.1408	16	20	8	50	1	20	38.31
29	0.35	60	55.14	0	0	0.2816	0	0	0.1408	16	20	8	50	1	20	56.53
30	0.26	60	88.25	0	0	0.2816	0	0	0.1408	16	20	8	50	1	20	89.27
31	0.53	60	37.14	0	0	1.408	0	0	0.7042	16	20	8	50	1	20	39.36
32	0.35	60	54.35	0	0	1.408	0	0	0.7042	16	20	8	50	1	20	55.92
33	0.26	60	86.14	0	0	1.408	0	0	0.7042	16	20	8	50	1	20	87.31
34	0.53	60	35.36	0	0	2.296	0	0	1.148	16	20	8	50	1	20	36.58
35	0.35	60	56.13	0	0	2.296	0	0	1.148	16	20	8	50	1	20	58.02
36	0.26	60	87.45	0	0	2.296	0	0	1.148	16	20	8	50	1	20	89.11
37	0.53	60	36.51	0	0	0.2816	0	0	0.1408	16	20	8	50	1	25	38.66
38	0.35	60	55.14	0	0	0.2816	0	0	0.1408	16	20	8	50	1	25	56.86
39	0.26	60	88.25	0	0	0.2816	0	0	0.1408	16	20	8	50	1	25	89.53
40	0.53	60	37.14	0	0	1.408	0	0	0.7042	16	20	8	50	1	25	38.53
41	0.35	60	54.35	0	0	1.408	0	0	0.7042	16	20	8	50	1	25	56.27
42	0.26	60	86.14	0	0	1.408	0	0	0.7042	16	20	8	50	1	25	87.58
43	0.53	60	35.36	0	0	2.296	0	0	1.148	16	20	8	50	1	25	35.82
44	0.35	60	56.13	0	0	2.296	0	0	1.148	16	20	8	50	1	25	58.51
45	0.26	60	87.45	0	0	2.296	0	0	1.148	16	20	8	50	1	25	89.51

A: w/c; B: specific surface area (m^{-1}); C: initial strength (MPa); D: FA dosage; E: slag dosage; F: Na^+ (mol/L); G: Mg^{2+} (mol/L); H: Cl^- (mol/L); I: SO_4^{2+} (mol/L); J: wetting time (°C); K: wetting temperature (°C); L: drying time (hours); M: drying temperature (°C); N: cycle period (days); O: exposure age (days).

From the errors of the prediction values and true values in Table 10, the strength can be accurately predicted by this BP neural network model. However, because this BP neural network model is a four-layer structure, it may cost a long time to do the calculation. Besides, the influencing factors may not be considered totally, and this will have an effect on the prediction results. Last but not least, weighing coefficients could be calculated by some new method, such as Grey relational theory, in advance. It could shorten the running time and simplify the model structure. And this method is under exploration.

5. Conclusions

The BP artificial neural network was developed for compressive strength of concrete in wet-dry environment prediction. The following conclusions are obtained in this study:

(1) Data collected to establish a prediction model are relatively representative in this research. However, data of some particular factors need more experiments to supplement, such as chloride ion and magnesium ion concentrations and slag dosage.

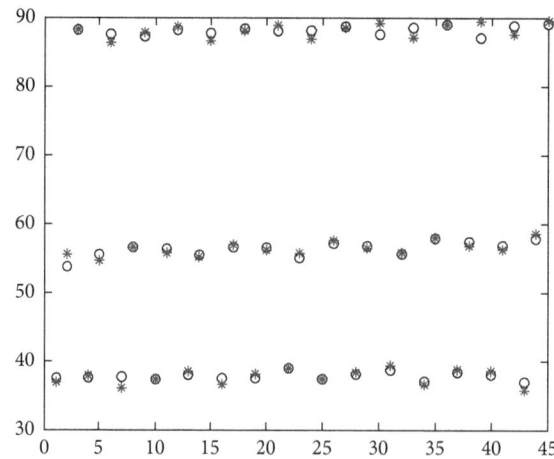

FIGURE 10: Distributions of the prediction value and actual value.

TABLE 9: Error percentage of actual values and prediction values.

Actual values (MPa)	36.93	55.53	88.43	37.72	54.76	86.45	36.04	56.54	87.86
Prediction values (MPa)	37.44	53.71	88.22	37.45	55.50	87.64	37.46	56.65	87.26
Error percentage (%)	1.39	−3.27	−0.24	−0.72	1.34	1.37	3.93	0.19	−0.68
Actual values (MPa)	37.26	55.89	88.79	38.29	55.18	86.77	36.72	57.02	88.28
Prediction values (MPa)	37.41	56.31	88.18	37.98	55.45	87.84	37.51	56.51	88.50
Error percentage (%)	0.40	0.75	−0.69	−0.81	0.49	1.24	2.14	−0.89	0.24
Actual values (MPa)	37.88	56.12	89.01	38.85	55.54	87.02	37.39	57.53	88.69
Prediction values (MPa)	37.55	56.55	88.04	39.10	55.06	88.26	37.15	57.15	88.80
Error percentage (%)	−0.87	0.76	−1.09	0.65	−0.87	1.42	−0.63	−0.67	0.12
Actual values (MPa)	38.31	56.53	89.27	39.36	55.92	87.31	36.58	58.02	89.11
Prediction values (MPa)	37.99	56.68	87.73	38.62	55.56	88.63	36.97	57.83	89.04
Error percentage (%)	−0.82	0.27	−1.72	−1.87	−0.64	1.52	1.06	−0.33	−0.08
Actual values (MPa)	38.66	56.86	89.53	38.53	56.27	87.58	35.82	58.51	89.51
Prediction values (MPa)	38.27	57.45	87.10	38.08	56.86	88.97	36.90	57.82	89.25
Error percentage (%)	−1.00	1.04	−2.71	−1.18	1.05	1.58	3.02	−1.18	−0.28

TABLE 10: Errors of the prediction values and true values.

Indicators	MAE (MPa)	RMSE (MPa)	MAPE (%)
Error	0.6356	0.8144	1.09%

And new data are continuously added into the database.

(2) Through multiple attempts, a four-hidden layer architecture BP artificial neural network is more appropriate to predict a multifactorial target because the combined effect of fifteen factors is quite complex. It turns out that the model is appropriate to predict concrete compressive strength in the wet-dry environment.

(3) The generalization of this model is validated by experiments. Although long-term strength prediction accuracy is not available, errors of short-term experimental values and prediction values are subtle. The BP artificial neural network is practical for concrete compressive strength prediction.

Conflicts of Interest

The authors declare that they have no conflicts of interest.

Acknowledgments

The research was financially supported by the National Program on Key Basic Research Project of China (Grant no. 2015CB6551002). The authors would like to express their thanks to the Key Basic Research Project of China for finance grant.

References

[1] P. S. Mangat and K. Gurusamy, "Long-term properties of steel fibre reinforced marine concrete," *Materials and Structures*, vol. 20, no. 4, pp. 273–282, 1987.

[2] Z. Kuhail and S. Shihada, "Effect of Gaza seawater on concrete strength for different exposures," *IUG Journal of Natural Studies*, vol. 11, no. 2, 2015.

[3] H. Toutanji, N. Delatte, S. Aggoun, R. Duval, and A. Danson, "Effect of supplementary cementitious materials on the compressive strength and durability of short-term cured concrete," *Cement and Concrete Research*, vol. 34, no. 2, pp. 311–319, 2004.

[4] T. Aye and C. T. Oguchi, "Resistance of plain and blended cement mortars exposed to severe sulfate attacks," *Construction and Building Materials*, vol. 25, no. 6, pp. 2988–2996, 2011.

[5] L. Jiang and D. Niu, "Study of deterioration of concrete exposed to different types of sulfate solutions under drying-wetting cycles," *Construction and Building Materials*, vol. 117, pp. 88–98, 2016.

[6] D. Chen, C. Du, X. Feng, and F. Ouyang, "An elastoplastic damage constitutive model for cementitious materials under wet-dry cyclic sulfate attack," *Mathematical Problems in Engineering*, vol. 2013, Article ID 562410, 7 pages, 2013.

[7] S. K. Kachigan, *Multivariate Statistical Analysis: A Conceptual Introduction*, Radius Press, Ann Arbor, MI, USA, 1991.

[8] P. Hajela and L. Berke, "Neurobiological computational modes in structural analysis and design," *Computers & Structures*, vol. 41, no. 4, pp. 657–67, 1991.

[9] J. Ghaboussi and A. Joghataie, "Active control of structures using neural networks," *Journal of Engineering Mechanics*, vol. 121, no. 4, pp. 555–567, 1995.

[10] H. R. Tavakoli, O. L. Omran, and S. S. Kutanaei, "Prediction of energy absorption capability in fiber reinforced self-compacting concrete containing nano-silica particles using artificial neural network," *Latin American Journal of Solids and Structures*, vol. 11, no. 6, pp. 966–979, 2014.

[11] H. R. Tavakoli, O. L. Omran, M. F. Shiade, and S. S. Kutanaei, "Prediction of combined effects of fibers and nanosilica on the mechanical properties of self-compacting concrete using artificial neural network," *Latin American Journal of Solids and Structures*, vol. 11, no. 11, pp. 1906–1923, 2014.

[12] H. G. Ni and J. Z. Wang, "Prediction of compressive strength of concrete by neural networks," *Cement and Concrete Research*, vol. 30, no. 8, pp. 1245–1250, 2000.

[13] S. C. Lee, "Prediction of concrete strength using artificial neural networks," *Engineering Structures*, vol. 25, no. 7, pp. 849–857, 2003.

[14] M. M. Alshihri, A. M. Azmy, and M. S. El-Bisy, "Neural networks for predicting compressive strength of structural light weight concrete," *Construction and Building Materials*, vol. 23, no. 6, pp. 2214–2219, 2009.

[15] A. Öztaş, M. Pala, E. Özbay, E. Kanca, N. Çaglar, and M. Asghar Bhatti, "Predicting the compressive strength and slump of high strength concrete using neural network," *Construction and Building Materials*, vol. 20, no. 9, pp. 769–775, 2006.

[16] N. D. Hoang, A. D. Pham, Q. L. Nguyen, and Q.-N. Pham, "Estimating compressive strength of high performance concrete with Gaussian process regression model," *Advances in Civil Engineering*, vol. 2016, Article ID 2861380, 8 pages, 2016.

[17] D. E. Rumelhart, J. L. McClelland, and PDP Research Group, *Parallel Distributed Processing*, MIT Press, Cambridge, MA, USA, 1987.

[18] D. Svozil, V. Kvasnicka, and J. Pospichal, "Introduction to multi-layer feed-forward neural networks," *Chemometrics and Intelligent Laboratory Systems*, vol. 39, no. 1, pp. 43–62, 1997.

[19] G. Li and J. Shi, "On comparing three artificial neural networks for wind speed forecasting," *Applied Energy*, vol. 87, no. 7, pp. 2313–2320, 2010.

[20] A. D. Ho and C. C. Yu, "Descriptive statistics for modern test score distributions: skewness, kurtosis, discreteness, and ceiling effects," *Educational and Psychological Measurement*, vol. 75, no. 3, pp. 365–388, 2015.

[21] J. A. Hanley, A. Negassa, and J. E. Forrester, "Statistical analysis of correlated data using generalized estimating equations: an orientation," *American Journal of Epidemiology*, vol. 157, no. 4, pp. 364–375, 2003.

[22] J. Bai and S. Ng, "Tests for skewness, kurtosis, and normality for time series data," *Journal of Business & Economic Statistics*, vol. 23, no. 1, pp. 49–60, 2005.

[23] M. S. Srivastava, "A measure of skewness and kurtosis and a graphical method for assessing multivariate normality," *Statistics & Probability Letters*, vol. 2, no. 5, pp. 263–267, 1984.

[24] J. C. F. de Winter, S. D. Gosling, and J. Potter, "Comparing the Pearson and Spearman correlation coefficients across distributions and sample sizes: a tutorial using simulations and empirical data," *Psychological Methods*, vol. 21, no. 3, pp. 273–290, 2016.

[25] A. K. Skidmore, "A comparison of techniques for calculating gradient and aspect from a gridded digital elevation model," *International Journal of Geographical Information System*, vol. 3, no. 4, pp. 323–334, 1989.

[26] D. G. Bonett and T. A. Wright, "Sample size requirements for estimating Pearson, Kendall and Spearman correlations," *Psychometrika*, vol. 65, no. 1, pp. 23–28, 2000.

[27] L. P. Prechelt, "A set of neural network benchmark problems and benchmarking rules," Universität Karlsruhe, Karlsruhe, Germany, Tech. Rep. 21/94, 1994.

[28] A. Mohemmed, S. Schliebs, S. Matsuda, and N. Kasabov, "Training spiking neural networks to associate spatio-temporal input–output spike patterns," *Neurocomputing*, vol. 107, pp. 3–10, 2013.

[29] A. Blais and D. Mertz, *An Introduction to Neural Networks-Pattern Learning with Back Propagation Algorithm*, Gnosis Software, Inc, 2001.

[30] H. Chen, C. Qian, C. Liang, and W. Kang, "An approach for predicting the compressive strength of cement-based materials exposed to sulfate attack," *PLoS One*, vol. 13, no. 1, article e0191370, 2018.

[31] K. Gurney, *An Introduction to Neural Networks*, CRC Press, Boca Raton, FL. USA, 1997.

[32] L. Prechelt, "Automatic early stopping using cross validation: quantifying the criteria," *Neural Networks*, vol. 11, no. 4, pp. 761–767, 1998.

[33] T. Ash, "Dynamic node creation in backpropagation networks," *Connection Science*, vol. 1, no. 4, pp. 365–375, 1989.

[34] J. Jiang and P. Lahiri, "Mixed model prediction and small area estimation," *Test*, vol. 15, no. 1, p. 1, 2006.

[35] T. H. Kim, I. Maruta, T. Sugie, S. Chun, and M. Chae, "Identification of multiple-mode linear models based on particle swarm optimizer with cyclic network mechanism," *Mathematical Problems in Engineering*, vol. 2017, Article ID 4321539, 10 pages, 2017.

[36] C. A. Gotway and W. W. Stroup, "A generalized linear model approach to spatial data analysis and prediction," *Journal of Agricultural, Biological, and Environmental Statistics*, vol. 2, no. 2, pp. 157–178, 1997.

6

A Mix Design Procedure for Alkali-Activated High-Calcium Fly Ash Concrete Cured at Ambient Temperature

Tanakorn Phoo-ngernkham ⓘ,[1] **Chattarika Phiangphimai,**[1]
Nattapong Damrongwiriyanupap,[2] **Sakonwan Hanjitsuwan** ⓘ,[3] **Jaksada Thumrongvut,**[1]
and Prinya Chindaprasirt ⓘ[4,5]

[1]*Department of Civil Engineering, Faculty of Engineering and Architecture, Rajamangala University of Technology Isan, Nakhon Ratchasima 30000, Thailand*
[2]*Civil Engineering Program, School of Engineering, University of Phayao, Phayao 56000, Thailand*
[3]*Program of Civil Technology, Faculty of Industrial Technology, Lampang Rajabhat University, Lampang 52100, Thailand*
[4]*Sustainable Infrastructure Research and Development Center, Department of Civil Engineering, Faculty of Engineering, Khon Kaen University, Khon Kaen 40002, Thailand*
[5]*Academy of Science, The Royal Society of Thailand, Dusit, Bangkok 10300, Thailand*

Correspondence should be addressed to Tanakorn Phoo-ngernkham; tanakorn.ph@rmuti.ac.th

Academic Editor: Marino Lavorgna

This research focuses on developing a mix design methodology for alkali-activated high-calcium fly ash concrete (AAHFAC). High-calcium fly ash (FA) from the Mae Moh power plant in northern Thailand was used as a starting material. Sodium hydroxide and sodium silicate were used as alkaline activator solutions (AAS). Many parameters, namely, NaOH concentration, alkaline activator solution-to-fly ash (AAS/FA) ratio, and coarse aggregate size, were investigated. The 28-day compressive strength was tested to validate the mix design proposed. The mix design methodology of the proposed AAHFAC mixes was given step by step, and it was modified from ACI standards. Test results showed that the 28-day compressive strength of 15–35 MPa was obtained. After modifying mix design of the AAHFAC mixes by updating the AAS/FA ratio from laboratory experiments, it was found that they met the strength requirement.

1. Introduction

Recently, alkali-activated binders have been widely studied to be used as a substitute for Portland cement. This is because they show great promise as an environmentally friendly binder, have high strength, are stable at high temperatures, and have high durability which are similar to those of Portland cement [1, 2]. From the past, alkali-activated binders are certainly emerged as a novel construction material and have a huge potential to become a prominent construction product of good environmental sustainability [3]. Alkali-activated binders are normally obtained from amorphous aluminosilicate materials such as fly ash, calcined kaolin, or metakaolin activated with high alkali solutions [4–8]. The sodium aluminosilicate hydrate (N-A-S-H) gel is the main reaction product for the low-calcium system, while calcium silicate hydrate (C-S-H) and calcium aluminosilicate hydrate (C-A-S-H) gels coexisted with sodium aluminosilicate hydrate (N-A-S-H) gel are the main reaction products for the high-calcium system [7, 9]. In Thailand, high-calcium fly ash (FA) from the Mae Moh power plant in the north of Thailand is a widely used starting material for making alkali-activated binders. High CaO content from this FA is very attractive for making alkali-activated binders because it can enhance the strength development when cured at ambient temperature [10–15]. This is why alkali-activated binders are needed to be used in practical works.

TABLE 1: Chemical compositions of FA (by weight).

Materials	SiO$_2$	Al$_2$O$_3$	Fe$_2$O$_3$	CaO	MgO	K$_2$O	Na$_2$O	SO$_3$	LOI
FA	31.32	13.96	15.64	25.79	2.94	2.93	2.83	3.29	1.30

TABLE 2: Material properties of the AAHFAC ingredients.

Materials	Specific gravity	Fineness modulus	Absorption capacity (%)	Dry density (kg/m^3)
FA	2.65	—	—	—
RS	2.52	2.20	0.85	1585
7 mm·LS	2.64	6.04	1.55	1420
10 mm·LS	2.65	7.00	1.50	1405
16 mm·LS	2.67	7.09	1.00	1400

It is well known that alkali activator solution is one of the most important factors influencing the strength development of alkali-activated binders. From previous study on this area by Pimraksa et al. [16], they reported that sodium hydroxide and sodium silicate solutions were widely used as liquid activators because of availability and good mechanical properties. Panias et al. [17] claimed that sodium hydroxide solution is commonly used for the dissolution of Si^{4+} and Al^{3+} ions from FA to form aluminosilicate materials, whereas sodium silicate solution contains soluble silicate species and thus is used to promote the condensation process of alkali-activated binders [13]. Many researchers [6, 13, 18–20] reported that a combination of sodium silicate and sodium hydroxide solutions showed the best mechanical performance of alkali-activated binders.

To be useful in practice, a mix design methodology of alkali-activated binders has been studied. For example, Lloyd et al. [21] conducted a mix design methodology for alkali-activated low-calcium fly ash concrete. Ananda Kumar and Sankara Narayanan. [22] studied a design procedure for different grades of alkali-activated concrete by using Indian standards. In this method, fly ash content and activator solution-to-fly ash ratio were selected based on the strength requirement and by keeping the fine aggregate percentage as constant. Ferdous et al. [23] proposed a mix design for fly ash-based alkali-activated binders concrete by considering the concrete density variability, specific gravity of the materials, air content, workability, and the strength requirement. Also, Lahoti et al. [24] studied the mix design factor and strength prediction of metakaolin-based geopolymer, but it was just a study on the basic properties of the geopolymer made from metakaolin which were not applicable for other raw materials such as fly ash and slag. There were few research studies conducted by Anuradha et al. [22] and Ferdous et al. [23] that used the trial-and-error approach for considering a mix design of alkali-activated binders. However, mix design and proportion of containing binders of alkali-activated concrete seem to be complex because more variables are being involved in it. Therefore, there is no standard mix design method available for designing alkali-activated binders concrete to date. According to the recent study, Pavithra et al. [25] studied a mix design procedure for alkali-activated low-calcium FA concrete. It is shown that its strength follows a similar trend to

that of Portland cement concrete as per ACI standards. However, this work still used the temperature curing for enhancing the strength development. From the literature review, there is no research investigated on a mix design methodology for alkali-activated high-calcium FA concrete. Therefore, in this research, we attempt to make a new mix design methodology for alkali-activated high-calcium FA concrete. Many parameters, namely, NaOH concentration, alkaline activator solution-to-binder ratio, and coarse aggregate size, have been investigated. Engineering properties, that is, setting time, compressive strength, flexural strength, modulus of elasticity, and Poisson's ratio, of the AAHFAC have been tested to understand behaviors for utilizing this material in the future. The step-by-step procedure of the mix design for alkali-activated high-calcium FA concrete will be explained in this paper. The outcome of this study would lay a foundation for the future use of alkali-activated high-calcium FA concrete for manufacturing this material in construction work.

2. Experimental Details

2.1. Materials. The precursor used in this study is fly ash (FA) from lignite coal combustion. The FA is the byproduct from the Mae Moh power plant in northern Thailand with a specific gravity of 2.65, a mean particle size of 15.6 μm, and a Blaine fineness of 4400 cm^2/g, respectively. Table 1 summarizes the chemical compositions of the FA used in the present experimental work. Note that the FA had a sum of SiO$_2$ + Al$_2$O$_3$ + Fe$_2$O$_3$ at 60.96% and CaO at 25.79%. Therefore, this FA was classified as class C FA as per ASTM C618 [26]. The fine aggregate is the local river sand (RS) with a specific gravity of 2.52 and a fineness modulus of 2.20, while coarse aggregates are the crushed lime stone (LS) with various different average sizes of 7, 10, and 16 mm, respectively. Material properties of alkali-activated high-calcium fly ash concrete (AAHFAC) ingredients are illustrated in Table 2.

Sodium hydroxide solution (NaOH) and sodium silicate solution (Na$_2$SiO$_3$) with 11.67% Na$_2$O, 28.66% SiO$_2$, and 59.67% H$_2$O were used as liquid activators. For example, for the preparation of 10 M·NaOH, sodium hydroxide pellets of 400 gram were dissolved by distilled water of 1 liter and then

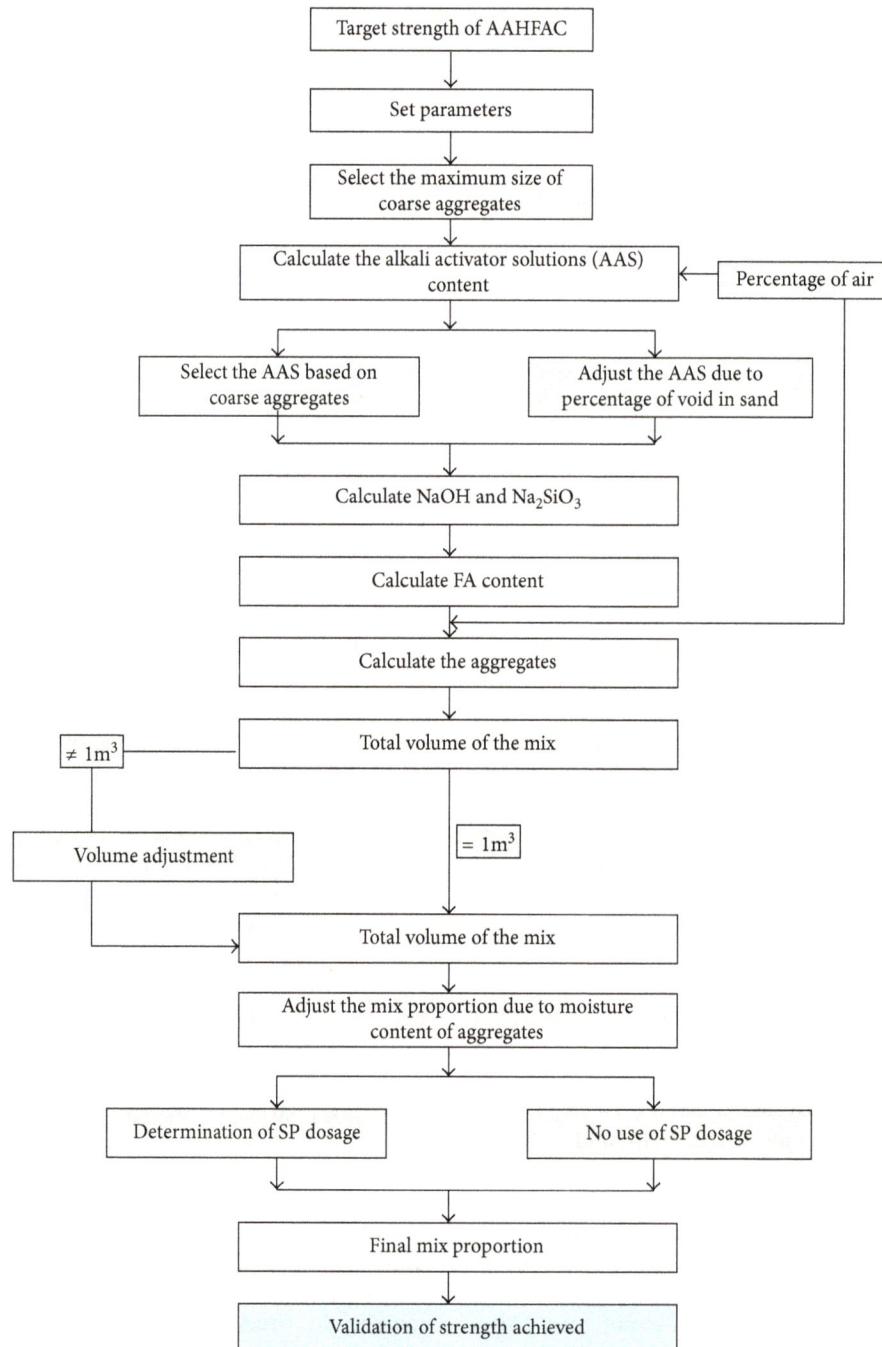

FIGURE 1: Flow chart for the mix design procedure.

allowed to cool down for 24 hours before use to avoid the uncontrolled acceleration of setting of alkali-activated binders [27–29].

2.2. Proposed Method for Designing Alkali-Activated High-Calcium Fly Ash Concrete. In this paper, we propose a novel mix design methodology for alkali-activated high-calcium fly ash concrete (AAHFAC) in a rational approach. It should be noted that the method is easy to use because it is based on the ACI 211.4R-93 [30] with some modification. The parametric studies are composed of different NaOH concentrations, alkaline activator solution- (AAS-) to-binder ratios, and coarse aggregate sizes. The flow chart for the mix design procedure in this study is illustrated in Figure 1. The step-by-step procedure is summarized as follows:

(1) Step 1: selection of the maximum size of the coarse aggregate

This step is to select the maximum sizes of coarse aggregates for mixing the AAHFAC. Three different sizes of coarse aggregates have been investigated, namely, 4.5–9.5 mm or average 7 mm, 9.5–12.5 mm or average 10 mm, and 12.5–20.0 mm or average 16 mm.

TABLE 3: Maximum water content and percentage of air per cubic meter of concrete [31].

Normal maximum size of the aggregate (mm)	Maximum water content (kg/m³)	Percentage of void (%)
10	225	3.0
12.5	215	2.5
20	200	2.0

(2) Step 2: selection of the alkaline activator solution (AAS) content and air content

AAS content and air content were based on the maximum coarse aggregate size as per ACI standards. Maximum AAS and percentage of air per cubic meter of concrete in this study were selected by using the slump condition of around 20 mm as per ACI standards. This condition is summarized in Table 3.

(3) Step 3: adjustment of the alkaline activator solution (AAS) content due to percentage of void in the fine aggregate

As per ACI 211.4R-93 [30], a mixture of concrete has been recommended to use the fine aggregate with fineness modulus values from 2.4 to 3.2. However, particle shape and surface texture of the fine aggregate have an effect on its voids content; therefore, mixing water requirements may be different from the values given. As mentioned, the values for the required mixing water given are applicable when the fine aggregate is used that has a void content of 35%. If not, an adjustment of water content must be added into the required water content. Therefore, this study will calculate the AAS content due to percentage of void in the fine aggregate in a similar way to Portland cement concrete. This adjustment can be calculated using the following equation:

$$AAS_{adjustment} = \left| \left\{ \left[1 - \left(\frac{\rho_{RS}}{S_G \rho_w} \right) \right] \times 100 \right\} - 35 \right| \times 4.75, \quad (1)$$

where $AAS_{adjustment}$ is an adjustment of the AAS content (kg/m³), ρ_{RS} is the density of the fine aggregate in SSD condition (kg/m³), S_G is the specific gravity of the fine aggregate, and ρ_w is the density of water (kg/m³).

(4) Step 4: selection of alkaline activator solution-to-fly ash (AAS/FA) ratio

This research attempts in adopting the standard AAS-to-FA ratio curve of the AAHFAC before use. The minimum compressive strength could be determined from the relationship between 28-day compressive strength and AAS-to-FA ratio. Only the compressive strength of AAHFAC was considered as it is the requirement as per ACI 211.4R-93 [30].

(5) Step 5: calculation of binder content

The weight of the binder required per cubic meter of the AAHFAC could be determined by dividing the values of mixing the AAS content after an adjustment of the AAS content due to percentage of void in the fine aggregate.

(6) Step 6: calculation of individual mass of AAS content (NaOH and Na₂SiO₃ solutions)

TABLE 4: Density of NaOH solution with different concentrations.

NaOH (molar)	5 M	10 M	15 M
Density (kg/m³)	1200	1413	1430

From the literature, NaOH and Na₂SiO₃ were found to be the commonly used alkali activators [21]. In this study, NaOH and Na₂SiO₃ have been selected as alkaline activator solutions. According to Table 4, the density of NaOH with different concentrations has been used for calculating the volume of AAS as per the volume method. The individual mass of alkaline activator solutions content could be calculated using the following equation:

$$Na_2SiO_3 = \frac{AAS}{[1 + (1/(Na_2SiO_3/NaOH))]},$$

$$NaOH = AAS - \frac{AAS}{[1 + (1/(Na_2SiO_3/NaOH))]}. \quad (2)$$

(7) Step 7: calculation of fine and coarse aggregates

The mass of fine and coarse aggregates content is determined as per the absolute volume method. Let the percentage of the fine aggregate in the total aggregate be 30% and that of the coarse aggregate be 70%. Fine and coarse aggregates content are determined using the following equation:

$$M_{RS} = 0.3 S_{G(RS)} \left[1 - V_{FA} - V_{NaOH} - V_{Na_2SiO_3} - V_{air} \right] \times 1000,$$

$$M_{LS} = 0.7 S_{G(LS)} \left[1 - V_{FA} - V_{NaOH} - V_{Na_2SiO_3} - V_{air} \right] \times 1000, \quad (3)$$

where M_{RS} is the mass of the fine aggregate (kg), M_{LS} is the mass of the coarse aggregate (kg), $S_{G(RS)}$ is the specific gravity of the fine aggregate, $S_{G(LS)}$ is the specific gravity of the coarse aggregate, V_{FA} is the volume of high-calcium fly ash, V_{NaOH} is the volume of NaOH, $V_{Na_2SiO_3}$ is the volume of Na₂SiO₃, and V_{air} is the volume of entrapped air.

(8) Step 8: calculation of superplasticizer dosage

The AAS has the higher viscosity than tap water when used for making the AAHFAC. Hardjito et al. [32] reported that the dosage of the superplasticizer was effective for the range between 0.8 and 2% of binder content to improve the workability of alkali-activated binder concrete. Pavithra et al. [25] also claimed that the use of superplasticizer dosage was found to have impact on behavior of fresh alkali-activated binder concrete; however, it had a little effect on strength and other properties. Therefore, to improve the workability of the AAHFAC, a small amount of the superplasticizer was incorporated in the mixture.

(9) Step 9: validation of strength attained with the proposed mix design

The 28-day compressive strength obtained from testing will be verified with the target strength.

(10) Step 10: recalculation of Step 4 by the strength obtained from Step 9

TABLE 5: Mix proportion used in this study based on the mix design procedure.

Mix	Symbol	FA (kg/m^3)	NaOH (kg/m^3)		Na$_2$SiO$_3$ (kg/m^3)	Aggregates (kg/m^3)			RS	SP (kg/m^3)
			10 M	15 M		7 mm·LS	10 mm·LS	16 mm·LS		
1	0.45AAS10M7mmLS	523	118	—	118	1124	—	—	459	5.2
2	0.45AAS10M10mmLS	500	113	—	113	—	1166	—	475	5.0
3	0.45AAS10M16mmLS	478	108	—	108	—	—	1211	490	4.8
4	0.50AAS10M7mmLS	470	118	—	118	1161	—	—	474	4.7
5	0.50AAS10M10mmLS	450	113	—	113	—	1201	—	489	4.5
6	0.50AAS10M16mmLS	430	108	—	108	—	—	1245	504	4.3
7	0.55AAS10M7mmLS	428	118	—	118	1191	—	—	487	4.3
8	0.55AAS10M10mmLS	409	113	—	113	—	1231	—	501	4.1
9	0.55AAS10M16mmLS	391	108	—	108	—	—	1273	515	3.9
10	0.60AAS10M7mmLS	392	118	—	118	1216	—	—	497	3.9
11	0.60AAS10M10mmLS	375	113	—	113	—	1255	—	511	3.8
12	0.60AAS10M16mmLS	359	108	—	108	—	—	1296	525	3.6
13	0.45AAS15M7mmLS	523	—	118	118	1126	—	—	460	5.2
14	0.45AAS150M10mmLS	500	—	113	113	—	1168	—	475	5.0
15	0.45AAS15M16mmLS	478	—	108	108	—	—	1212	491	4.8
16	0.50AAS15M7mmLS	470	—	118	118	1163	—	—	475	4.7
17	0.50AAS15M10mmLS	450	—	113	113	—	1203	—	490	4.5
18	0.50AAS15M16mmLS	430	—	108	108	—	—	1246	505	4.3
19	0.55AASB15M7mmLS	428	—	118	118	1193	—	—	487	4.3
20	0.55AAS15M10mmLS	409	—	113	113	—	1232	—	502	4.1
21	0.55AAS15M16mmLS	391	—	108	108	—	—	1274	516	3.9
22	0.60AASB150M7mmLS	392	—	118	118	1218	—	—	498	3.9
23	0.60AAS15M10mmLS	375	—	113	113	—	1257	—	512	3.8
24	0.60AAS15M16mmLS	359	—	108	108	—	—	1298	525	3.6

After the 28-day compressive strength has been obtained from testing, all strengths will be recalculated in Step 4 for determining the strength requirement of the AAHFAC with various AAS-to-FA ratios.

(11) Step 11: validation of strength achieved

Compressive strength tests will be conducted in the laboratory using the mix design proposed above. When the designed mix satisfies the strength requirement, the final development of the AAHFAC can be made by employing the above design steps.

2.3. Manufacturing and Testing of the AAHFAC. The laboratory experiments have been conducted to validate the proposed mix design. Based on the AAHFAC trial mix, high-calcium fly ash (FA) from the Mae Moh power plant in northern Thailand is used as a starting material for making the AAHFAC. Constant NaOH/Na$_2$SiO$_3$ ratio is fixed at 1.0 in all mixes. The AAHFAC has been manufactured with different NaOH concentrations of 10 M and 15 M. Both fine and coarse aggregates in saturated surface dry (SSD) condition have been used for making the AAHFAC. Crushed lime stone with different average sizes of 7, 10, and 16 mm, respectively, is investigated. In this present work, mix design of the AAHFAC with slump at around 17.5–22.5 mm or average 20 mm has been controlled in order to ensure the workability of the AAHFAC.

For the mixing of the AAHFAC, fine and coarse aggregates were mixed together first for 60 s. After that, NaOH solution was added, and then, they were mixed again for 30 s. After 30 s, FA was added, and then, the mixture was mixed for 60 s. Afterward, Na$_2$SiO$_3$ solution and superplasticizer were added into the mixture, and the mixture was mixed again for a further 60 s until becoming homogeneous. The mix proportions of the AAHFAC are illustrated in Table 5.

After mixing, the workability of the AAHFAC was tested using the slump cone test as per ASTM C143 [33] as shown in Figure 2. Setting time of the AAHFAC was evaluated using the method of penetration resistance as per ASTM C403/C403M-16 [34]. After that, a fresh AAHFAC was placed into a cylinder mold with 100 mm diameter and 200 mm height to measure the compressive strength, modulus of elasticity, and Poisson's ratio. Tests for the determination of the static chord modulus of elasticity and Poisson's ratio of the samples have been carried out as per ASTM C469 [35]. The 75 × 75 × 300 mm3 long beam was used to measure the flexural strength of the AAHFAC, and the flexural strength was calculated as per ASTM C78 [36]. The test setup for the compressive strength, flexural strength, modulus of elasticity, and Poisson's ratio of the AAHFAC is illustrated in Figures 3 and 4. After curing, the AAHFAC samples were demolded with the aid of slight tapping at the side of the mold at the age of 1 day and immediately wrapped with vinyl sheet to protect moisture loss and kept at the ambient room temperature. All samples of the compressive strength, flexural strength, modulus of elasticity, and Poisson's ratio were tested at the age of 28 days of curing of the AAHFAC. Five identical samples were tested for each mix, and the average value was used as the test result.

(a)   (b)

FIGURE 2: Fresh AAHFAC (a) and slump test of the AAHFAC (b).

FIGURE 3: Test setup for compressive strength, modulus of elasticity, and Poisson's ratio.

FIGURE 4: Test setup for flexural strength.

3. Experimental Results and Discussion

Figures 5–10 show the test results of fresh AAHFAC and mechanical properties of the AAHFAC. According to Figure 5, the slump values are between 18 and 22 mm; therefore, mix design of the AAHFAC from Table 5 is in line with the slump value at around 17.5–22.5 mm. As mentioned, the AAS and air contents based on the maximum size of the aggregate as per the ACI standard could be taken for this study. The final setting time of the AAHFAC tends to obviously increase with the increase of AAS/FA ratio and NaOH concentration as illustrated in Figure 6. This result conforms to the previous studies [37, 38] that an increase of fluid medium content resulted in less particle interaction and increased the workability of the mixture. Hanjitsuwan et al. [39] and Rattanasak and Chindaprasirt [40] also explained that NaOH concentration is a main reason for leaching out of Si^{4+} and Al^{3+} ions; therefore, the time of setting tends to increase. However, different sizes of the coarse aggregate are marginal changed in the setting time of the AAHFAC.

Test results of the compressive strength, modulus of elasticity, and Poisson's ratio have been shown in Figures 7 and 8. It is found that the compressive and flexural strengths tend to increase with the increase of AAS/FA ratio and NaOH concentration; however, they are decreased with increasing coarse aggregate sizes. Sinsiri et al. [38] claimed that the excess OH^- concentration in the mixture at high AAS/FA ratio causes the decrease of strength development of the AAHFAC similar to that in case of increasing water-to-cement ratios for Portland cement concrete. Also, the excess liquid solution could disrupt the polymerization process [37]. For the effect of NaOH concentration, it is found that the increase in the leaching out of Si^{4+} and Al^{3+} ions from FA particles at high NaOH concentration could improve the N-A-S-H gel, and thus, it gives high strength development of the AAHFAC [40]. Also, CaO oxide could react with silica and alumina from FA to form C-(A)-S-H gel which coexisted with N-A-S-H gel [11, 12, 14, 15, 20], resulting in an enhancement of strength. The short setting time of less than 60 minutes at ambient temperature with relatively high 28-day compressive strengths of 16.0–36.0 MPa indicated the role of calcium in the system. Therefore, the AAHFAC could be used as an alternative repair material as reported by several publications [13, 14, 20, 29, 41–43]. There are many reasons such as high compressive strength [44], negligible drying shrinkage [45], low creep [46], good bond with reinforcing steel [47, 48], good resistance to acid and sulfate [3, 27], fire resistance [49], and excellent bond with old concrete [14, 20, 28]. For the effect of coarse aggregate size, the strength development tends to decrease with the increase of coarse aggregate size. Generally,

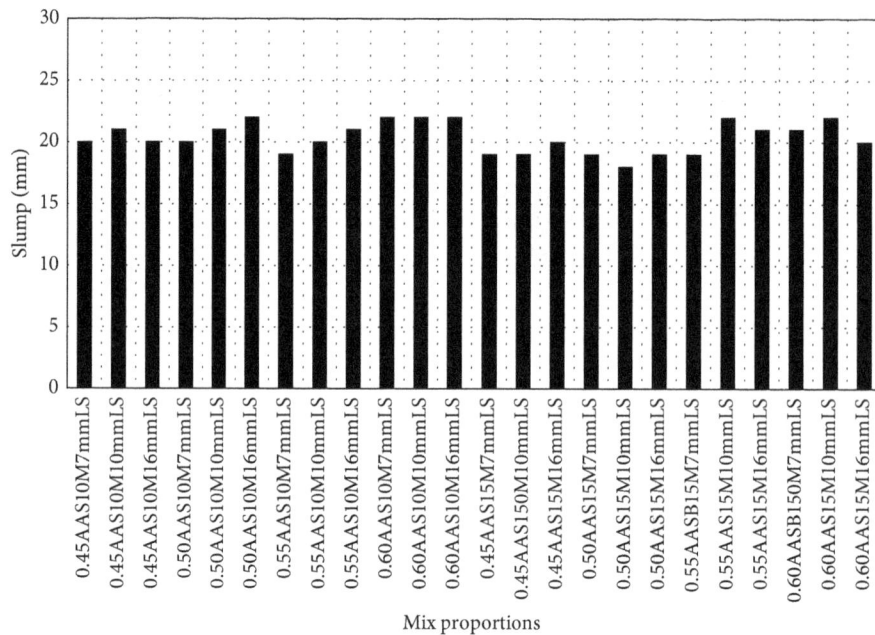

FIGURE 5: Slump of the AAHFAC

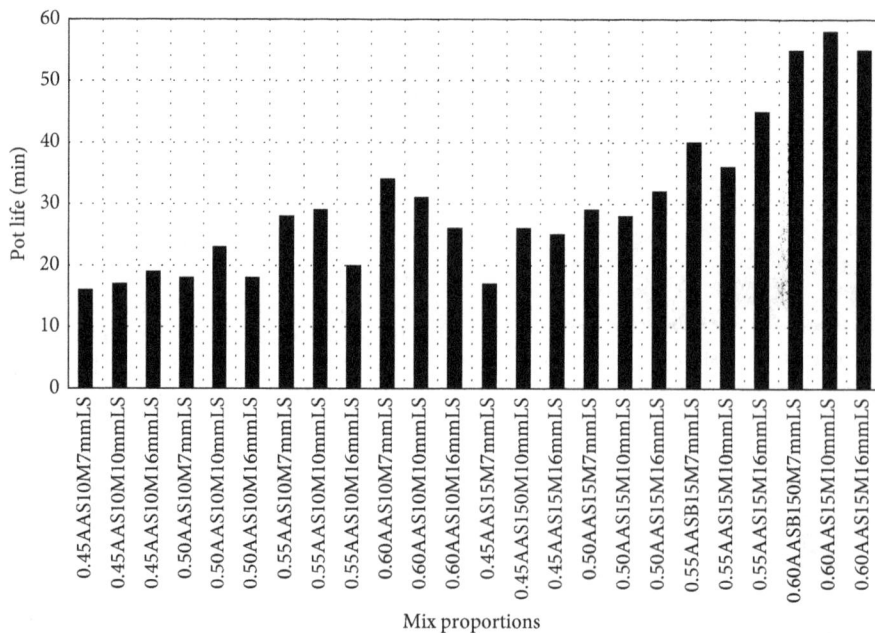

FIGURE 6: Pot life of the AAHFAC.

the binder within the AAHFAC is an important factor on the strength development of the matrix. This agrees with the findings of Pavithra et al. [25] that the increase in total aggregate corresponding to the decrease of paste has an adverse effect on the strength development of low-calcium fly ash geopolymer concrete.

The experimental results obtained for the modulus of elasticity and Poisson's ratio of the AAHFAC mixes show an overall increase with the increase of compressive strength as shown in Figures 9 and 10; however, the coarse aggregate

size influenced the modulus of elasticity and Poisson's ratio. Normally, it is well known that the values of modulus of elasticity and Poisson's ratio of concrete depend on the stiffness of paste and aggregates [50]. The modulus of elasticity obtained from this study agrees with Nath and Sarker [51] that the modulus of elasticity of geopolymer concrete with 28-day compressive strength of 25–45 MPa was between 17.4 and 24.6 GPa, while the modulus of elasticity of Portland cement concrete was between 25 and 35 GPa [51, 52]. According to Figure 10, Poisson's ratio for

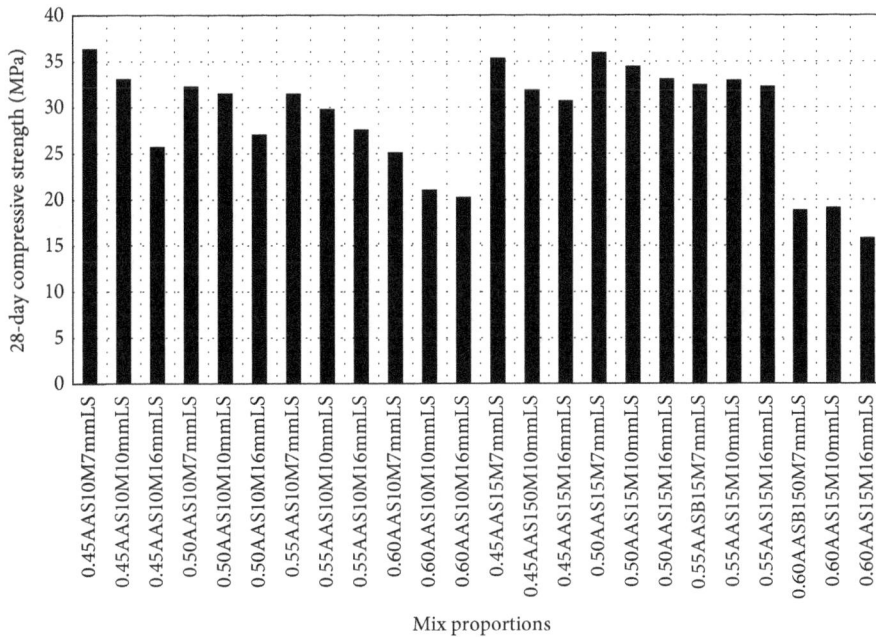

FIGURE 7: 28-day compressive strength of the AAHFAC.

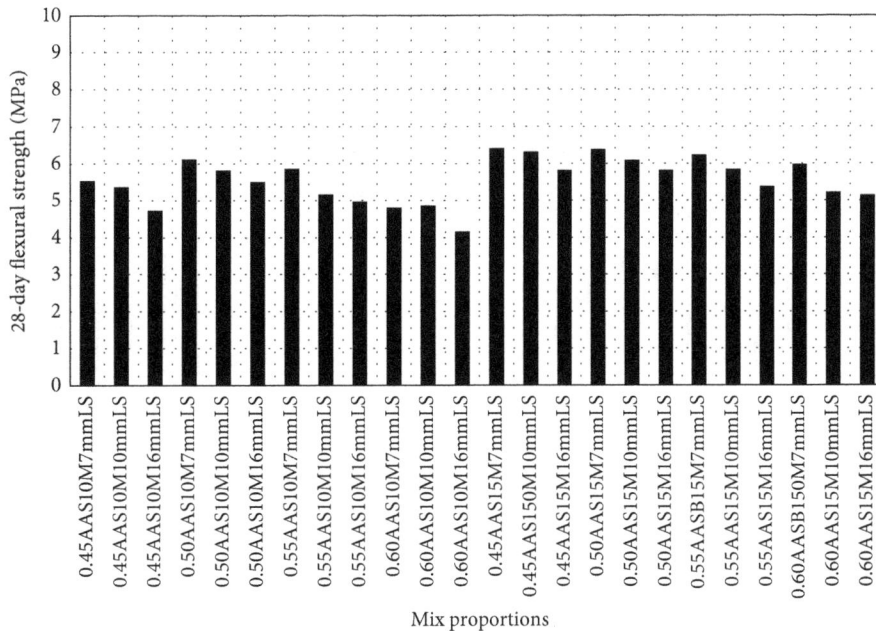

FIGURE 8: 28-day flexural strength of the AAHFAC.

all of the AAHFAC mixes is between 0.21 and 0.31. They show values slightly higher than the values assigned for normal strength of Portland cement concrete which are between 0.11 and 0.21 [50]. The high Poisson's ratios were associated with the mixes with large aggregates of 16 mm. For the mixes with smaller aggregates of 7 and 11 mm, Poisson's ratios are between 0.21 and 0.27 similar to the previously reported values of inorganic polymer concrete between 0.23 and 0.26 and also in the same range as those of high strength concrete between 0.20 and 0.25 [50].

4. Verification of the Mix Design Methodology Using Laboratory Experiments

To develop the mix design methodology of the AAHFAC, some parameters from laboratory experiments should be updated before use. In order to support the basic principles of mix design, the relationship between 28-day compressive strength and AAS/FA ratio from laboratory experiments has been produced as shown in Figure 11. The AAS/FA ratio curve can be determined for the minimum 28-day

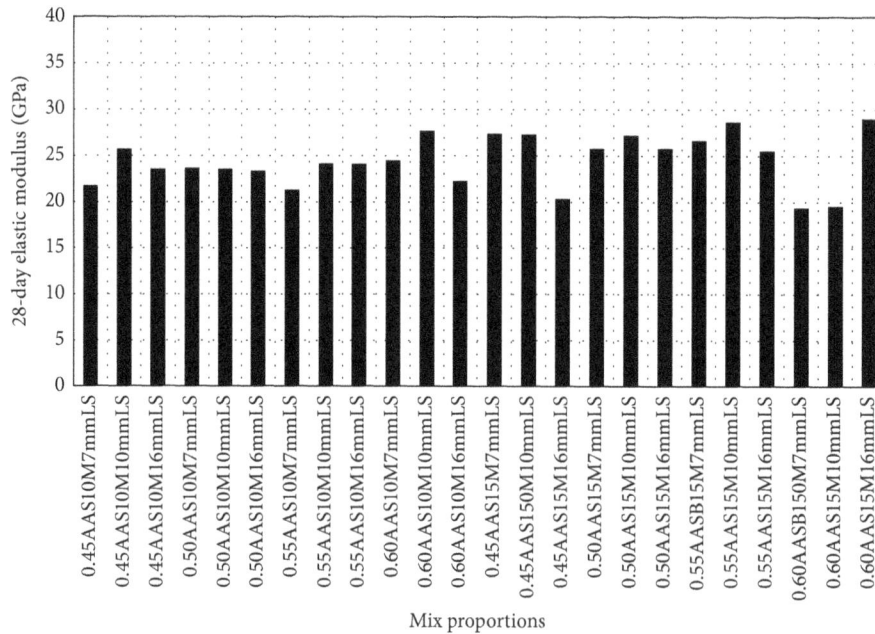

FIGURE 9: 28-day elastic modulus of the AAHFAC.

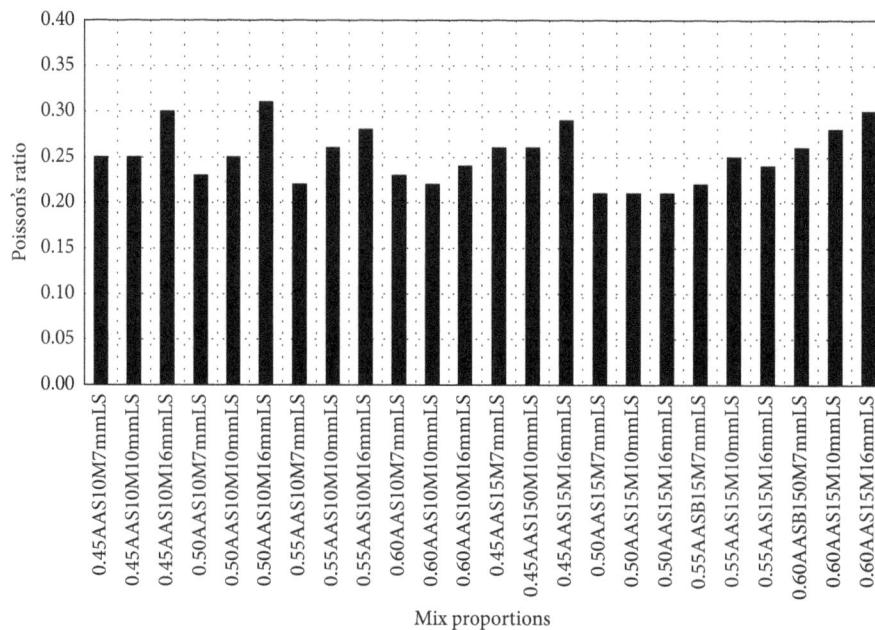

FIGURE 10: 28-day Poisson's ratio of the AAHFAC.

compressive strength of the AAHFAC similar to that of the water-to-cement ratio curve of Portland cement concrete.

According to Table 5 and Figures 5–10, the 24 different mixes of the AAHFAC were conducted to develop the mix design methodology. The 28-day compressive strength of the AAHFAC cured at ambient temperature ranging from 15 to 35 MPa was obtained by using the proposed mix design methodology. All obtained compressive strengths of the AAHFAC mixes can be separated into 5 strength requirements, namely, 15, 20, 25, 30, and 35 MPa, respectively.

These obtained strength requirements of the AAHFAC will be used to recalculate in Step 4 for determining the strength requirement of the AAHFAC with various AAS/FA ratios (Figure 11).

In order to verify the mix design procedure, an example design of the AAHFAC mix with the target 28-day compressive strength of 30 MPa has been considered. The parameters are NaOH concentration of 10 M, maximum aggregate size of 10 mm, and $Na_2SiO_3/NaOH$ ratio of 1.0. Material properties of the AAHFAC ingredients are

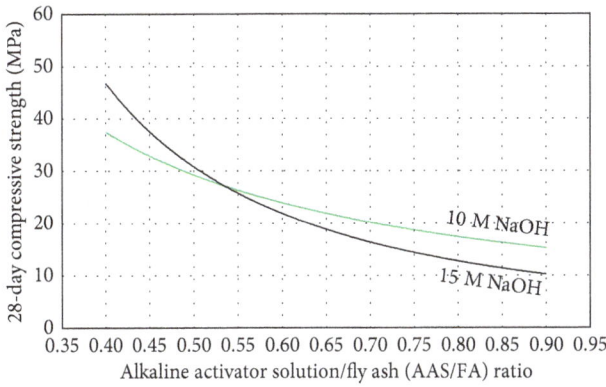

FIGURE 11: Twenty-eight-day compressive strength versus the AAS/FA ratio curve.

illustrated in Table 2. The step-by-step procedure of the mix design is explained as follows:

Step 1: selection of the maximum coarse aggregate size

This mix design of the AAHFAC has been selected the maximum coarse aggregate size of 10 mm for mixing the AAHFAC.

Step 2: selection of AAS and air contents

AAS and air contents are based on the maximum size of the coarse aggregate; therefore, AAS content of 225 kg/m^3 and air content of 3.0% have been used as given in Table 3.

Step 3: adjustment of AAS content due to percentage of void in the fine aggregate

The fine aggregate used in this example has a void of 37%; therefore, the value of adjustment could be calculated as follows:

$$\text{AAS}_{\text{adjustment}} = \left| \left\{ \left[1 - \left(\frac{1585}{2.52 * 1000} \right) \right] \times 100 \right\} - 35 \right| \\ \times 4.75 = 10 \text{ kg/m}^3. \tag{4}$$

Therefore, the AAS content after an adjustment of AAS content due to percentage of void in the fine aggregate is 235 kg/m^3.

Step 4: selection of AAS/FA ratio

From Figure 11, for the minimum 28-day compressive strength of 30 MPa, it is found that the AAS/FA ratio of 0.50 is obtained when 10 M·NaOH was used as alkaline activator solution.

Step 5: calculation of binder content

$$\text{FA content} = \frac{\text{AAS content}}{\text{AAS/FA ratio}} = \frac{235}{0.50} = 470 \text{ kg/m}^3. \tag{5}$$

Step 6: calculation of NaOH and Na$_2$SiO$_3$ content

The individual mass of NaOH and Na$_2$SiO$_3$ content could be calculated as follows:

$$\text{Na}_2\text{SiO}_3 = \frac{\text{AAS}}{\left[1 + (1/(\text{Na}_2\text{SiO}_3/\text{NaOH})) \right]} \\ = \frac{235}{\left[1 + (1/1) \right]} = 117.50 \text{ kg/m}^3,$$

$$\text{NaOH} = \text{AAS} - \frac{\text{AAS}}{\left[1 + (1/(\text{Na}_2\text{SiO}_3/\text{NaOH})) \right]} \\ = 117.50 - \frac{235}{\left[1 + (1/1) \right]} = 117.50 \text{ kg/m}^3. \tag{6}$$

Step 7: calculation of fine and coarse aggregates

$$M_{\text{RS}} = 0.3(2.52) \left[1 - \frac{470}{2.65 \times 1000} - \frac{117.5}{1413} - \frac{117.5}{1485} - \frac{3}{100} \right] \\ \times 1000 = 477 \text{ kg/m}^3,$$

$$M_{\text{LS}} = 0.7(2.64) \left[1 - \frac{470}{2.65 \times 1000} - \frac{117.5}{1413} - \frac{117.5}{1485} - \frac{3}{100} \right] \\ \times 1000 = 1164 \text{ kg/m}^3. \tag{7}$$

Step 8: calculation of superplasticizer dosage

$$\text{SP dosage} = \left(\frac{1}{100} \right) \times 470 = 4.7 \text{ kg/m}^3. \tag{8}$$

Step 9: summarization of mix design

Based on the mix design methodology of the AAHFAC mix, mix proportions of ingredients are concluded as follows:

$$\begin{aligned} \text{FA} &= 470 \text{ kg/m}^3, \\ \text{RS} &= 477 \text{ kg/m}^3, \\ \text{LS} &= 1164 \text{ kg/m}^3, \\ \text{NaOH} &= 117.5 \text{ kg/m}^3, \\ \text{Na}_2\text{SiO}_3 &= 117.5 \text{ kg/m}^3, \\ \text{SP} &= 4.7 \text{ kg/m}^3. \end{aligned} \tag{9}$$

Step 10: validation of strength achieved

After following the mix design of the AAHFAC mix, the AAHFAC samples were tested for compressive strength on cylinder molds with 100 mm diameter and 200 mm height. After testing, it is found that the 28-day compressive strength of the AAHFAC is 31.15 MPa. Therefore, the mix design of the AAHFAC above meets the strength requirement.

5. Conclusion

In this study, the novel mix design methodology for alkali-activated high-calcium fly ash concrete (AAHFAC) cured at ambient temperature in a rational way was proposed. From the review of mix design of alkali-activated binders concrete, there is no standard mix design method available for designing the AAHFAC. Some works have been investigated in this area in an attempt to develop mix design methodology for alkali-activated low-calcium fly ash. However, it still used the temperature curing for enhancing the strength development of alkali-activated low-calcium fly ash, but this is limited for construction work. To be useful in practice, alkali-activated binders concrete cured at ambient temperature has been used to solve this problem. In this study, a new mix design methodology for AAHFAC was modified from ACI standards. The step-by-step procedure of the mix design for the AAHFAC mixes has been explained in the earlier section. From laboratory experiments, the 28-day compressive strength of the AAHFAC cured at ambient temperature ranging from 15 to 35 MPa was obtained. After compressive strength was obtained, the alkaline activator solution-to-binder ratios were used to modify the mix design of the AAHFAC. Using the modified mix design of the AAHFAC mix, it is found that the proposed mix design of the AAHFAC in this study meets the strength requirement. This mix design would lay a foundation for the future use of AAHFAC in construction industry.

Conflicts of Interest

The authors declare that there are no conflicts of interest.

Acknowledgments

This work was financially supported by the TRF New Research Scholar Grant no. MRG6080174 and the Thailand Research Fund (TRF) under the TRF Distinguished Research Professor Grant no. DPG6180002. The part of the present work was also supported by the European Commission Research Executive Agency via the Marie Skłodowska-Curie Research and Innovation Staff Exchange Project (689857-PRIGeoC-RISE-2015) and the Industry Academia Partnership Programme-2 (IAPP1617\16) "Development of Sustainable Geopolymer Concrete." The authors also would like to acknowledge the support of the Department of Civil Engineering, Faculty of Engineering and Architecture, Rajamangala University of Technology Isan, Thailand.

References

[1] F. Pacheco-Torgal, Z. Abdollahnejad, S. Miraldo, S. Baklouti, and Y. Ding, "An overview on the potential of geopolymers for concrete infrastructure rehabilitation," *Construction and Building Materials*, vol. 36, pp. 1053–1058, 2012.

[2] F. Xu, X. Deng, C. Peng, J. Zhu, and J. P. Chen, "Mix design and flexural toughness of PVA fiber reinforced fly ash-geopolymer composites," *Construction and Building Materials*, vol. 150, pp. 179–189, 2017.

[3] P. Chindaprasirt, P. Paisitsrisawat, and U. Rattanasak, "Strength and resistance to sulfate and sulfuric acid of ground fluidized bed combustion fly ash–silica fume alkali-activated composite," *Advanced Powder Technology*, vol. 25, no. 3, pp. 1087–1093, 2014.

[4] M. Albitar, P. Visintin, M. S. Mohamed Ali, and M. Drechsler, "Assessing behaviour of fresh and hardened geopolymer concrete mixed with class-F fly ash," *KSCE Journal of Civil Engineering*, vol. 19, pp. 1445–1455, 2015.

[5] Y. Jun and J. E. Oh, "Microstructural characterization of alkali-activation of six Korean class F fly ashes with different geopolymeric reactivity and their zeolitic precursors with various mixture designs," *KSCE Journal of Civil Engineering*, vol. 19, no. 6, pp. 1775–1786, 2015.

[6] S. Kumar, R. Kumar, and S. P. Mehrotra, "Influence of granulated blast furnace slag on the reaction, structure and properties of fly ash based geopolymer," *Journal of Materials Science*, vol. 45, no. 3, pp. 607–615, 2010.

[7] C. Li, H. Sun, and L. Li, "A review: the comparison between alkali-activated slag (Si + Ca) and metakaolin (Si + Al) cements," *Cement and Concrete Research*, vol. 40, no. 9, pp. 1341–1349, 2010.

[8] D. Ravikumar, S. Peethamparan, and N. Neithalath, "Structure and strength of NaOH activated concretes containing fly ash or GGBFS as the sole binder," *Cement and Concrete Composites*, vol. 32, no. 6, pp. 399–410, 2010.

[9] A. Palomo, M. W. Grutzeck, and M. T. Blanco, "Alkali-activated fly ashes: a cement for the future," *Cement and Concrete Research*, vol. 29, no. 8, pp. 1323–1329, 1999.

[10] T. Phoo-ngernkham, P. Chindaprasirt, V. Sata, S. Hanjitsuwan, and S. Hatanaka, "The effect of adding nano-SiO$_2$ and nano-Al$_2$O$_3$ on properties of high calcium fly ash geopolymer cured at ambient temperature," *Materials & Design*, vol. 55, pp. 58–65, 2014.

[11] T. Phoo-ngernkham, P. Chindaprasirt, V. Sata, S. Pangdaeng, and T. Sinsiri, "Properties of high calcium fly ash geopolymer pastes containing Portland cement as additive," *International Journal of Minerals, metallurgy and Materials*, vol. 20, no. 2, pp. 214–220, 2013.

[12] T. Phoo-ngernkham, P. Chindaprasirt, V. Sata, and T. Sinsiri, "High calcium fly ash geopolymer containing diatomite as additive," *Indian Journal of Engineering & Materials Sciences*, vol. 20, no. 2, pp. 214–220, 2013.

[13] T. Phoo-ngernkham, A. Maegawa, N. Mishima, S. Hatanaka, and P. Chindaprasirt, "Effects of sodium hydroxide and sodium silicate solutions on compressive and shear bond strengths of FA–GBFS geopolymer," *Construction and Building Materials*, vol. 91, pp. 1–8, 2015.

[14] T. Phoo-ngernkham, V. Sata, S. Hanjitsuwan, C. Ridtirud, S. Hatanaka, and P. Chindaprasirt, "High calcium fly ash geopolymer mortar containing Portland cement for use as repair material," *Construction and Building Materials*, vol. 98, pp. 482–488, 2015.

[15] T. Phoo-ngernkham, V. Sata, S. Hanjitsuwan, C. Ridtirud, S. Hatanaka, and P. Chindaprasirt, "Compressive strength, bending and fracture characteristics of high calcium fly ash geopolymer mortar containing Portland cement cured at ambient temperature," *Arabian Journal for Science and Engineering*, vol. 41, no. 4, pp. 1263–1271, 2016.

[16] K. Pimraksa, P. Chindaprasirt, A. Rungchet, K. Sagoe-Crentsil, and T. Sato, "Lightweight geopolymer made of highly porous siliceous materials with various Na$_2$O/Al$_2$O$_3$ and SiO$_2$/Al$_2$O$_3$ ratios," *Materials Science and Engineering A*, vol. 528, no. 21, pp. 6616–6623, 2011.

[17] D. Panias, I. P. Giannopoulou, and T. Perraki, "Effect of synthesis parameters on the mechanical properties of fly ash-based geopolymers," *Colloids and Surfaces A: Physicochemical and Engineering Aspects*, vol. 301, no. 1–3, pp. 246–254, 2007.

[18] A. M. Rashad, "Properties of alkali-activated fly ash concrete blended with slag," *Iranian Journal of Materials Science and Engineering*, vol. 10, no. 1, pp. 57–64, 2013.

[19] D. Ravikumar and N. Neithalath, "Effects of activator characteristics on the reaction product formation in slag binders activated using alkali silicate powder and NaOH," *Cement and Concrete Composites*, vol. 34, no. 7, pp. 809–818, 2012.

[20] T. Phoo-ngernkham, S. Hanjitsuwan, N. Damrongwiriyanupap, and P. Chindaprasirt, "Effect of sodium hydroxide and sodium silicate solutions on strengths of alkali activated high calcium fly ash containing Portland cement," *KSCE Journal of Civil Engineering*, vol. 21, no. 6, pp. 2202–2210, 2017.

[21] R. R. Lloyd, J. L. Provis, and J. S. J. van Deventer, "Pore solution composition and alkali diffusion in inorganic polymer cement," *Cement and Concrete Research*, vol. 40, no. 9, pp. 1386–1392, 2010.

[22] S. Ananda Kumar and T. S. N. Sankara Narayanan, "Thermal properties of siliconized epoxy interpenetrating coatings," *Progress in Organic Coatings*, vol. 45, no. 4, pp. 323–330, 2002.

[23] W. Ferdous, A. Manalo, A. Khennane, and O. Kayali, "Geopolymer concrete-filled pultruded composite beams–concrete mix design and application," *Cement and Concrete Composites*, vol. 58, pp. 1–13, 2015.

[24] M. Lahoti, P. Narang, K. H. Tan, and E. H. Yang, "Mix design factors and strength prediction of metakaolin-based geopolymer," *Ceramics International*, vol. 43, no. 14, pp. 11433–11441, 2017.

[25] P. Pavithra, M. Srinivasula Reddy, P. Dinakar, B. Hanumantha Rao, B. K. Satpathy, and A. N. Mohanty, "A mix design procedure for geopolymer concrete with fly ash," *Journal of Cleaner Production*, vol. 133, pp. 117–125, 2016.

[26] ASTM C618-15, "Standard specification for coal fly ash and raw or calcined natural pozzolan for use in concrete," in *Annual Book of ASTM Standard*ASTM International, West Conshohocken, PA, USA, 2015.

[27] S. Hanjitsuwan, T. Phoo-ngernkham, L. Y. Li, N. Damrongwiriyanupap, and P. Chindaprasirt, "Strength development and durability of alkali-activated fly ash mortar with calcium carbide residue as additive," *Construction and Building Materials*, vol. 162, pp. 714–723, 2018.

[28] T. Phoo-ngernkham, S. Hanjitsuwan, L. Y. Li, N. Damrongwiriyanupap, and P. Chindaprasirt, "Adhesion characterization of Portland cement concrete and alkali-activated binders under different types of calcium promoters," *Advances in Cement Research*, pp. 1–11, 2018.

[29] T. Phoo-ngernkham, S. Hanjitsuwan, C. Suksiripattanapong, J. Thumrongvut, J. Suebsuk, and S. Sookasem, "Flexural strength of notched concrete beam filled with alkali-activated binders under different types of alkali solutions," *Construction and Building Materials*, vol. 127, pp. 673–678, 2016.

[30] ACI 211.4R-93, *Guide for Selecting Proportions for High-Strength Concrete with Portland Cement and Fly Ash*, American Concrete Institute, Farmington Hills, MI, USA, 1993.

[31] ACI 211.1-91, *Standard Practice for Selecting Proportions for Normal, Heavyweight, and Mass Concrete*, American Concrete Institute, Farmington Hills, MI, USA, 1991.

[32] D. Hardjito, S. E. Wallah, D. M. J. Sumajouw, and B. V. Rangan, "On the development of fly ash-based geopolymer concrete," *ACI Materials Journal*, vol. 101, no. 6, pp. 467–472, 2004.

[33] ASTM C143/C143M-15a, "Standard test method for slump of hydraulic-cement concrete," *Annual Book of ASTM Standard*, ASTM International, West Conshohocken, PA, USA, 2015.

[34] ASTM C403/C403M-16, "Standard test method for time of setting of concrete mixtures by penetration resistance," in *Annual Book of ASTM Standard*ASTM International, West Conshohocken, PA, USA, 2016.

[35] ASTM C469, "Standard test method for static modulus of elasticity and Poisson's ratio of concrete in compression," in *Annual Book of ASTM Standard*ASTM International, West Conshohocken, PA, USA, 2002.

[36] ASTM C78, "Standard test method for flexural strength of concrete (using simple beam with third-point loading)," in *Annual Book of ASTM Standard*ASTM International, West Conshohocken, PA, USA, 2002.

[37] A. Sathonsaowaphak, P. Chindaprasirt, and K. Pimraksa, "Workability and strength of lignite bottom ash geopolymer mortar," *Journal of Hazardous Materials*, vol. 168, no. 1, pp. 44–50, 2009.

[38] T. Sinsiri, T. Phoo-ngernkham, V. Sata, and P. Chindaprasirt, "The effects of replacement fly ash with diatomite in geopolymer mortar," *Computers and Concrete*, vol. 9, no. 6, pp. 427–437, 2012.

[39] S. Hanjitsuwan, S. Hunpratub, P. Thongbai, S. Maensiri, V. Sata, and P. Chindaprasirt, "Effects of NaOH concentrations on physical and electrical properties of high calcium fly ash geopolymer paste," *Cement and Concrete Composites*, vol. 45, pp. 9–14, 2014.

[40] U. Rattanasak and P. Chindaprasirt, "Influence of NaOH solution on the synthesis of fly ash geopolymer," *Minerals Engineering*, vol. 22, no. 12, pp. 1073–1078, 2009.

[41] A. Hawa, D. Tonnayopas, W. Prachasaree, and P. Taneerananon, "Development and performance evaluation of very high early strength geopolymer for rapid road repair," *Advances in Materials Science and Engineering*, vol. 2013, Article ID 764180, 9 pages, 2013.

[42] F. Pacheco-Torgal, J. P. Castro-Gomes, and S. Jalali, "Adhesion characterization of tungsten mine waste geopolymeric binder. Influence of OPC concrete substrate surface treatment," *Construction and Building Materials*, vol. 22, no. 3, pp. 154–161, 2008.

[43] F. Pacheco-Torgal, J. A. Labrincha, C. Leonelli, A. Palomo, and P. Chindaprasirt, *Handbook of Alkali-Activated Cements, Mortars and Concretes*, WoodHead Publishing Limited-Elsevier Science and Technology, Cambridge, UK, 2014.

[44] P. Chindaprasirt, T. Chareerat, S. Hatanaka, and T. Cao, "High strength geopolymer using fine high calcium fly ash," *Journal of Materials in Civil Engineering*, vol. 23, no. 3, pp. 264–270, 2011.

[45] C. Ridtirud, P. Chindaprasirt, and K. Pimraksa, "Factors affecting the shrinkage of fly ash geopolymers," *International Journal of Minerals, Metallurgy, and Materials*, vol. 18, no. 1, pp. 100–104, 2011.

[46] K. Sagoe-Crentsil, T. Brown, and A. Taylor, "Drying shrinkage and creep performance of geopolymer concrete," *Journal of Sustainable Cement-Based Materials*, vol. 2, no. 1, pp. 35–42, 2013.

[47] S. Songpiriyakij, T. Pulngern, P. Pungpremtrakul, and C. Jaturapitakkul, "Anchorage of steel bars in concrete by geopolymer paste," *Materials & Design*, vol. 32, no. 5, pp. 3021–3028, 2011.

[48] P. K. Sarker, "Bond strength of reinforcing steel embedded in fly ash-based geopolymer concrete," *Materials and Structures*, vol. 44, no. 5, pp. 1021–1030, 2011.

[49] P. K. Sarker, S. Kelly, and Z. Yao, "Effect of fire exposure on cracking, spalling and residual strength of fly ash geopolymer concrete," *Materials & Design*, vol. 63, pp. 584–592, 2014.

[50] M. Sofi, J. S. J. van Deventer, P. A. Mendis, and G. C. Lukey, "Engineering properties of inorganic polymer concretes (IPCs)," *Cement and Concrete Research*, vol. 37, no. 2, pp. 251–257, 2007.

7

Comparisons of Tensile Fracturing Behaviors of Hydraulic Fully Graded and Wet-Screened Concretes: A Mesoscale Study

Lei Xu [iD],[1] Yongmiao Jin [iD],[1] Shuaizhao Jing [iD],[1] Jie Liu [iD],[2] Yefei Huang [iD],[1] and Changqiao Zhou [iD][1]

[1]College of Water Conservancy and Hydropower Engineering, Hohai University, Nanjing 210098, China
[2]Chongqing Surveying and Design Institute of Water Resources, Electric Power and Architecture, Chongqing 400020, China

Correspondence should be addressed to Lei Xu; leixu@hhu.edu.cn

Academic Editor: Jun Liu

The widely used wet-screening method in the experimental testing of hydraulic fully graded concrete inevitably results in a gap between the real mechanical parameters of hydraulic fully graded concrete specimens and those of the corresponding wet-screened specimens and therefore necessitates the comparative study on their mechanical behaviors. To this end, a two-dimensional mesoscale modeling methodology is developed for simulating the tensile fracturing behaviors of hydraulic fully graded and wet-screened concretes, and extensive Monte Carlo simulations are performed. The individual effects of specimen size variation, variation of gradation and volume fraction of coarse aggregates, and the weaker interfacial transition zones surrounding the large coarse aggregates to be removed by wet-screening are detailed followed by the discussion on the combined effect of these three main factors. All the mean values of the macroscopic mechanical parameters related to tensile fracturing behaviors are found to show significant change in response to wet-screening, and the underlying differentiation mechanism and governing factor(s) are identified. Furthermore, it is shown that the randomness of the investigated parameters can be roughly described by the Gaussian distribution, and the dispersion of each of the investigated parameters of hydraulic fully graded concrete is higher than that of the corresponding wet-screened concrete.

1. Introduction

Fully graded concrete is widely used in the construction of hydraulic structures such as gravity dam and arch dam [1]. Compared to ordinary concrete with the maximum size of aggregate (MSA) no more than 40 mm, hydraulic fully graded concrete (hereinafter referred to as HFGC) employs larger coarse aggregates. Typically, the MSA reaches to 80 mm in the case of hydraulic three-graded concrete, while for the four-graded case, the MSA is even larger and increases up to 150 mm (or 120 mm). Moreover, HFGC is also featured by high volume fraction of aggregates composed of fine aggregates and coarse aggregates with the cutoff size between them taken to be 5 mm, low cement content, large admixture dosage, and high water-to-cement ratio [2]. As a result, the mechanical behaviors of HFGC are different from those of the extensively studied ordinary concrete [3]. Thus, to obtain the fundamental mechanical parameters especially the parameters related to fracturing behaviors,

which are needed by the failure analysis of hydraulic mass concrete structures, a detailed investigation on the mechanical behaviors of HFGC is required [4, 5].

With regard to identifying the mechanical parameters of concrete, the larger aggregate size implies that a larger size of concrete specimens, typically at least three times of MSA, is needed to fulfill the requirement of statistical representations [6]. According to the Chinese code SL 352-2006 (test code for hydraulic concrete), the length of specimens L with cube shape for hydraulic three-graded concrete should be set to 300 mm, while for the four-graded case, L is further increased to 450 mm. Compared to the standard specimen with L equal to 150 mm prepared for ordinary concrete, the larger specimen for HFGC causes particular challenges when conducting experimental testing in the usual concrete laboratory, including large size of the testing machine, inconvenience of testing operation, and high testing expense, which further hampers to a large extent the direct experimental testing on HFGC specimens [7].

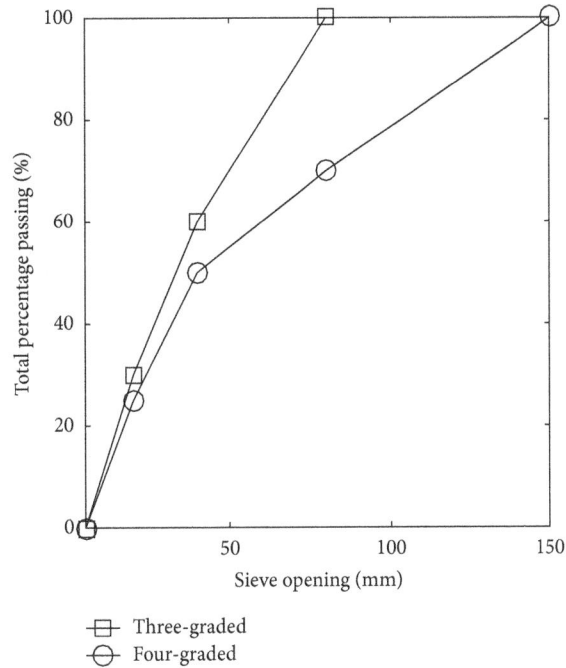

FIGURE 1: Typical coarse aggregate grading curves for hydraulic fully graded concrete.

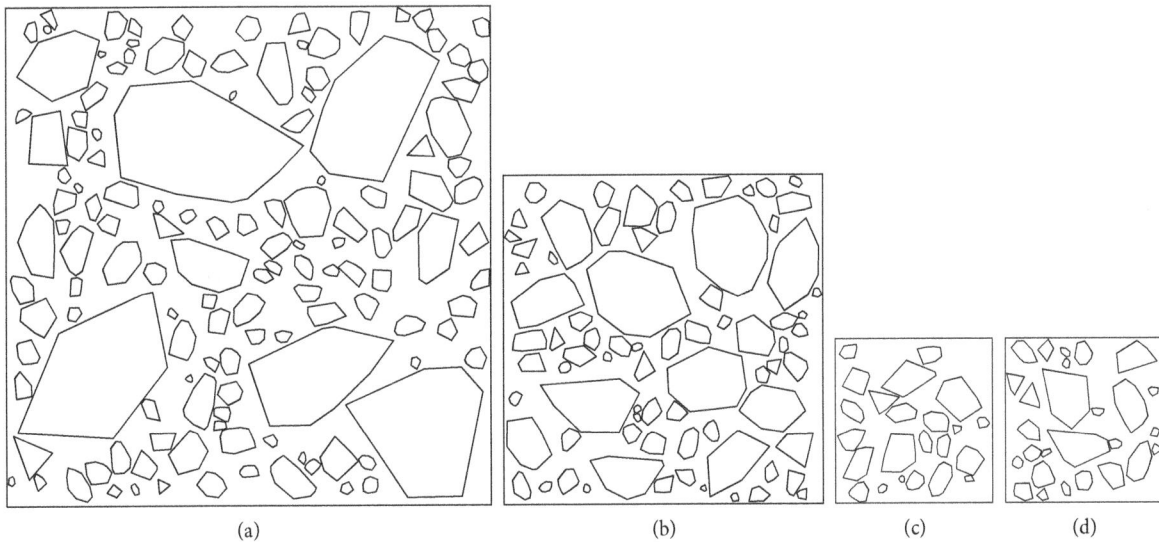

FIGURE 2: Hydraulic fully graded and wet-screened concrete specimens. (a) Four-graded specimen (A_F = 50%, L = 450 mm). (b) Three-graded specimen (A_F = 50%, L = 300 mm). (c) Wet-screened specimen corresponding to the four-graded specimen (A_F = 33.3%, L = 150 mm). (d) Wet-screened specimen corresponding to the three-graded specimen (A_F = 37.5%, L = 150 mm).

As an alternative to carrying out mechanical experiments on large specimens, the well-known wet-screening method is commonly employed in practice [8]. Following this method, experimental testing is performed on standard specimens, which are cast using the wet-screened concrete obtained by removing the coarse aggregates with the size larger than 40 mm from the original fully graded concrete through sieving, enabling the testing under usual laboratory conditions. Although the wet-screening method can avoid the direct testing on large specimens, a gap inevitably exists [9] between the real mechanical parameters of HFGC specimens and those of the corresponding wet-screened specimens;

that is, the standard specimens cast using the wet-screened concrete. Therefore, an extrapolation procedure should be executed with the aim of acquiring the accurate characterization of the mechanical properties of HFGC [10], which necessitates a detailed comparative study on the mechanical parameters of hydraulic fully graded and wet-screened concretes [11].

In spite of the awareness of the differences between the mechanical parameters of HFGC and those of the corresponding wet-screened concrete, related experimental comparisons are still sparse due to the practical difficulties of performing extensive large-scale tests. Pioneered by Blanks

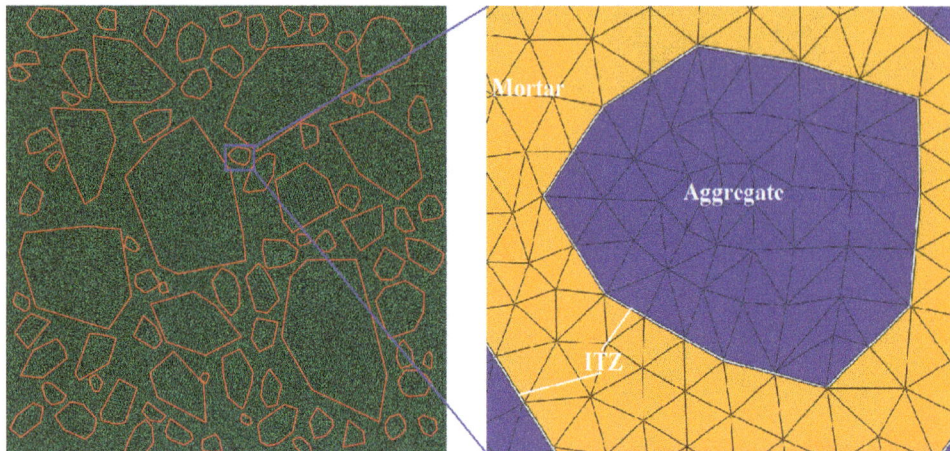

FIGURE 3: FE mesh discretization.

and McNamara [12], limited comparative experimental results can be found in [13–17], and the most recent representative investigations were carried out by Deng et al. [18] who presented the behaviors of dam and sieved concretes in uniaxial tension and compression together with the effects of curing age, by Serra et al. [19] who studied the influence of the wet-screening of the dam concrete on the development of the modulus of elasticity based on a specific experimental program using in situ creep cells, and by Shi et al. [20] who performed experimental study on uniaxial compression properties of large aggregates and wet-screened concrete at different strain rates. On the whole, although it has been well accepted that the differences mentioned above are mainly attributed to specimen size variation, variation of gradation and volume fraction of coarse aggregate, and the weaker interfacial transition zones (ITZs) surrounding large coarse aggregate particles in HFGC [21], the individual effect of these three factors, which can provide fundamental knowledge of the underlying differentiation mechanism, has been rarely studied in the past experimental research works. Furthermore, being aware of the fact that the mechanical parameters of concrete are intrinsically random, it is necessary to compare the mechanical parameters of HFGC and the corresponding wet-screened concrete in the statistical sense, which, to the best of our knowledge, has not been conducted.

On the contrary, since the mechanical behaviors of concrete on the structural scale (macroscale) are greatly controlled by its components and their interactions taking place on a finer scale (mesoscale) [22, 23], several mesoscale models have been developed to provide tools for a better understanding of the extremely complicated mechanical behaviors of concrete, especially fracturing [24]. Roughly, there are two types of mesoscale models, namely, the continuum model [25–28] and the lattice model [29–31]. In the continuum model, concrete is usually characterized by a continuum composite material with each component discretized by finite elements, while for the lattice model, a discrete system composed of lattice elements is used to represent concrete. On the whole, both of them are capable of realizing reasonable simulations of microcracking,

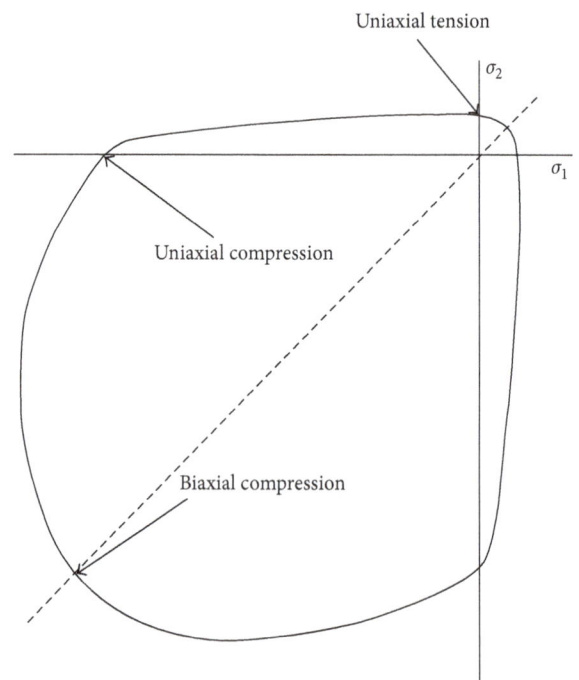

FIGURE 4: Yield surface in plane stress.

coalescence of multiple distributed cracks into localized cracks, and fracture propagation, provided each component in concrete is modeled with use of a well-suited mechanical model. In contrast to the continuum model, the lattice model is considered computationally less demanding as concrete mesostructure is roughly represented by a discrete system with relatively less degrees of freedom and meanwhile can still possess the ability to capture the most important aspects of concrete fracturing. However, it is hard to investigate the interactions of concrete components in a real sense as the actual concrete mesostructure is not fully taken into account. Consequently, mesoscale numerical simulation using the continuum model can be considered as a promising complement to experimental testing in the comparative study on the mechanical parameters of HFGC and wet-screened concrete.

TABLE 1: Mechanical properties of concrete components.

Material	E_0 (GPa)	Poisson's ratio (—)	σ_{t0} (MPa)	ϵ (—)	ψ (°)	α (—)	γ (—)
Aggregate	50	0.2	—	—	—	—	—
Mortar	20	0.2	1.94	0.1	35	0.12	2.0
ITZ	15	0.2	1.46	0.1	35	0.12	2.0

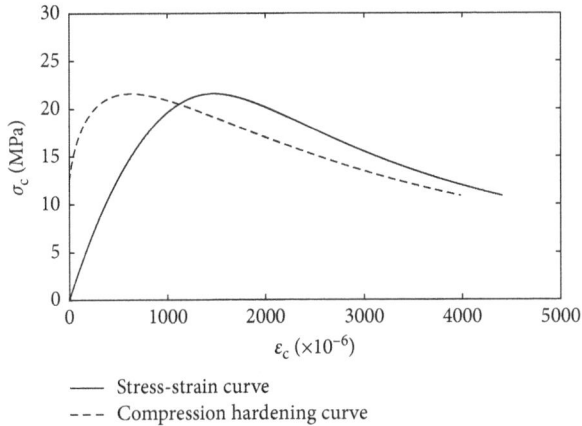

FIGURE 5: Total stress-strain relation under uniaxial compression and compression hardening curve of mortar.

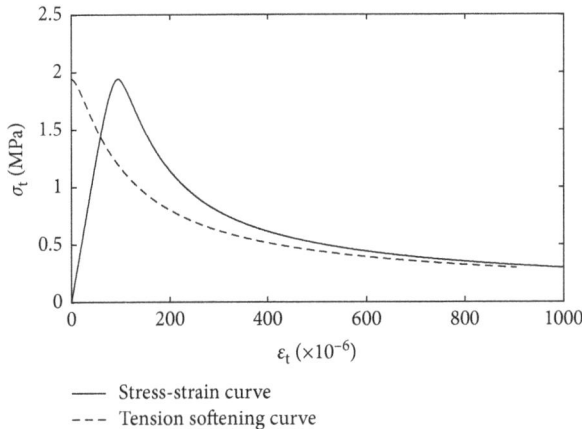

FIGURE 6: Total stress-strain relation under uniaxial tension and tension softening curve of mortar.

Considering this, a two-dimensional (2D) finite element (FE) mesoscale modeling framework for hydraulic fully graded and wet-screened concretes is proposed in this study, in which concrete is considered as a three-phase composite composed of coarse aggregate, mortar, and ITZ, and comprehensive comparisons of tensile fracturing behaviors of HFGC and the wet-screened concrete are performed based on extensive Monte Carlo simulations (MCS).

2. Mesostructure Generation of HFGC and Wet-Screened Concrete

To explicitly model the components and their interactions of both HFGC and the wet-screened concrete on the mesoscale, the internal material structure of concrete should be generated. In this regard, concrete is considered to be

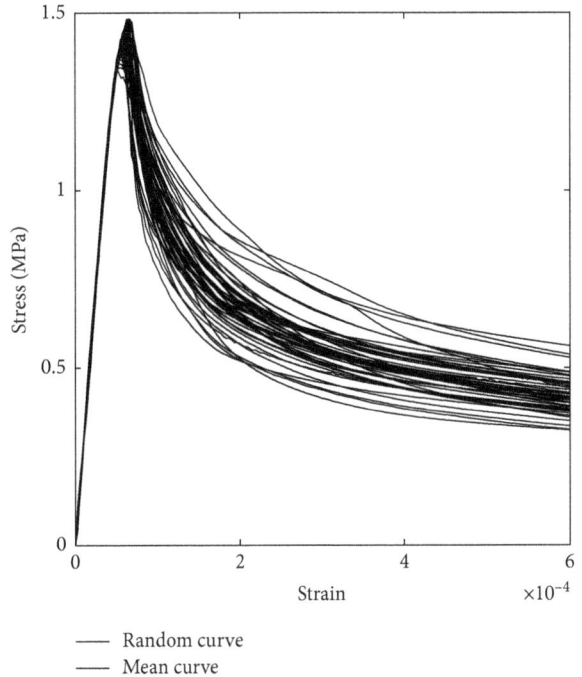

FIGURE 7: Macroscopic stress-strain curves and their mean curve of Type I.

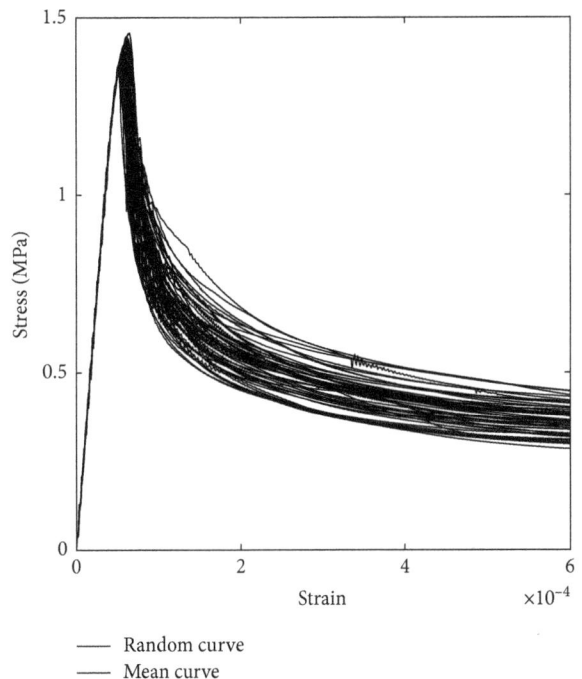

FIGURE 8: Macroscopic stress-strain curves and their mean curve of Type II.

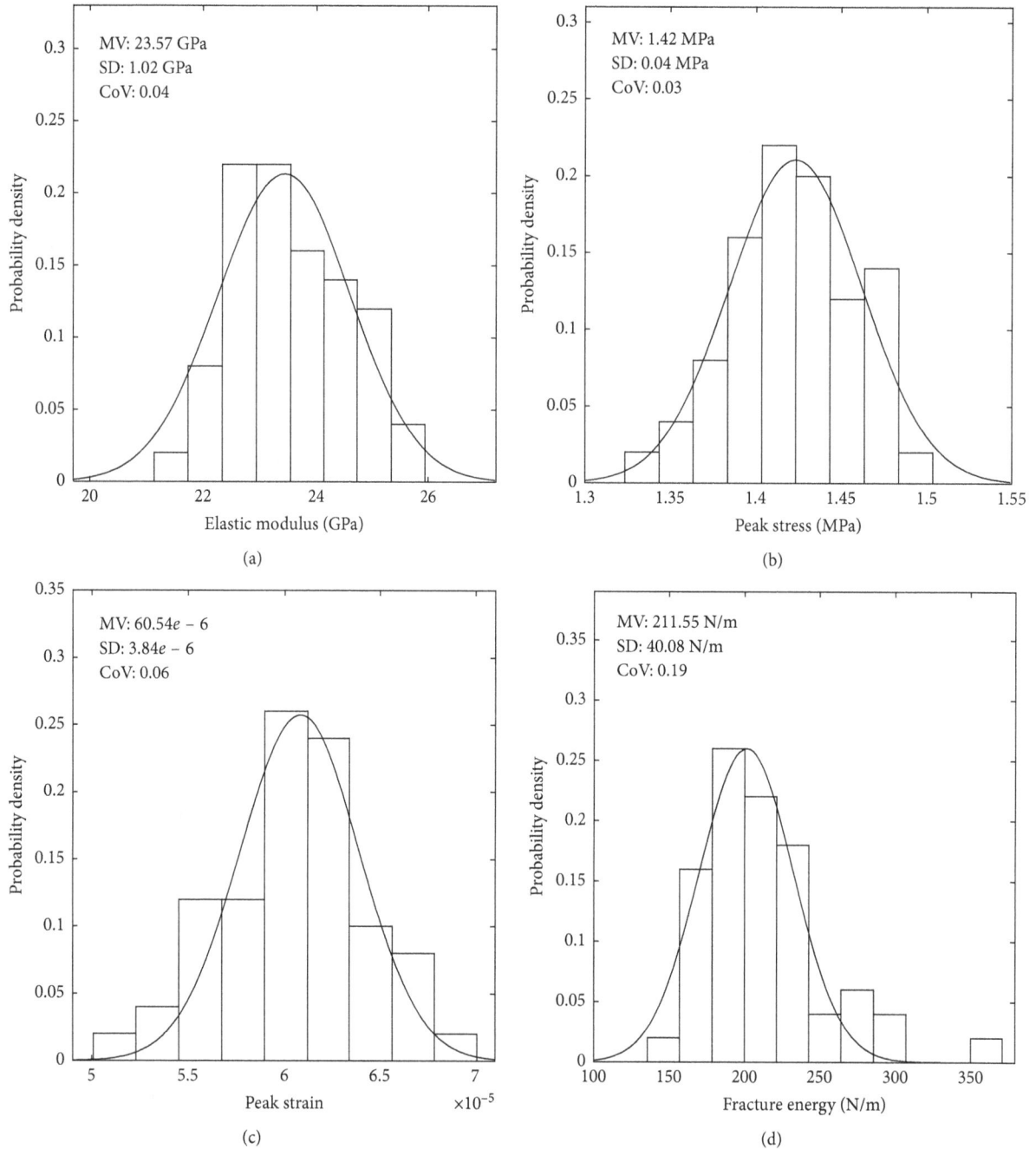

FIGURE 9: Probability density of macroscopic mechanical parameters related to tensile fracturing behaviors of Type I. (a) Tensile elastic modulus. (b) Peak stress. (c) Peak strain. (d) Fracture energy.

a composite consisting of the homogeneous mortar matrix representing cement paste and fine aggregates, randomly distributed coarse aggregates, and ITZs surrounding coarse aggregates. For a specimen, the key issue in generating its internal material structure is to determine the configuration of coarse aggregates, given the size and spatial distributions, volume fraction, and geometric characteristics of the coarse aggregate.

Unlike ordinary concrete usually adopting the well-known Fuller's curve, the size distribution of coarse aggregates of HFGC is commonly described by three or four grading segments with fixed maximum and minimum aggregate sizes for each segment. With respect to hydraulic three-graded concrete, coarse aggregates are divided into three groups: the small coarse aggregate with size ranging from 5 to 20 mm, the medium coarse aggregate with size ranging from 20 to 40 mm, and the large coarse aggregate with size ranging from 40 to 80 mm, while regarding the four-graded case, another group of coarse aggregate with even larger size ranging from 80 to 150 mm (or 120 mm), namely, the extra-large coarse aggregate, is used besides the above three groups. In addition, it is usually assumed that

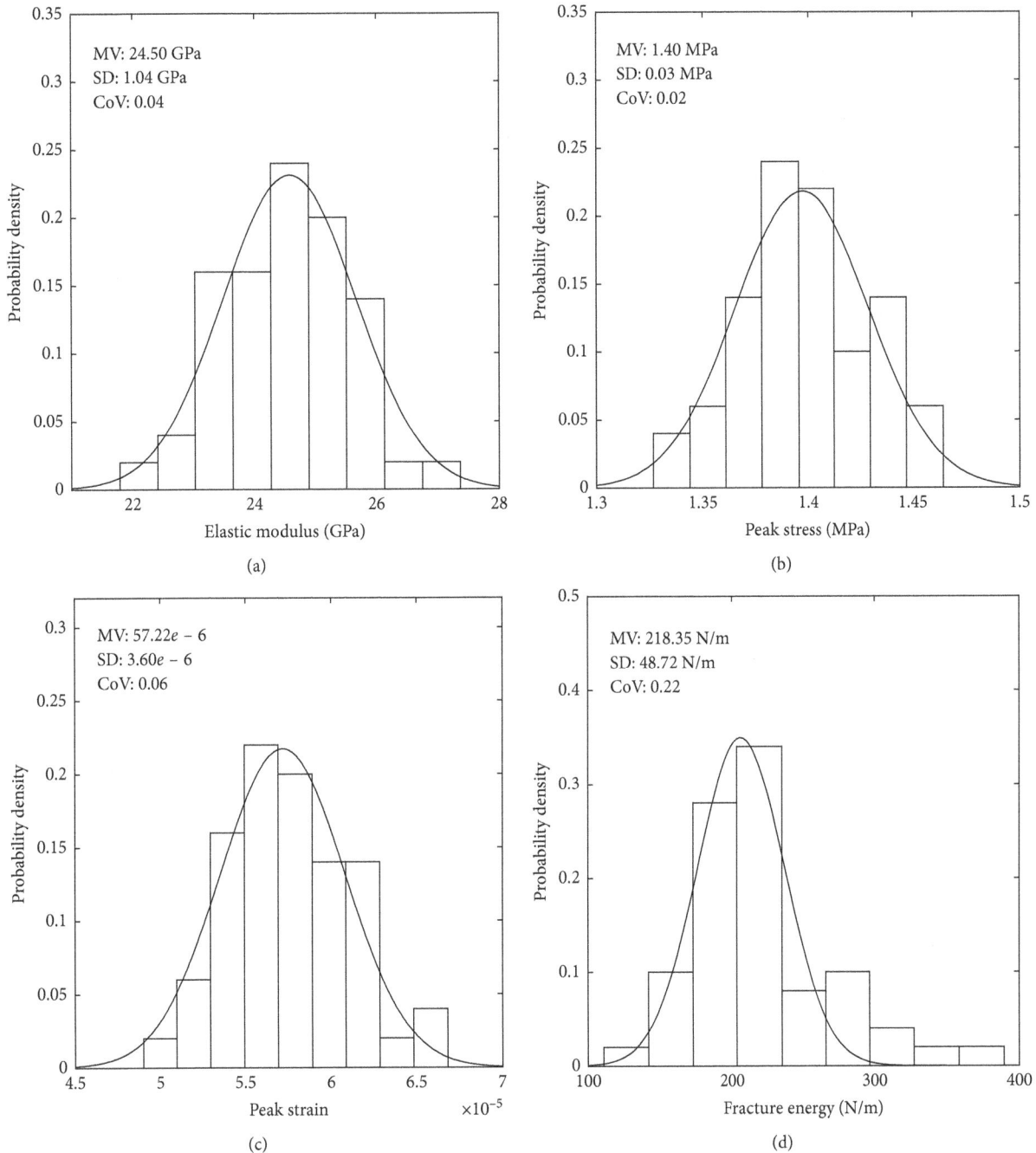

FIGURE 10: Probability density of macroscopic mechanical parameters related to tensile fracturing behaviors of Type II. (a) Tensile elastic modulus. (b) Peak stress. (c) Peak strain. (d) Fracture energy.

the size of the coarse aggregate in a certain segment follows the uniform distribution as the size distribution of the coarse aggregate for each segment is commonly not precisely controlled during the production of HFGC. Hence, on the basis of the given coarse aggregate weight of each grading segment, the size distribution of coarse aggregate can be determined with ease. Figure 1 depicts two typical coarse aggregate grading curves corresponding to hydraulic three- and four-graded concretes, respectively. While for the wet-screened concrete, the coarse aggregate gradation can be calculated in a straightforward way, by removing grading

segment(s) with size larger than 40 mm from the coarse aggregate gradation of HFGC.

Concerning the spatial distribution of coarse aggregates, it is well accepted that coarse aggregates can be considered to be randomly distributed in a certain specimen. Consequently, the location of an individual coarse aggregate needed to be placed in the specimen, which can be represented by the coordinates of its geometric center, is assumed to be uniformly distributed throughout the specimen in this study.

For a given concrete mix design, the volume fraction of coarse aggregate in the three-dimensional (3D) sense can be

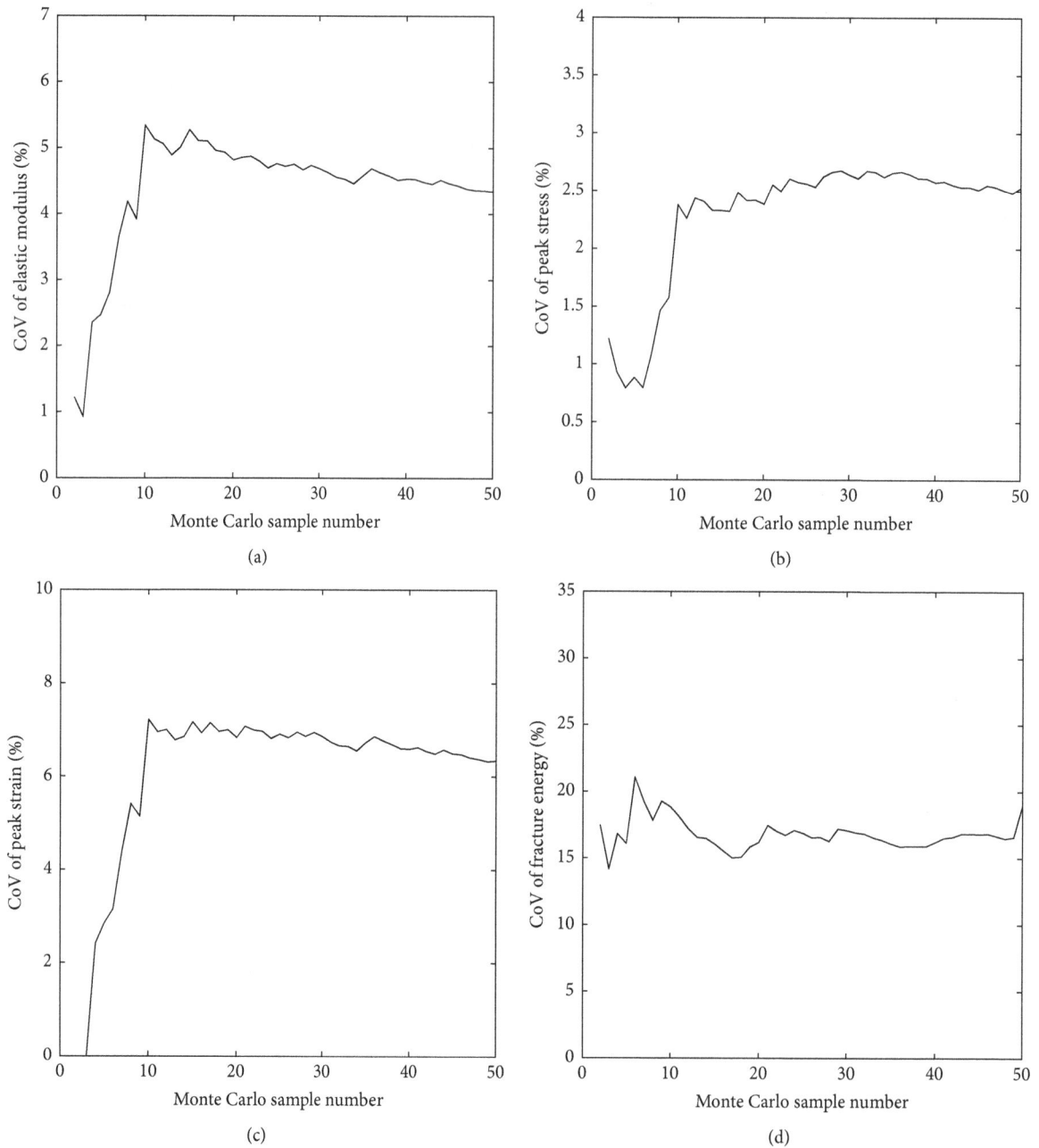

FIGURE 11: Influence of the sample number on CoVs of macroscopic mechanical parameters of Type I. (a) Tensile elastic modulus. (b) Peak stress. (c) Peak strain. (d) Fracture energy.

directly calculated through dividing the total weight of coarse aggregate used in a unit volume of concrete by its unit weight. However, for the simplified two-dimensional case, a conversion of the volume fraction of coarse aggregate from 3D to 2D should be conducted. To this end, the well-known Walraven's conversion equation used by several researchers is employed in the present work [32].

The existing studies indicate that the shape of coarse aggregate affects the mechanical behaviors of concrete, especially the mesoscale fracture mechanism [24]. Regarding HFGC, there are mainly two types of coarse aggregates: the gravel stone and the crushed stone. In the 2D case, an ellipse

can be utilized to represent the gravel stone, while the crushed stone can be simplified to a polygon. Due to the fact that the crushed stone is more commonly used in practice compared to the gravel one, the shape of the coarse aggregate in this study is modeled by the polygon.

After determining the gradation and volume fraction of the coarse aggregate along with the shape and size of the concrete specimen, the required concrete mesostructure can be randomly generated by employing the widely used parameterized modeling approach based on the take-and-place method [33]. In this study, the versatile mesostructure generator for concrete (MGC), developed using MATLAB, is

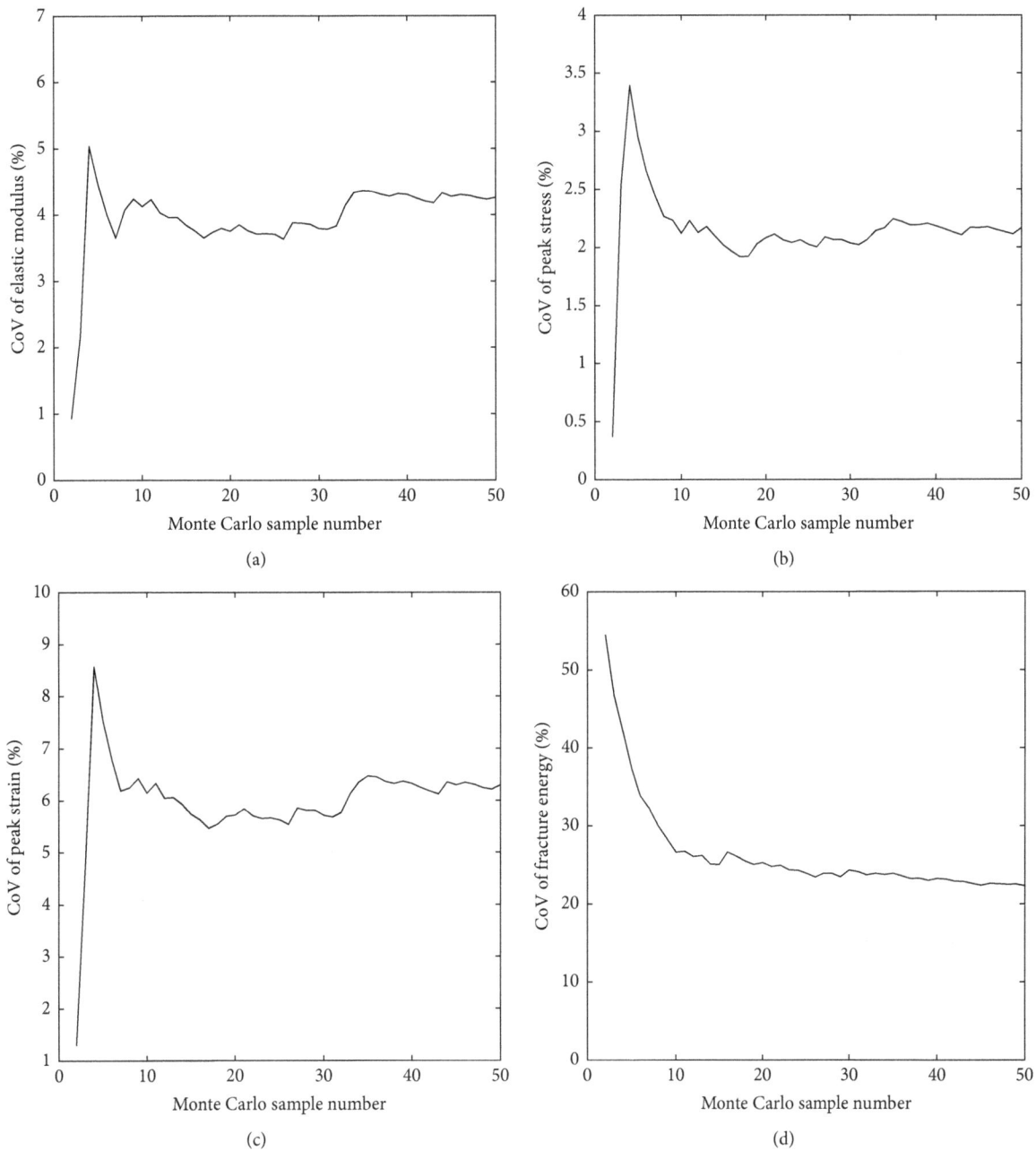

FIGURE 12: Influence of sample number on CoVs of macroscopic mechanical parameters of Type II. (a) Tensile elastic modulus. (b) Peak stress. (c) Peak strain. (d) Fracture energy.

used to generate the following required HFGC specimens and the wet-screened concrete specimens, and the detailed procedure and implementation of MGC can be referred to our previous work [34]. Figure 2(a) sketches an example of a hydraulic four-graded concrete specimen with the coarse aggregate volume fraction $A_F = 50\%$ and the proportions of the small, medium, large, and extra-large coarse aggregates set to 0.25, 0.25, 0.2, and 0.3, respectively, while an example of a hydraulic three-graded concrete specimen with the same coarse aggregate volume fraction and the proportions of the small, medium, and large coarse aggregates set to 0.3, 0.3, and 0.4, respectively, is illustrated in Figure 2(b). The wet-screened

concrete specimens corresponding to the above two examples are shown in Figures 2(c) and 2(d), respectively.

3. Finite Element Modeling Methodology and Monte Carlo Simulations

Provided the mesostructure of a concrete specimen, the corresponding computational model is needed for performing the following mesoscale study. A 2D mesoscale FE modeling methodology for tensile fracturing simulations is developed in this section. Moreover, the Monte Carlo method is employed to take into account the randomness of

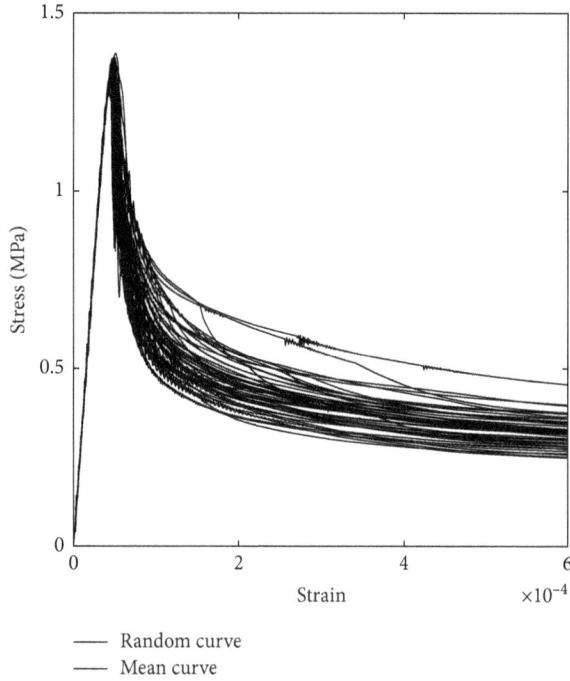

— Random curve
— Mean curve

FIGURE 13: Macroscopic stress-strain curves and their mean curve of Type III.

concrete mesostructure and to obtain statistical characteristics of the macroscopic mechanical parameters related to tensile fracturing behaviors, serving as the basis of the subsequent comparative study.

3.1. Mesh Discretization.

In order to realize the precise mesh discretization of the generated HFGC and wet-screened concrete specimens, a two-step meshing approach is proposed, by which the accurate representation of concrete mesostructure in the FE model can be achieved. Firstly, an original FE mesh including coarse aggregates and mortar, rigidly obeying the given arrangement of coarse aggregates, is generated by exploiting the powerful preprocessing modules provided by the commercial finite element software ABAQUS. Then, the thin-layer four-node elements with a uniform thickness approximately set to $100 \, \mu m$ according to the experimental observation [35], which are used to model ITZs, are automatically inserted between coarse aggregate elements and their surrounding mortar elements. Following this strategy, a mesh generator is developed using MATLAB and Python, with the detailed procedure and implementation described in the previous study [34]. Moreover, in order to ensure the accurate representation of the geometric characteristics of coarse aggregates and meanwhile to generate a computationally less demanding FE model, the distance between two neighbouring seeds along the edges of coarse aggregates and the boundary of the specimen, which controls the average element size (or mesh density) to a great extent, is set to 0.4 times of the minimum size of the coarse aggregate. An example of the final FE mesh discretization with ITZ elements highlighted and the zooming image of a part of the FE mesh are depicted in Figure 3.

3.2. Constitutive Modeling.

Concerning concrete fracturing, it is well recognized that mesoscale cracking under loading first appears in ITZs owing to their weaker properties. Afterwards, the existing cracks propagate into mortar and additional cracks may initiate within mortar during the process of further loading, while coarse aggregates commonly behave elastically. Thus, the mechanical behavior of coarse aggregates is simulated herein by the isotropic linear elastic model, whereas a continuum damaged plasticity (CDP) model implemented in ABAQUS [36] is used to describe the mechanical behaviors of both mortar and ITZ, which is briefly summarized below.

In the CDP model, two independent hardening variables, that is, equivalent compressive and tensile plastic strains ($\tilde{\varepsilon}_c^p$ and $\tilde{\varepsilon}_t^p$), are employed in order to consider compressive crushing and tensile cracking, respectively. Then, two independent damage variables $d_c(\tilde{\varepsilon}_c^p)$ and $d_t(\tilde{\varepsilon}_t^p)$ are introduced to characterize the compressive and tensile damage states. In addition, to represent the overall damage in an isotropic manner, a scale variable d is defined as

$$d = 1 - (1 - s_t d_c)(1 - s_c d_t), \tag{1}$$

where s_t and s_c are the functions of the stress state which are used to represent stiffness recovery effects associated with stress reversals [37].

Thus, the damaged elastic modulus E related to different failure mechanisms under tension and compression can be calculated by

$$E = (1 - d)E_0, \tag{2}$$

where E_0 represents the initial elastic modulus.

Based on the concept of damage mechanics, the effective stress $\bar{\sigma}$ can be obtained as

$$\bar{\sigma} = \frac{\sigma}{1 - d}, \tag{3}$$

where σ is the Cauchy stress.

The yield function of the CDP model is given in the effective stress space as

$$F(\bar{\sigma}, \tilde{\varepsilon}^p) = \frac{1}{1 - \alpha}\left(\bar{q} - 3\alpha\bar{p} + \beta(\tilde{\varepsilon}^p)\langle\hat{\bar{\sigma}}_{max}\rangle - \gamma\langle-\hat{\bar{\sigma}}_{max}\rangle\right)$$
$$- \bar{\sigma}_c(\tilde{\varepsilon}_c^p), \tag{4}$$

where $\tilde{\varepsilon}^p = [\tilde{\varepsilon}_t^p \cdot \tilde{\varepsilon}_c^p]^T$; $\bar{p}$ and $\bar{q}$ are the effective hydrostatic pressure and the effective Mises equivalent deviatoric stress, respectively; $\hat{\bar{\sigma}}_{max}$ is the algebraically maximum eigenvalue of $\bar{\sigma}$; the brackets $\langle\rangle$ are used in Macaulay sense; $\bar{\sigma}_c(\tilde{\varepsilon}_c^p)$ is the uniaxial compressive effective strength; α and γ are the dimensionless material constants, which can be determined by comparing the initial equibiaxial and uniaxial compressive yield stress and by comparing the yield conditions along the tensile and compressive meridians, respectively; and $\beta(\tilde{\varepsilon}^p)$ can be calculated by

$$\beta(\tilde{\varepsilon}^p) = \frac{\bar{\sigma}_c(\tilde{\varepsilon}_c^p)}{\bar{\sigma}_t(\tilde{\varepsilon}_t^p)}(1 - \alpha) - (1 + \alpha), \tag{5}$$

where $\bar{\sigma}_t(\tilde{\varepsilon}_t^p)$ is the uniaxial tensile effective strength. Figure 4 illustrates the yield surface in the case of plane stress.

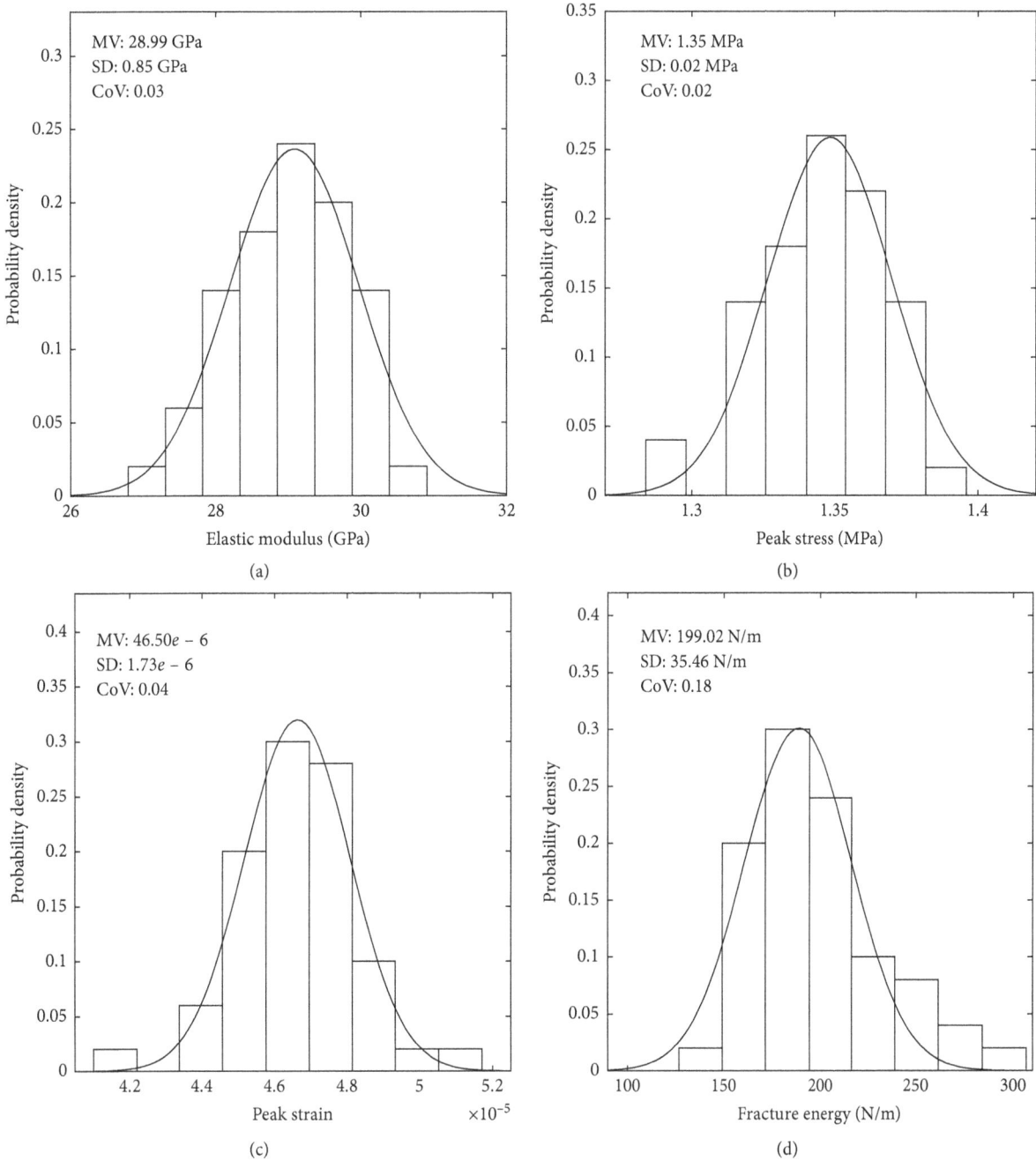

FIGURE 14: Probability density of macroscopic mechanical parameters related to tensile fracturing behaviors of Type III. (a) Tensile elastic modulus. (b) Peak stress. (c) Peak strain. (d) Fracture energy.

In order to describe the dilatancy reasonably, the non-associated flow rule is adopted in the CDP model, and the flow potential is formulated as

$$G(\overline{\sigma}) = \sqrt{\left(\epsilon\sigma_{t0} \tan \psi\right)^2 + \overline{q}^2} - \overline{p} \tan(\psi), \quad (6)$$

where ϵ is the parameter defining the rate at which the function approaches the asymptote; σ_{t0} is the uniaxial tensile stress at failure; ψ is the dilation angle measured in the $\overline{p} - \overline{q}$ plane at high confining pressure.

As stated earlier, the material softening under tension is defined by the relationship between the uniaxial tensile

effective strength and equivalent tensile plastic strain (5), which means mesh sensitivity will be encountered when applying the CDP model in FE simulations. Therefore, a stress-displacement relation is used in this study to define the tensile softening behavior for alleviating the influence of mesh sensitivity on the simulation results.

3.3. Numerical Solution Algorithm. Owing to the highly nonlinear and softening behavior of concrete in the process of tensile fracturing, the ABAQUS/Explicit solver is employed in the present work in order to capture the entire fracturing process.

(a)

(b)

(c)

(d)

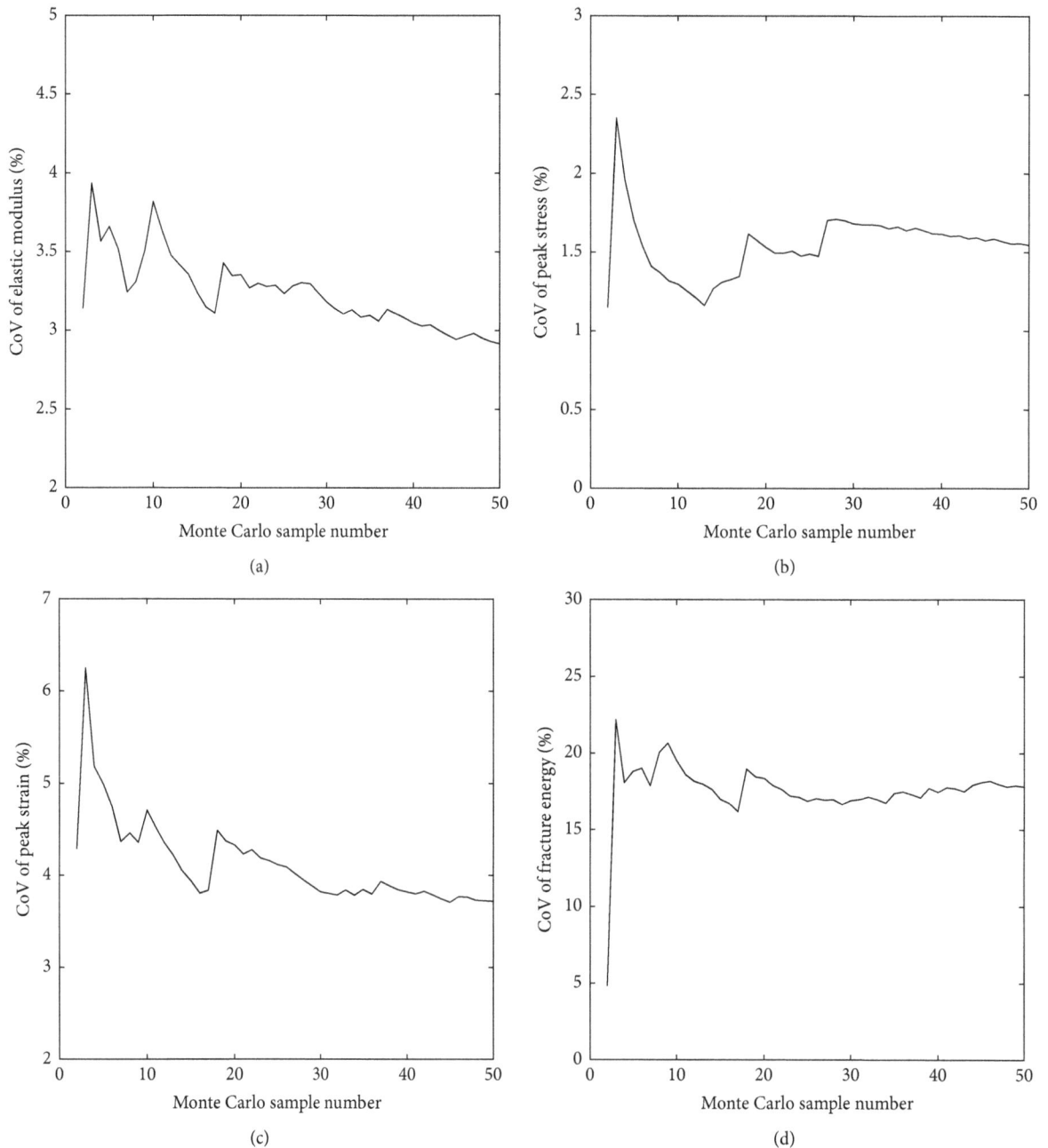

FIGURE 15: Influence of thw sample number on CoV of macroscopic mechanical parameters of Type III. (a) Tensile elastic modulus. (b) Peak stress. (c) Peak strain. (d) Fracture energy.

Regarding the explicit FE modeling, the dynamic effect inevitably exists, and therefore, its influence on the solution of a quasistatic problem should be small enough to be neglected. In order to minimise the dynamic effect, the loading duration should be large enough, while on the other hand, the computational effort increases proportionally with the increase of loading duration. Hence, a balance has to be made between the computational efficiency and simulation accuracy, which can be achieved through comparing the results under different loading durations (or loading rates).

3.4. Monte Carlo Simulations and Statistical Analysis. Due to the random arrangement of coarse aggregates, the mechanical behaviors of concrete obtained from both experimental testing and numerical modeling vary with specimens having the same size, especially in the nonlinear regime, which consequently necessitates the analysis of concrete mechanical behaviors in a statistical sense in order to get more conclusive and accurate results. Hence, the Monte Carlo simulations are conducted in this study aiming at obtaining the statistical characteristics of the mechanical parameters related to tensile fracturing behaviors of both HFGC and wet-screened concrete.

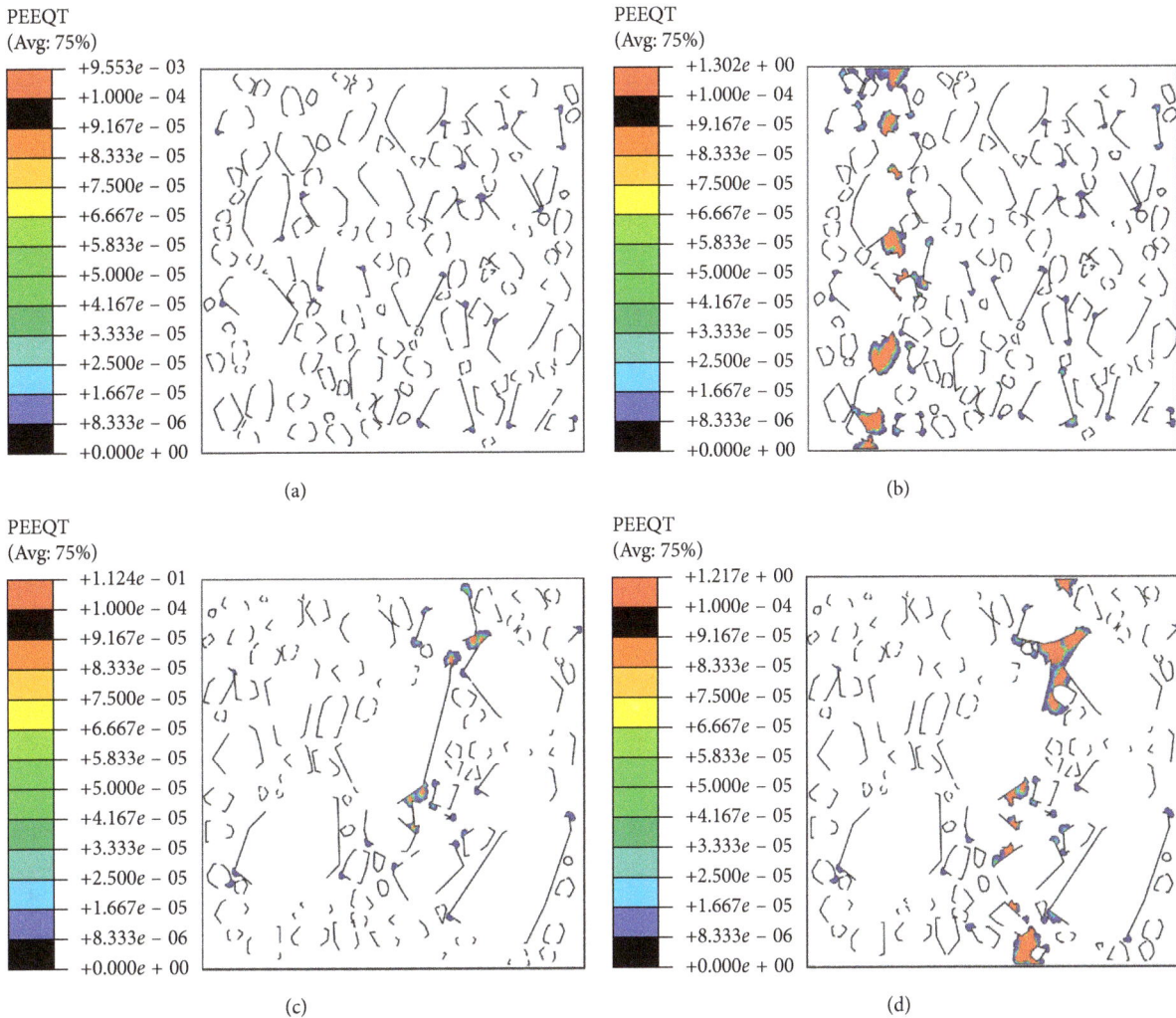

FIGURE 16: Distribution of tensile plastic strain on the mesoscale. At peak stress: (a) Type II; (c) Type III. At the maximum displacement: (b) Type II; (d) Type III.

Within the context of MCS, the sample number (the number of concrete specimens), which controls the accuracy of statistical characteristics and therefore should be large enough in order to meet the requirement of statistical convergence, can be taken as the smallest one needed to stabilize the values of statistical parameters with respect to the number of concrete specimens. On the contrary, for each specimen in the MCS, the corresponding macroscopic mechanical parameters related to concrete tensile fracturing behaviors, including tensile elastic modulus, peak stress (tensile strength), peak strain, and fracture energy, can be determined with ease based on the uniaxial tensile stress-strain (or displacement) curve obtained from the mesoscale FE modeling, and then, statistical analysis can be performed for each parameter mentioned above, providing the statistical characteristics of these parameters, including mean value (MV), standard deviation (SD), and coefficient of variation (CoV), which can be used as the basis of the subsequent comparative study.

4. Results and Discussion

The uniaxial tensile fracturing behaviors of hydraulic three-graded concrete with $A_F = 50\%$ in the sense of 2D and the proportions of small, medium, and large coarse aggregate set to 0.3, 0.3, and 0.4, respectively, and the corresponding wet-screened concrete are compared in detail in this section.

4.1. Numerical Specimens and Mechanical Properties. As stated earlier, the differences in macroscopic mechanical parameters related to tensile fracturing behaviors of HFGC and the wet-screened concrete specimens are mainly attributed to three factors: specimen size variation, variation of gradation and volume fraction of coarse aggregate, and the weaker ITZs surrounding large coarse aggregates. In order to study the effects of these three factors, six types of numerical concrete specimens are generated in this study. For each type of numerical specimens, MCS is performed, and the specimen number is taken as 50 to satisfy the requirement of statistical convergence.

Numerical specimens in Type I with standard dimensions of 150 mm × 150 mm are used to represent the wet-screened concrete specimens, while numerical specimens in Type II with dimensions of 300 mm × 300 mm are generated to represent the wet-screened concrete specimens with the same size as the hydraulic three-graded concrete

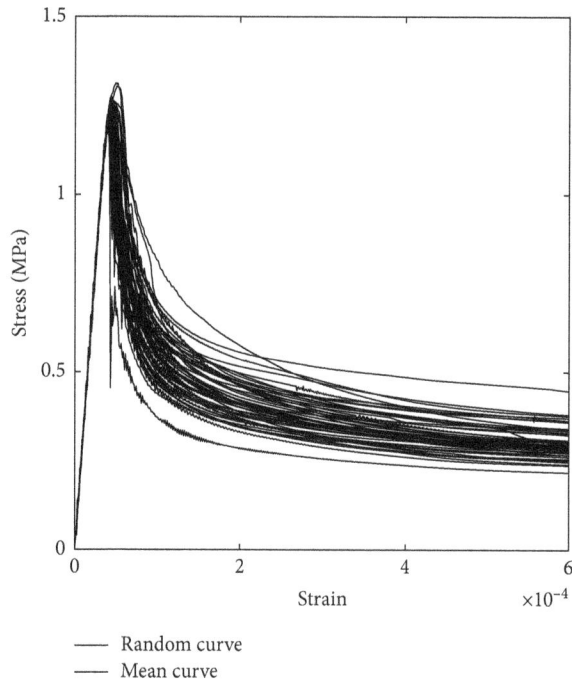

FIGURE 17: Macroscopic stress-strain curves and their mean curve of Type IV.

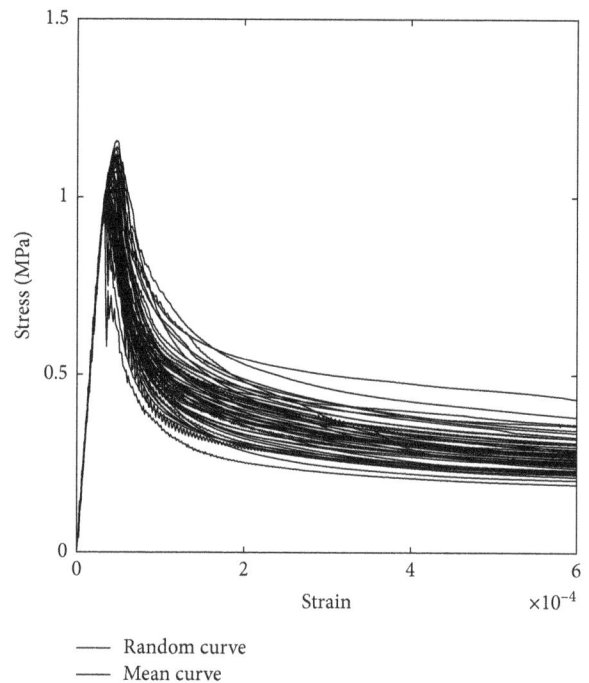

FIGURE 18: Macroscopic stress-strain curves and their mean curve of Type V.

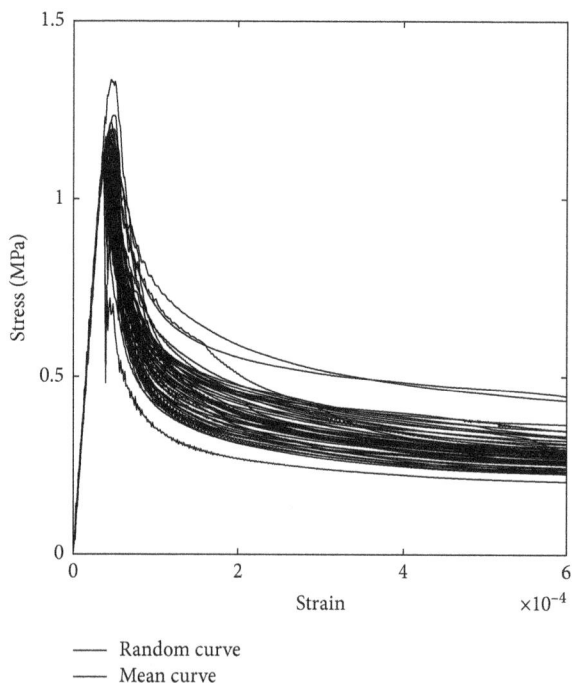

FIGURE 19: Macroscopic stress-strain curves and their mean curve of Type VI.

MCS results of Types I and II, the effect of variation of gradation and volume fraction of coarse aggregate can be individually discussed. The numerical specimens in Types IV, V, and VI employ the same mesostructures as those of the numerical specimens in Type III, and the only difference lies in the mechanical properties of the ITZs surrounding large coarse aggregates (to be removed in the wet-screening process). For Type IV, the ratio of the mechanical properties of ITZs surrounding coarse aggregates with size larger than 40 mm to those of ITZs surrounding coarse aggregates with size no more than 40 mm is set to 90%, while for Types V and VI, the ratios are set to 80% and 70%, respectively. Thus, the effect of the long weaker ITZs can be roughly evaluated by comparing the MCS results of Types III, IV, V, and VI. Finally, by comparing the MCS results of Types I, II, III, IV, V, and VI, the combined effect of three factors on the differences of the investigated macroscopic mechanical parameters can be analyzed, and the governing factor(s) corresponding to each of these parameters can also be identified.

For each numerical specimen, uniaxial tensile fracturing simulation is performed. In all FE simulations, the left end of the specimen is fixed in the horizontal direction, while the opposite end is subjected to a uniformly distributed horizontal displacement up to 0.18 mm corresponding to 1200 microstrains for Type I or 600 microstrains for Types II, III, IV, V, and VI, namely, a displacement-controlled loading scheme is used. Following the strategy discussed in Section 3.3, the loading time is set to 0.036 s, which corresponds to a loading rate 5 mm/s.

According to the setup of numerical specimen type, only the mechanical properties of coarse aggregate, mortar, and

specimens. Thus, by comparing the MCS results of Types I and II, the effect of specimen size variation can be extracted. Furthermore, by conducting the MCS of Type III consisting of the hydraulic three-graded concrete specimens with dimensions of 300 mm × 300 mm and with all ITZs assumed to having the same mechanical parameters and comparing the

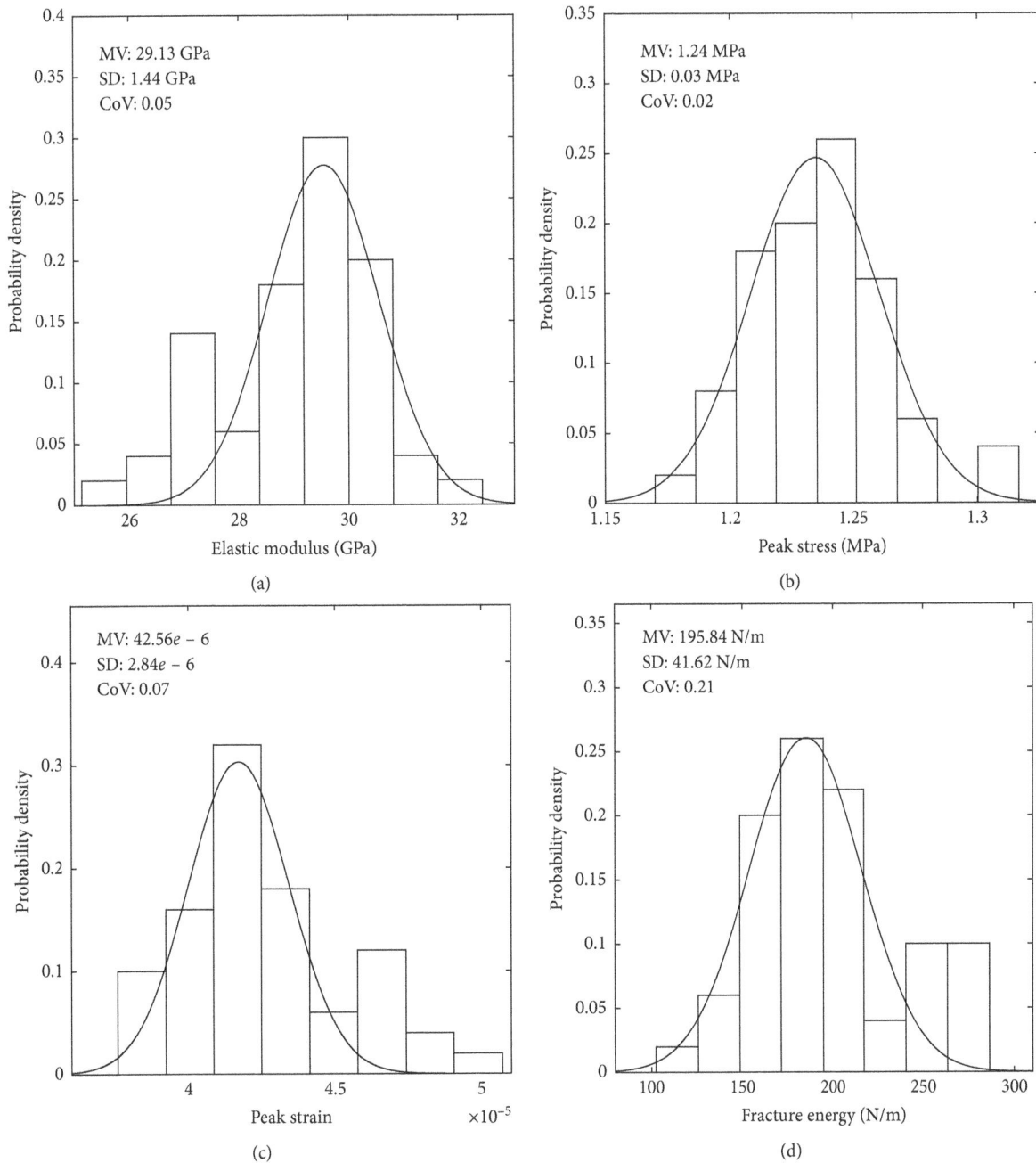

FIGURE 20: Probability density of macroscopic mechanical parameters related to tensile fracturing behaviors of Type IV. (a) Tensile elastic modulus. (b) Peak stress. (c) Peak strain. (d) Fracture energy.

ITZ in the wet-screened concrete are independent and should be defined in the mesoscale FE simulations. However, due to the lack of experimental results for ITZ, the compressive and tensile strengths and elastic modulus of ITZ are assumed to be 75% of those of mortar since ITZ is considered to be weaker than mortar, while the other mechanical properties of ITZ are taken as the same as those of mortar. Table 1 lists the mechanical properties of coarse aggregate, mortar, and ITZ adopted in this study. It is noted that the mechanical properties of mortar is directly obtained from the Chinese code GB 50010-2002 (the code for designing concrete structures), and the compression hardening

curve and the tension softening curve are shown in Figures 5 and 6, respectively.

4.2. Effect of Specimen Size Variation.

The macroscopic stress-strain curves of all 50 specimens of Type I with the mean curve are plotted in Figure 7, and Figure 8 depicts the corresponding results of Type II. Overall, it can be observed that all the concrete specimens approximately exhibit linear elastic responses on the macroscale in the prepeak stage, whereas in the postpeak phase, the macroscopic stress decreases in a nonlinear way (softening) with the increase of

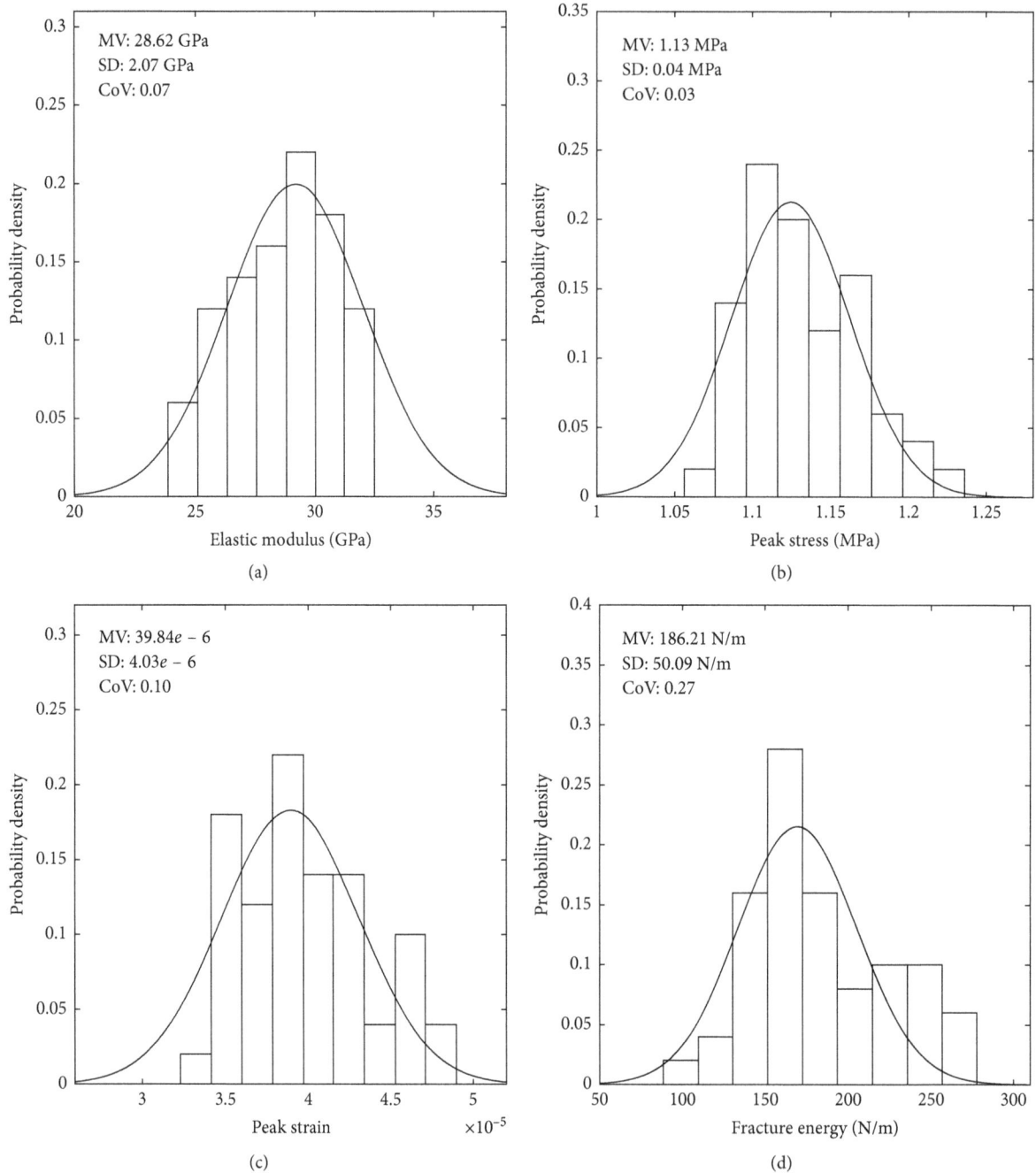

FIGURE 21: Probability density of macroscopic mechanical parameters related to tensile fracturing behaviors of Type V. (a) Tensile elastic modulus. (b) Peak stress. (c) Peak strain. (d) Fracture energy.

applied displacement, which can be attributed to the mesoscopic cracking taking place in ITZs and mortar. Furthermore, it is worth noting that although the elastic responses of all the specimens of Type I or Type II are quite close to each other, dispersion clearly appears during tensile softening, indicating the necessity of MCS.

With the peak stress, peak strain and fracture energy extracted from each stress-strain (displacement) curve, the statistical analysis of the investigated macroscopic parameters can be executed. Figure 9 presents the probability densities and the best fit Gaussian probability density

functions (PDFs) of the tensile elastic modulus, peak stress, peak strain, and fracture energy of Type I, together with MVs, SDs, and CoVs, whereas the statistical results of Type II are depicted in Figure 10. It is shown that the randomness of the investigated parameters can be roughly characterized by the Gaussian distribution. However, fracture energy is distinguished from elastic modulus, peak stress, and peak strain by its large CoV, which suggests that the random mesostructure has a bigger effect on fracture energy. With respect to each investigated macroscopic parameter, similar probability distributions can be found in Types I and II,

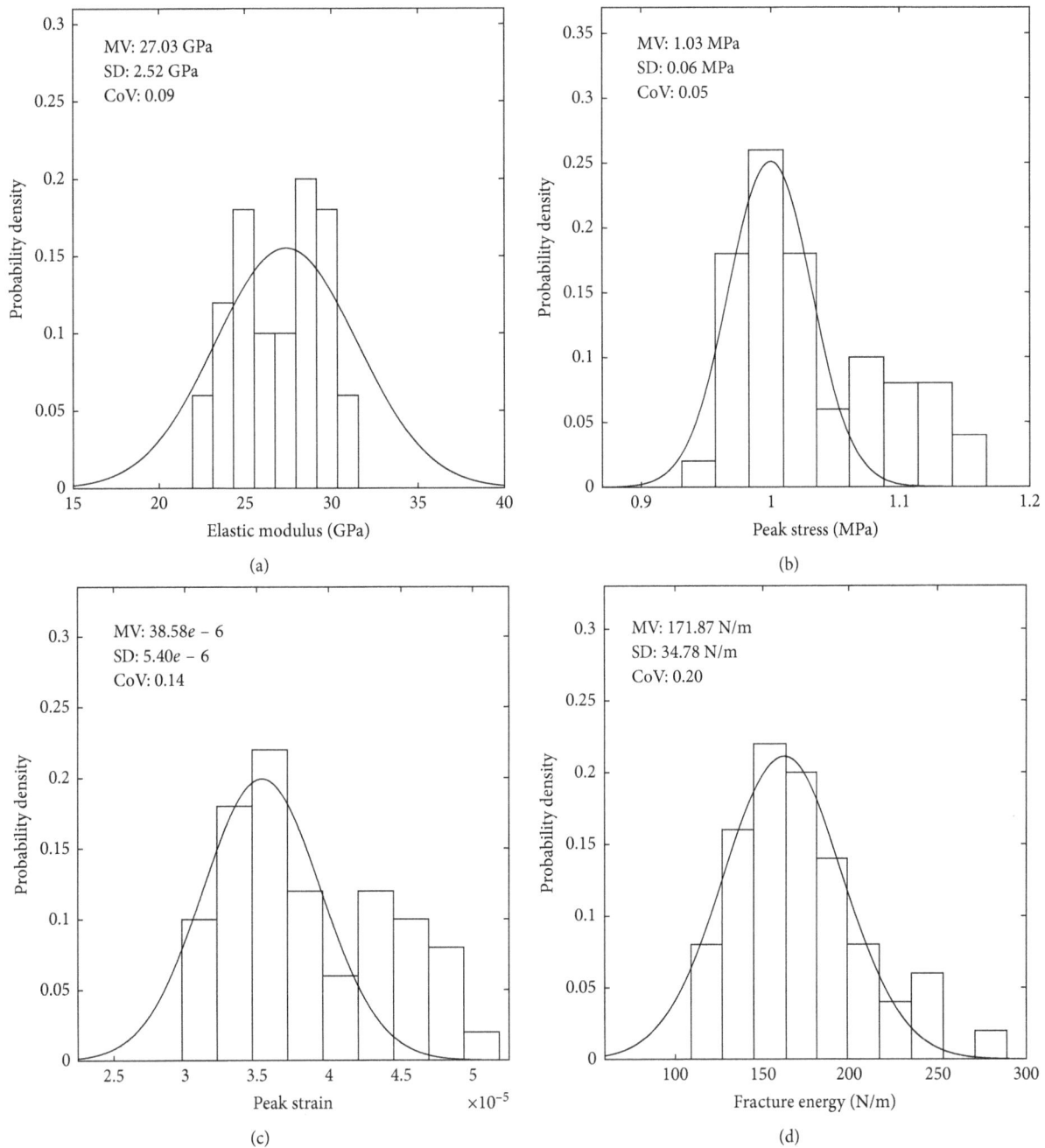

FIGURE 22: Probability density of macroscopic mechanical parameters related to tensile fracturing behaviors of Type VI. (a) Tensile elastic modulus. (b) Peak stress. (c) Peak strain. (d) Fracture energy.

which indicates the influence of the specimen size on the probability distributions of these parameters is not evident. Moreover, with the increase of the specimen size, different variations of the MVs of the investigated parameters can be observed. For the tensile elastic modulus and fracture energy, their MVs slightly increase with increasing specimen size (3.95% and 3.21%, respectively), whereas small MV reductions are shown for peak stress and peak strain (1.76% and 5.48%, respectively). On the whole, it can be concluded that the increase of the specimen size does result in the variations of the macroscopic mechanical parameters related to tensile fracturing behaviors, which is consistent with the

existing research results concerning the size effect of concrete [38]. However, in view of the large variation amplitudes of the investigated parameters obtained by limited experimental studies [11] (e.g., the typical ratio of the tensile strength of HFGC specimen to that of the corresponding wet-screened concrete specimen is ranging from 0.65 to 0.85), the variation of the specimen size is not a governing factor.

Figures 11 and 12 illustrate the variations of CoVs of tensile elastic modulus, peak stress, peak strain, and fracture energy with the sample number for Types I and II, respectively, from which it is noted that 50 random specimens are enough to achieve statistical convergence.

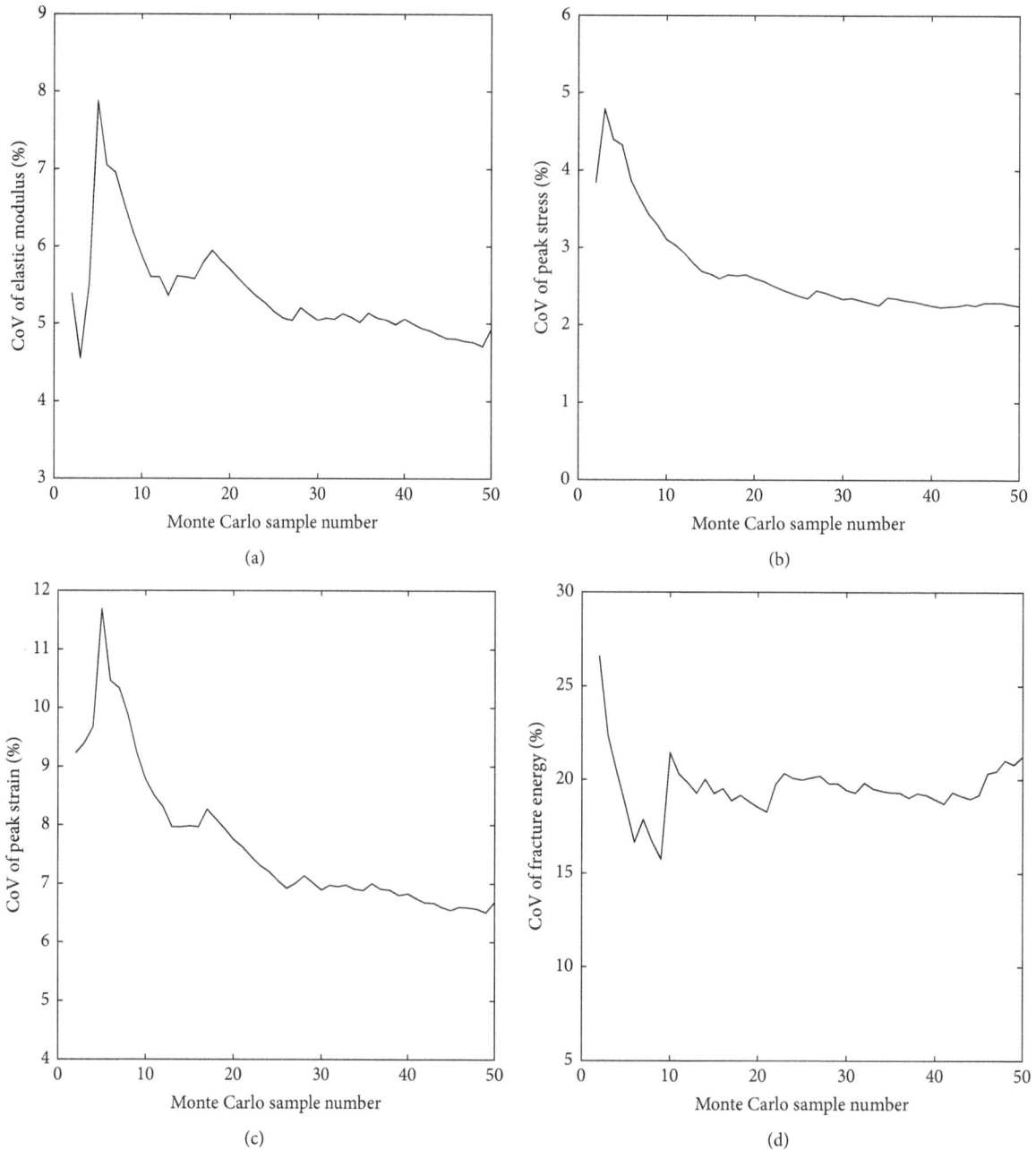

FIGURE 23: Influence of the sample number on CoV of macroscopic mechanical parameters of Type IV. (a) Tensile elastic modulus. (b) Peak stress. (c) Peak strain. (d) Fracture energy.

4.3. Effect of Variation of Gradation and Volume Fraction of Coarse Aggregate. The macroscopic stress-strain curves of all 50 specimens of Type III are drawn in Figure 13, along with the mean curve, and it is shown that the characteristic of the curves is similar to those of Type II.

Figure 14 demonstrates the probability densities and the best fit Gaussian probability density functions (PDFs) of the tensile elastic modulus, peak stress, peak strain, and fracture energy of Type III, with the calculated MVs, SDs, and CoVs. Also, it can be found that the randomness of the macroscopic parameters can be generally described by the Gaussian distribution, and the CoV of fracture energy is much larger than that of tensile elastic modulus, peak stress, and peak strain. Compared to Type II, all calculated CoVs of Type III tend to decrease, which may be due to the increase of coarse aggregate volume fraction. For tensile elastic modulus, it is shown that its MV of Type III is clearly bigger than that of Type II (increased by 18.32%), which can be attributed to the existence of large coarse aggregates (much stiffer than mortar and ITZ) in HFGC. On the contrary, the MVs of peak stress, peak strain, and fracture energy of Type III are smaller than those of Type II (reduced by 3.65%, 18.73%, and 8.85%, respectively). Overall, the variation of gradation and volume fraction of coarse aggregate is found

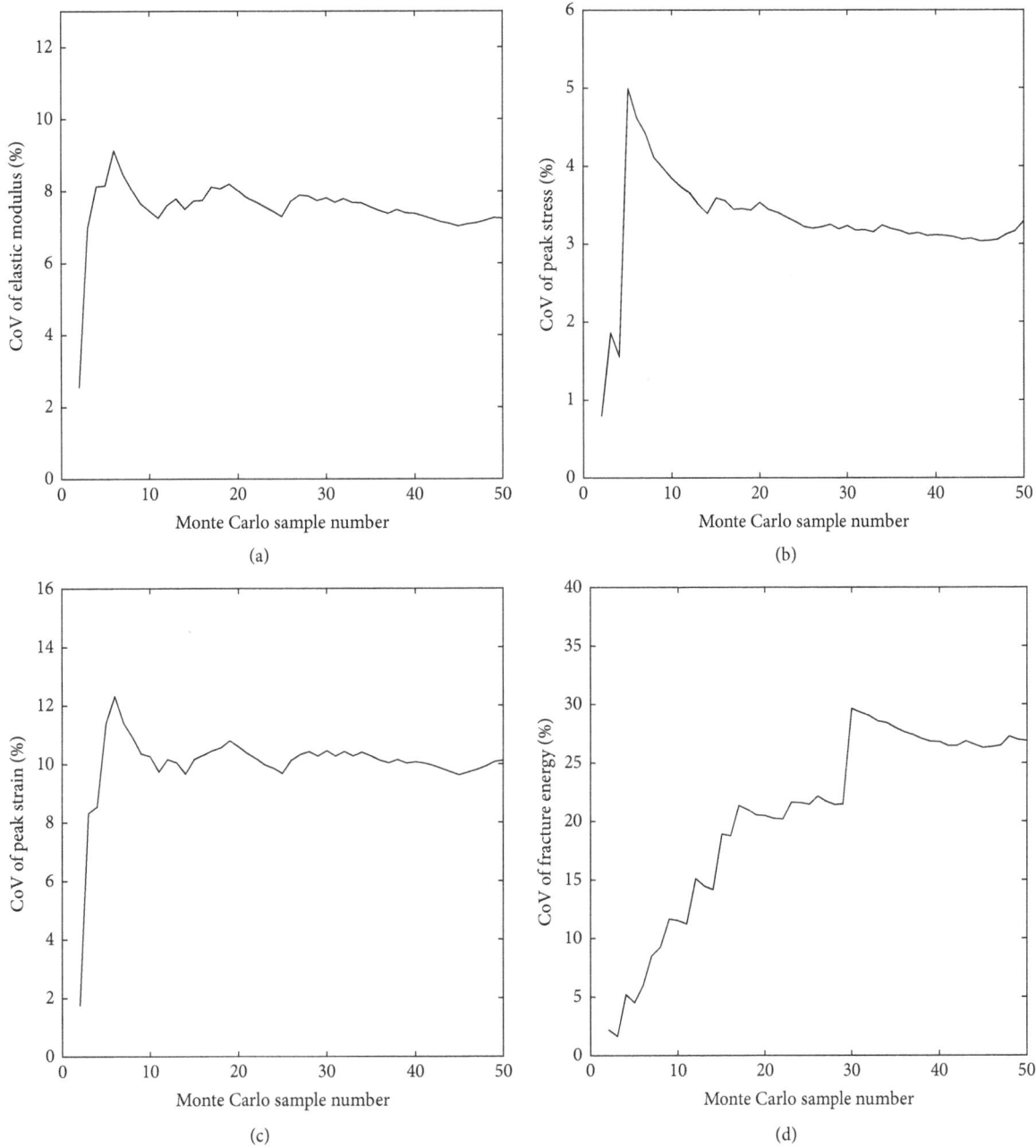

FIGURE 24: Influence of the sample number on CoV of macroscopic mechanical parameters of Type V. (a) Tensile elastic modulus. (b) Peak stress. (c) Peak strain. (d) Fracture Energy.

to give rise to big differences in the investigated parameters except peak stress and consequently can be considered to play an important role in the variations of tensile elastic modulus, peak strain, and fracture energy. Additionally, it is worth noting that specimen size variation and the variation of gradation and volume fraction of coarse aggregate have opposite effects on the variation of mean fracture energy.

The variations of CoVs of elastic modulus, peak stress, peak strain, and fracture energy with the number of samples for Type III are shown in Figure 15, from which it can also be concluded that the adopted sample number meets the requirement of statistical convergence.

To better understand the underlying differentiation mechanism on the mesoscale with respect to gradation and volume fraction, the elements with nonzero tensile plastic strain at peak stress and the maximum displacement of two typical specimens (one for Type II and the other for Type III) are visualized in Figure 16, in which the element with equivalent tensile plastic strain bigger than 100 microstrains is highlighted in red. It can be found that the majority of tensile cracks on the mesoscale is initiated in ITZs, and the subsequent propagation of mesoscale cracking is prone to take place and develop around the end of long ITZs (surrounding large coarse aggregates) due to their lower capacity

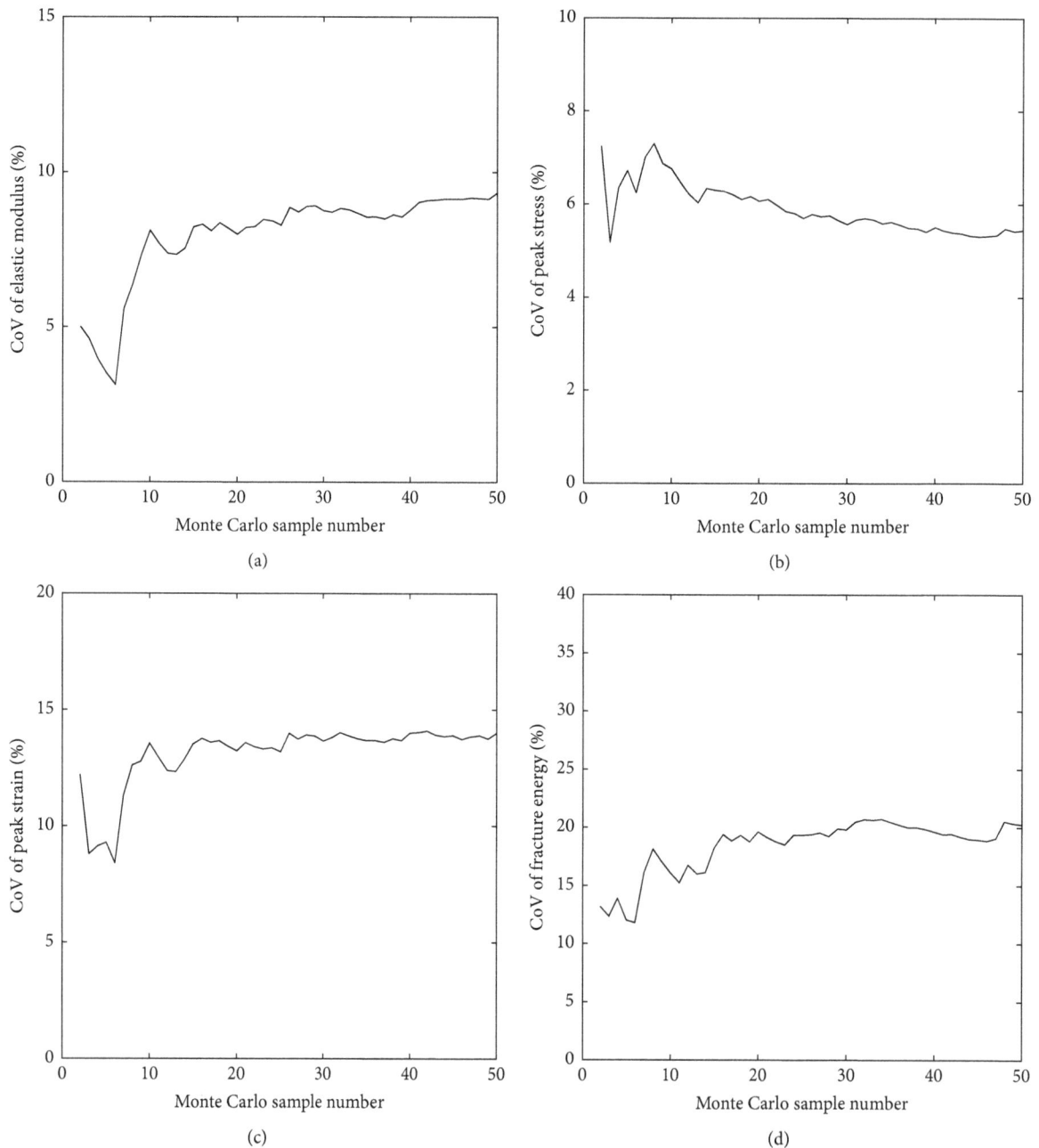

FIGURE 25: Influence of the sample number on CoV of macroscopic mechanical parameters of Type VI. (a) Tensile elastic modulus. (b) Peak stress. (c) Peak strain. (d) Fracture energy.

of resisting fracture compared to the short ITZs (around other coarse aggregates), leading to the lower peak stress of Type III. Moreover, it is shown that ITZs account for a bigger proportion of the macroscopic crack in the case of Type III compared to the Type II case, resulting in the lower fracture energy of Type III.

4.4. Effect of the Weaker ITZs Surrounding Large Coarse Aggregates. The macroscopic stress-strain curves with respect to Types IV, V, and VI, along with the corresponding mean curves, are illustrated in Figures 17–19, respectively. It

can be observed that the weakened mechanical properties of ITZs surrounding large coarse aggregates do not lead to significant change of characteristic of the macroscopic stress-strain relation, which can still be roughly divided into two stages: the prepeak linear stage and the postpeak softening stage.

The statistical results of the tensile elastic modulus, peak stress, peak strain, and fracture energy of Types IV, V, and VI are given in Figures 20–22, respectively. It is shown that the Gaussian distribution generally remains suitable for describing the randomness of these macroscopic parameters, and the dispersion of fracture energy is still much

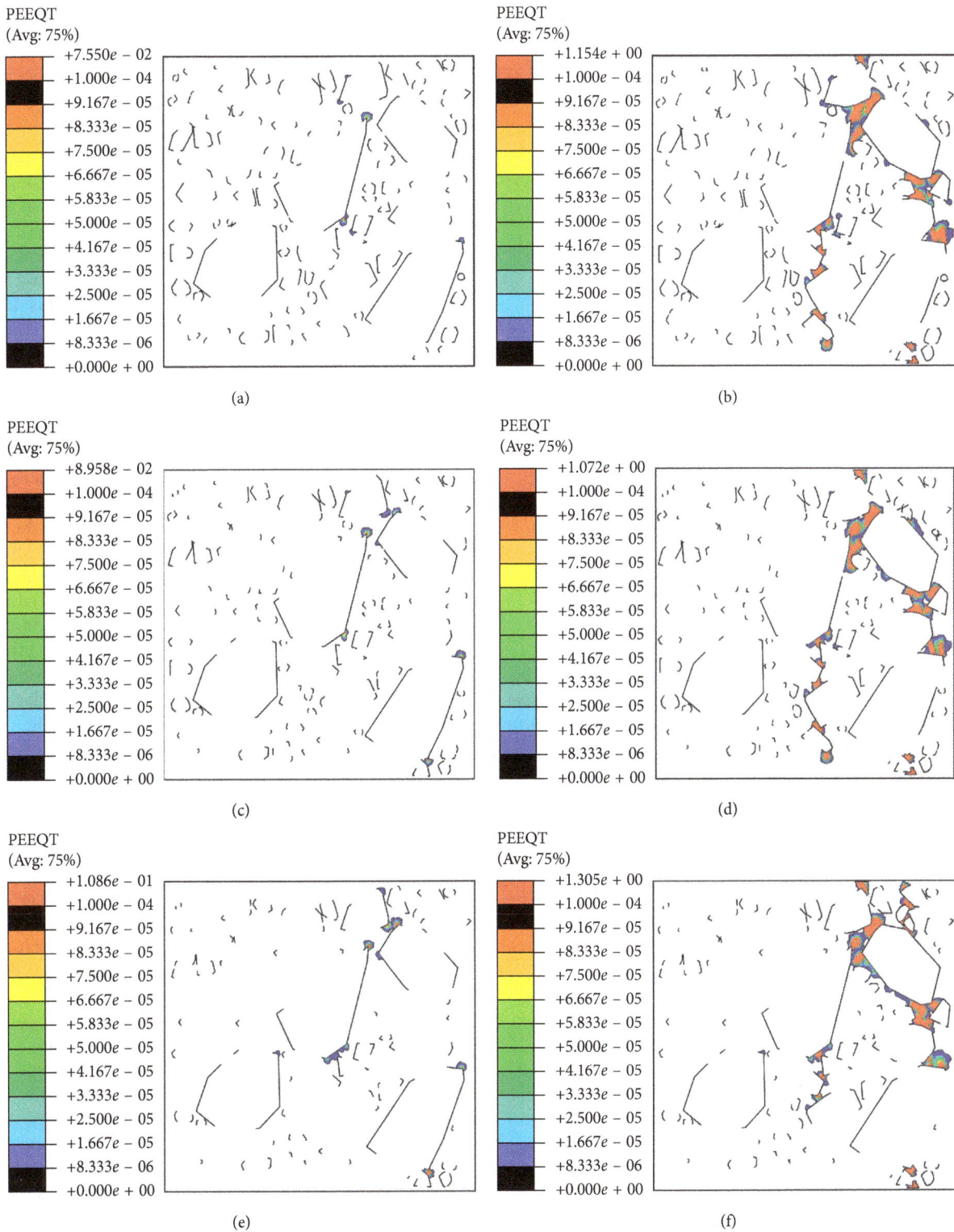

FIGURE 26: Distribution of tensile plastic strain on the mesoscale. At peak stress: (a) Type IV; (c) Type V; (e) Type VI. At the maximum displacement: (b) Type IV; (d) Type V; (f) Type VI.

higher than that of the other three parameters. Moreover, the dispersion of the investigated parameters tends to increase as the mechanical properties of the long ITZs gradually weakened, whereas tensile elastic modulus, peak stress, peak strain, and fracture energy all suffer decreased MV. Compared to Type III, the MVs of tensile elastic modulus, peak stress, peak strain, and fracture energy of Type VI are decreased by 6.76%, 23.53%, 17.03%, and

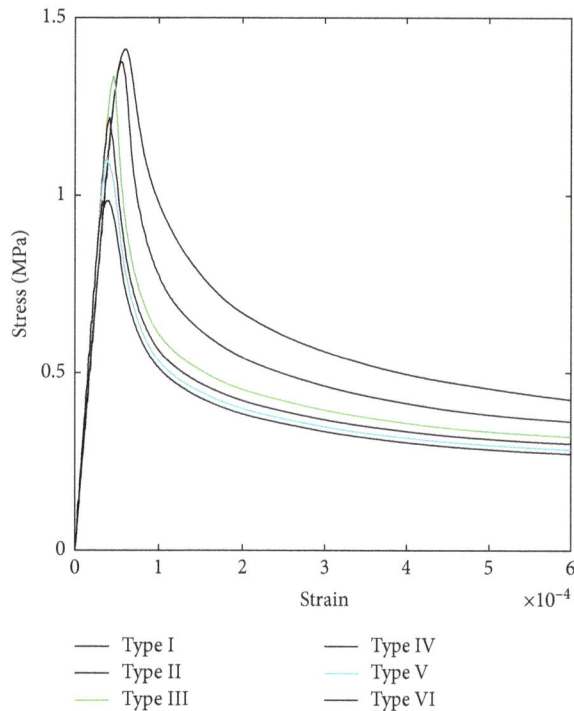

FIGURE 27: Macroscopic mean stress-strain curves of Types I, III, IV, V, and VI.

13.64%, respectively. On the whole, all the investigated parameters show significant change in response to the weaker ITZs surrounding large coarse aggregates and consequently should be considered to play a key role in the differences in the investigated parameters.

Again, it can be observed from Figures 23–25 that the number of specimens (50) is sufficient to reach converged results for Types IV, V, and VI.

Furthermore, to reveal the differentiation mechanism induced by the weaker ITZs surrounding large coarse aggregates, the elements with nonzero tensile plastic strain at peak stress and the maximum displacement of three specimens with the same mesostructure as the one illustrated in Figures 16(c) and 16(d) but different mechanical properties of ITZs surrounding large coarse aggregates, which belong to Types IV, V, and VI, respectively, are sketched in Figure 26. Although the long weaker ITZs definitely give rise to the decrease of tensile elastic modulus, the resulting lower load bearing capacity still reaches to its peak value (represented by peak stress) at a smaller macroscopic strain (peak strain), which is responsible for fewer aggregate-mortar interfacial cracks on the mesoscale (Figures 16(c), 26(a), 26(c), and 26(e)). On the contrary, as the long ITZs become weaker, the number of mesoscale cracks at the maximum displacement gradually decreases (Figures 16(d), 26(b), 26(d), and 26(f)), which means the existence of long weaker ITZs will suppress to a certain degree the development of mesoscale cracking in other regions and consequently leads to less energy dissipation.

4.5. Discussion on the Combined Effect and the Governing Factor(s). The mean stress-strain curves of Types I, II, III,

IV, V, and VI are plotted in Figure 27, and the PDFs of these types with respect to tensile elastic modulus, peak stress, peak strain, and fracture energy, with MVs, SDs, and CoVs, are shown in Figure 28.

Even though the consideration of the long weaker ITZs in HFGC results in the decrease of its tensile elastic modulus, it is shown in Figure 27 that the mean tensile elastic modulus of HFGC is still bigger than that of the corresponding wet-screened concrete. Furthermore, since the increase of tensile elastic modulus caused by increased specimen size is relatively small, the variation of gradation and volume fraction of coarse aggregate should be viewed as the governing factor of the difference in the mean tensile elastic modulus. Unlike tensile elastic modulus, the mean peak stress of HFGC is much smaller than that of the corresponding wet-screened concrete, and all the three factors are responsible for this observation in the same direction. However, the effect of the weaker ITZs surrounding large coarse aggregates is much more significant than that of the other two factors and therefore plays a governing role. Similarly, the mean peak strain of HFGC is much smaller compared to that of the corresponding wet-screened concrete, and each of the three factors plays a positive role in this regard. As the effect of increased specimen size on the decrease of mean peak strain is small, the other two factors, especially the variation of aggregate gradation and coarse aggregate content, dominate the difference in the mean peak strain. In addition, although the increased specimen size leads to a small increase in the mean fracture energy (obtained from the stress-displacement curve rather than the stress-strain curve shown in Figure 27), it is found that the mean fracture energy of HFGC is lower than that of the corresponding wet-screened concrete. Since either the variation of aggregate gradation and coarse aggregate content or the long weaker ITZs decrease the mean fracture energy to a relatively large extent, these two factors are viewed to be dominant in this respect. On the whole, although the MVs of tensile elastic modulus, peak stress, peak strain, and facture energy caused by the wet-screened method indeed show significant change in response to wet-screening, different governing factor(s) can be identified for different mechanical parameters.

Regarding the dispersion of the investigated parameters, it is shown in Figure 28 that the effects of these factors differ from each other. Specifically, the effect of specimen size variation is quite small, and the effects of the other two factors are relatively larger. The dispersions of the investigated parameters of HFGC are found to be lower than those of the corresponding wet-screened concrete when taking into consideration only the variation of gradation and volume fraction of coarse aggregate, while an opposite effect of the weaker ITZs surrounding large coarse aggregates on the dispersions can be observed. In view of the fact that the dispersion of each of the investigated parameters of HFGC is clearly higher than that of the corresponding wet-screened concrete due to the combined effect of three main factors, the weaker ITZs surrounding large coarse aggregates are considered to be the key factor controlling the dispersion variations of the investigated parameters.

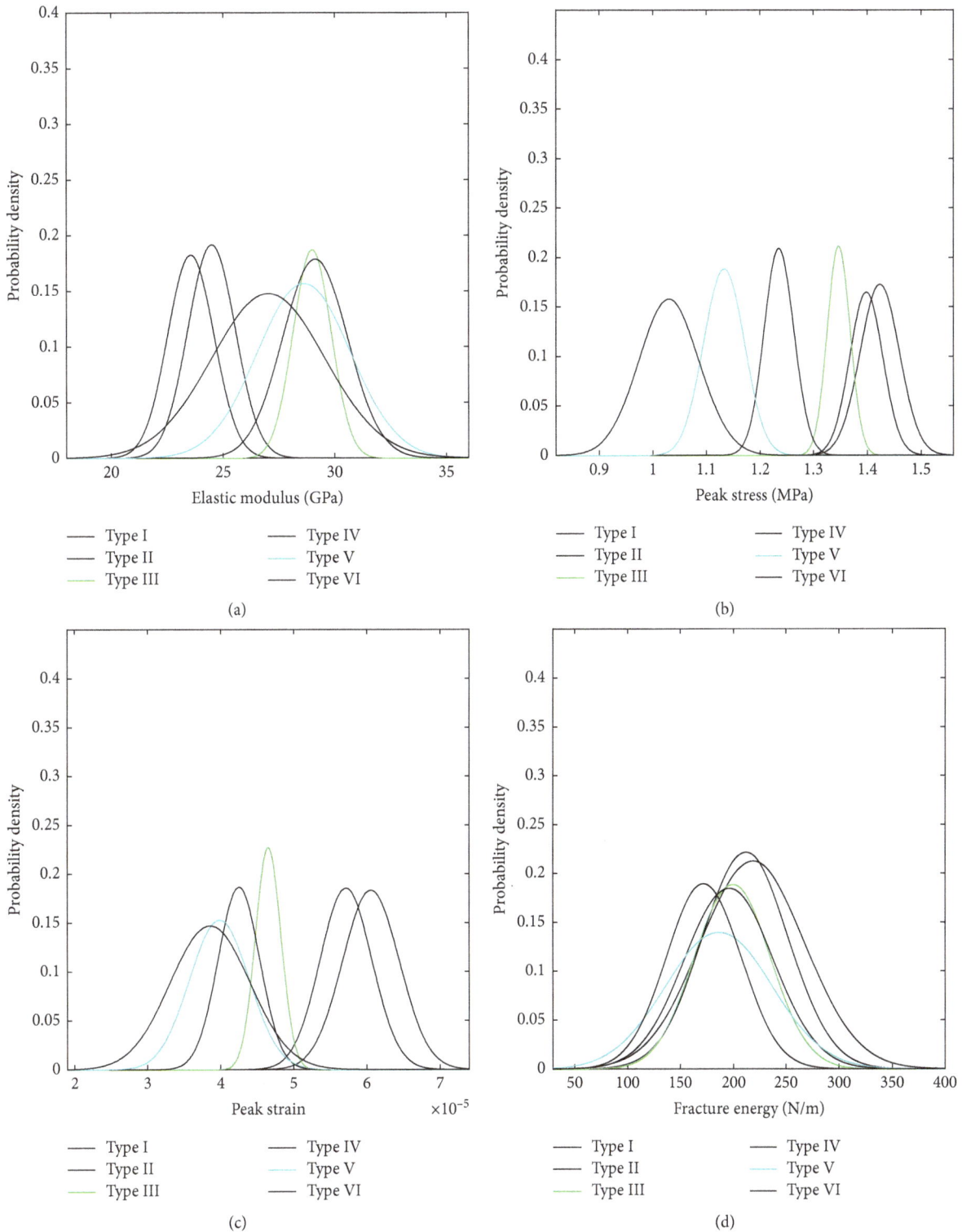

FIGURE 28: Probability densities of macroscopic mechanical parameters related to tensile fracturing behaviors of Types I, II, III, IV, V, and VI. (a) Tensile elastic modulus. (b) Peak stress. (c) Peak strain. (d) Fracture energy.

5. Conclusions

A finite element mesoscale modeling methodology for hydraulic fully graded and wet-screened concretes is first proposed in the present work. Then, based on the mesoscale FE modeling and extensive Monte Carlo simulations, a detailed comparative study on the tensile fracturing behaviors of fully graded and wet-screened concretes is carried out with prime attention placed on the individual and combined effects of three main factors, namely, specimen size variation,

variation of gradation and volume fraction of coarse aggregate, and the weaker ITZs surrounding large coarse aggregates. The work results in the following conclusions:

(i) It is shown that the tensile fracturing behaviors of hydraulic fully graded and wet-screened concretes are clearly sensitive to the internal material structure, which necessitates the extensive Monte Carlo simulations performed in this study.

(ii) All the investigated macroscopic mechanical parameters related to tensile fracturing behaviors, that is, tensile elastic modulus, peak stress, peak strain, and fracture energy, show significant change in response to wet-screening, even though the characteristics of the stress-strain (or displacement) relation of hydraulic fully graded concrete are similar to those of the wet-screened concrete.

(iii) The mean tensile elastic modulus of hydraulic fully graded concrete is found to be larger than that of the corresponding wet-screened concrete, which is mainly due to the variation of gradation and volume fraction of coarse aggregate.

(iv) Owing to the existence of the long weaker ITZs in hydraulic fully graded concrete, the mean peak stress of hydraulic fully graded concrete is smaller than that of the corresponding wet-screened concrete.

(v) Both the variation of gradation and volume fraction of coarse aggregate and the long weaker ITZs are considered to lead to a smaller mean peak strain of hydraulic fully graded concrete compared to the corresponding wet-screened concrete.

(vi) It is found that the mean fracture energy of hydraulic fully graded concrete is lower than that of the corresponding wet-screened concrete and the variation of gradation and volume fraction of coarse aggregate and the long weaker ITZs are identified to be the governing factors.

(vii) The degree of dispersion of the investigated parameters of HFGC is found to be higher than that of the corresponding wet-screened concrete, and the long weaker ITZs can be considered to play a dominant role in this respect.

Conflicts of Interest

The authors declare that there are no conflicts of interest regarding the publication of this paper.

Acknowledgments

This work was supported by projects of the National Natural Science Foundation of China (Grant nos. 51739006 and U1765204) and by "the Fundamental Research Funds for the Central Universities" (Grant no. 2018B11414).

References

[1] J. Lu, S. Huang, and L. Zp, "Statistical analysis of limit tensile strain of full graded concrete samples," *Journal of Hohai University*, vol. 26, pp. 101–104, 1998.

[2] Y. Dong, K. Xiao, and H. Yang, "Experimental studies on properties of fully-graded arch dam concrete with natural aggregate," *Chinese Journal of Concrete*, vol. 2013, pp. 140–143, 2013.

[3] D. Zheng, W. Han, and Y. Zhang, "Research on strength ratio of full grade and wet-sieving hydraulic concrete," *Chinese Journal of Water Resources and Architectural Engineering*, vol. 9, pp. 10–14, 2011.

[4] M. Paggi, G. Ferro, and F. Braga, "A multiscale approach for the seismic analysis of concrete gravity dams," *Computers and Structures*, vol. 122, pp. 230–238, 2013.

[5] L. Xu, S. Jing, J. Liu, and Y. Huang, "Cracking behavior of a concrete arch dam with weak upper abutment," *Mathematical Problems in Engineering*, vol. 2017, Article ID 6541975, 13 pages, 2017.

[6] Y. Xiao, "Study on the real compressive strength of high-arch dam concrete," *Chinese Journal of Water Power*, vol. 36, pp. 103–105, 2010.

[7] T. Tang, S. Duan, and D. Yl, "Analysis on full-graded and wet-screened tests for Jinping I Arch dam concrete," *Chinese Journal of Yellow River*, vol. 34, pp. 111-112, 2012.

[8] J. Xie, Y. Tang, and G. Chen, "Study on the properties of concrete for arch dam of Xiaowan Hydropower Station," *Chinese Journal of Water Power*, vol. 35, pp. 38–41, 2009.

[9] H. Yang, M. Rao, and Y. Dong, "Influence study of extra-large stone limited size and content on full-graded concrete properties," *Construction and Building Materials*, vol. 127, pp. 774–783, 2016.

[10] A. Sajna and H. Linsbauer, "Fracture mechanics of mass concrete–wet-screening procedure (FMWS)," *Fracture Mechanics of Concrete Structures*, vol. 1, pp. 101–110, 1998.

[11] C. Serra, A. Batista, and N. M. Azevedo, "Dam and wet-screened concrete creep in compression: in situ experimental results and creep strains prediction using model B3 and composite models," *Materials and Structures*, vol. 49, no. 11, pp. 4831–4851, 2016.

[12] R. Blanks and C. McNamara, "Mass concrete tests in large cylinders," *ACI Journal Proceedings*, vol. 31, pp. 280–303, 1935.

[13] W. Liu, W. Ye, H. Ge, and N. Xie, "Experimental study on mechanical properties of full graded concrete of Dongjiang dam," *Chinese Journal of Water Resources and Hydropower Engineering*, vol. 1986, pp. 8–14, 1986.

[14] G. Huang, "Experimental research of full-graded mass concrete strength," *Chinese Journal of Water Power*, vol. 1992, pp. 45–48, 1992.

[15] Z. Yang and R. Chen, "Experimental study on mechanical properties of fully graded aggregate concrete in TGP," *Chinese Journal of Yangtze River Scientific Research Institute*, vol. 15, pp. 1–5, 1998.

[16] G. Li, Z. Yang, and D. Yang, "Experimental research on mechanical properties of full graded mass concrete for high arch dams," *Chinese Journal of China Three Gorges University*, vol. 25, pp. 500–503, 2003.

[17] Z. Deng, Q. Li, and H. Fu, "Tensile and compressive behaviors of full grade concrete made of crushed coarse aggregate," *Chinese Journal of Hydraulic Engineering*, vol. 36, pp. 214–218, 2005.

[18] Z. Deng, Q. Li, and H. Fu, "Comparison between mechanical properties of dam and sieved concretes," *Chinese Journal of Materials in Civil Engineering*, vol. 20, no. 4, pp. 321–326, 2008.

[19] C. Serra, A. L. Batista, and N. Monteiro Azevedo, "Effect of wet screening in the elastic properties of dam concrete: experimental in situ test results and fit to composite models," *Journal of Materials in Civil Engineering*, vol. 28, no. 12, p. 04016146, 2016.

[20] L. Shi, L. Wang, Y. Song, and L. Shen, "Dynamic properties of large aggregate concrete under triaxial loading," *Magazine of Concrete Research*, vol. 67, no. 6, pp. 282–293, 2015.

[21] C. Serra, A. L. Batista, N. M. Azevedo, and J. Custódio, "Prediction of dam concrete compressive and splitting tensile strength based on wet-screened concrete test results," *Journal of Materials in Civil Engineering*, vol. 29, no. 10, p. 04017188, 2017.

[22] G. Cusatis, D. Pelessone, and A. Mencarelli, "Lattice discrete particle model (LDPM) for failure behavior of concrete. I: theory," *Cement and Concrete Composites*, vol. 33, no. 9, pp. 881–890, 2011.

[23] G. Cusatis, A. Mencarelli, D. Pelessone, and J. Baylot, "Lattice discrete particle model (LDPM) for failure behavior of concrete. II: calibration and validation," *Cement and Concrete composites*, vol. 33, no. 9, pp. 891–905, 2011.

[24] X. Wang, M. Zhang, and A. P. Jivkov, "Computational technology for analysis of 3D meso-structure effects on damage and failure of concrete," *International Journal of Solids and Structures*, vol. 80, pp. 310–333, 2016.

[25] P. Wriggers and S. Moftah, "Mesoscale models for concrete: homogenisation and damage behaviour," *Finite Elements in Analysis and Design*, vol. 42, no. 7, pp. 623–636, 2006.

[26] X. Du, L. Jin, and G. Ma, "Numerical simulation of dynamic tensile-failure of concrete at meso-scale," *International Journal of Impact Engineering*, vol. 66, pp. 5–17, 2014.

[27] Y. Huang, Z. Yang, W. Ren, G. Liu, and C. Zhang, "3D meso-scale fracture modelling and validation of concrete based on in-situ X-ray computed tomography images using damage plasticity model," *International Journal of Solids and Structures*, vol. 67-68, pp. 340–352, 2015.

[28] X. Wang, Z. Yang, J. Yates, A. Jivkov, and C. Zhang, "Monte Carlo simulations of mesoscale fracture modelling of concrete with random aggregates and pores," *Construction and Building Materials*, vol. 75, pp. 35–45, 2015.

[29] R. Rezakhani and G. Cusatis, "Asymptotic expansion homogenization of discrete fine-scale models with rotational degrees of freedom for the simulation of quasi-brittle materials," *Journal of the Mechanics and Physics of Solids*, vol. 88, pp. 320–345, 2016.

[30] K. Kim and Y. Lim, "Simulation of rate dependent fracture in concrete using an irregular lattice model," *Cement and Concrete Composites*, vol. 33, no. 9, pp. 949–955, 2011.

[31] Z. Bažant, M. Tabbara, M. Kazemi, and G. Pijaudier-Cabot, "Random particle model for fracture of aggregate or fiber composites," *Journal of Engineering Mechanics*, vol. 116, no. 8, pp. 1686–1705, 1990.

[32] J. Walraven and H. Reinhardt, "Theory and experiments on the mechanical behaviour of cracks in plain and reinforced concrete subjected to shear loading," *Heron*, vol. 26, no. 1, 1981.

[33] Z. Wang, A. Kwan, and H. Chan, "Mesoscopic study of concrete I: generation of random aggregate structure and finite element mesh," *Computers and Structures*, vol. 70, no. 5, pp. 533–544, 1999.

[34] L. Xu and Y. Huang, "Effects of voids on concrete tensile fracturing: a mesoscale study," *Advances in Materials Science and Engineering*, vol. 2017, pp. 1–14, 2017.

[35] J. Unger and S. Eckardt, "Multiscale modeling of concrete," *Archives of Computational Methods in Engineering*, vol. 18, no. 3, pp. 341–393, 2011.

[36] ABAQUS, *ABAQUS 6.5 User's Manual*, ABAQUS Inc., Johnston, RI, USA, 2004.

[37] J. Lee and G. L. Fenves, "Plastic-damage model for cyclic loading of concrete structures," *Journal of Engineering Mechanics*, vol. 124, no. 8, pp. 892–900, 1998.

[38] Z. P. Bažant, "Size effect in blunt fracture: concrete, rock, metal," *Journal of Engineering Mechanics*, vol. 110, no. 4, pp. 518–535, 1984.

8

Mechanical Properties and Conversion Relations of Strength Indexes for Stone/Sand-Lightweight Aggregate Concrete

Xianggang Zhang,[1,2] Dapeng Deng,[1] and Jianhui Yang ⓘ[1,2]

[1]*Henan Province Engineering Laboratory of Eco-architecture and the Built Environment, Henan Polytechnic University, Jiaozuo 454000, China*
[2]*School of Civil Engineering, Henan Polytechnic University, Jiaozuo 454000, China*

Correspondence should be addressed to Jianhui Yang; yangjianhui@hpu.edu.cn

Academic Editor: Barbara Liguori

This is a study of the basic mechanical properties of specified density shale aggregate concrete, which is based on different replacement rates in stone-lightweight aggregate concrete (stone-LAC) and sand-lightweight aggregate concrete (sand-LAC). They were prepared by replacing the ceramsite and pottery sand with stone and river sand, respectively. Many tests were performed regarding the basic mechanical property indexes, including tests of cube compressive strength, axial compressive strength, splitting tensile strength, flexural strength, elastic modulus and Poisson's ratio. The failure modes of specified density shale aggregate concrete were obtained. The effects of replacement rates on the mechanical property indexes of specified density shale aggregate concrete were analyzed. Calculation models were implemented for elastic modulus, for the conversion relations between the axial compressive strength and the cube compressive strength, and for the relations between the tension-compression ratio and Poisson's ratio. It was shown that when the replacement rate of stone or river sand increased from 0% to 100%, the cube compressive strength of stone-LAC and sand-LAC increased, respectively, by 55% and 25%, the axial compressive strength increased, respectively, by 91% and 72%, splitting tensile strength increased, respectively, by 99% and 44%, and the flexural strength increased, respectively, by 46% and 26%. Similarly, the elastic modulus of stone-LAC and sand-LAC increased, respectively, by 16% and 30%. However, Poisson's ratio for stone-LAC decreased first and then increased, eventually increased by 11%; Poisson's ratio for sand-LAC only reduced gradually, eventually reduced by 67%. After introducing the influence parameter for the replacement rate, the established calculation models become simple and practical, and the calculation accuracies are favorable.

1. Introduction

The rapid development of the construction industry in recent years has been marked by intensifying research and development for full-lightweight aggregate concrete (FLAC), where the aggregate consists of ceramsite and shale pottery. Compared with ordinary concrete, FLAC can effectively alleviate the environmental destruction resulting from the exploitation of normal stone and sand. FLAC also has the advantage of low density, as well as favorable thermal insulation, and fine frost resistance [1–3]. However, there are some weaknesses for FLAC, for example, FLAC can't be widely used because of its high cost [4]. In addition, the tensile strength of FLAC was about 0.8 times that of normal concrete under same condition [5]. In particularly, compared to ordinary concrete, the brittleness of FLAC reflected as proportional strain ratio is about 20% higher than of ordinary concrete [6], it also indicates that the tensile strength or shearing strength of FLAC is lower than of ordinary concrete. In order to improve the physical and mechanical performance of FLAC, some normal aggregate instead of partial lightweight aggregate is employed in FLAC to create a new kind of specified density lightweight aggregate concrete. In general, the density of specified density lightweight aggregate concrete varies in the range of 1840–2240 kg/m³ [7]. Thus, in this paper, two new specified density shale aggregate concretes are formulated. The concrete with ordinary stone instead of partial coarse lightweight aggregate is referred to as stone-LAC for simplicity; similarly, the concrete with some ordinary river sand instead of fine lightweight aggregate is referred to as sand-LAC. Compared to FLAC, stone-LAC and sand-LAC have higher strength and elasticity modulus, lower shrinkage deformation, and less pumping and construction difficulty. Moreover, it is worth mentioning that under this methodology, construction costs are observably reduced [8, 9].

In this work, failure processes and failure mechanisms of stone-LAC and sand-LAC will be revealed based on comprehensive test data. Furthermore, conversion relations among strength indexes will be established by regression analysis. The material parameters underlying the structural design of stone-LAC and sand-LAC will be provided with a scientific foundation through the aforementioned methods.

Without a doubt, lightweight aggregate assumes a leading role in LAC properties, which include mechanics, durability, thermal conductivity, etc. In other words, different kinds of lightweight aggregate can produce LACs with distinguishable properties. In previous studies, quantities of lightweight materials were selected as aggregate in lightweight concrete. The lightweight aggregates could be classified into the following categories: natural (such as pumice, diatomite, volcanic ash, etc.) and artificial (such as perlite, expanded shale, clay, slate, sintered pulverized-fuel ash, etc.) [10]. Onoue et al. [11] presented the result that the shock-absorbing capability of lightweight concrete utilizing volcanic pumice aggregate was superior to the control concrete using crushed limestone as coarse aggregate. Topçu and Işıkdağ [12] and Sengul et al. [13] studied the effect of expanded perlite aggregate on the properties of lightweight concrete; they proved that increased use of expanded perlite aggregate resulted in less strength and less weight in the concrete, while at the same time, thermal conductivity was substantially improved.

In contrast to LAC with other lightweight aggregates, LAC with shale aggregate originating from natural shale has not been as well researched. Natural shale can be manufactured into shale ceramsite and shale pottery sand via high temperature and calcination, which has been generally adopted as lightweight aggregate [14, 15]. Shale aggregate is suitable for wear resistance, corrosion resistance, and adsorption [16–18], it also has the advantages of weight, compressibility, heat retention, seismic resistance, and nonradioactivity. Thus, shale aggregate has been deemed as an appropriate material for reducing energy usage in buildings [19, 20]. These qualities, combined with its low price, have helped spur its growing application in agriculture and other industries [21–23]. However, the mechanical properties of shale LAC require more study for its continued development and application.

At present, many experimental and theoretical studies on the physical and mechanical properties of FLAC have been conducted. Tasdemire et al. [24] found that lightweight aggregates can reduce the thermal conductivity of FLAC and established a significant correlation between thermal conductivity and unit weight for the concrete. Zaetang et al. [25] showed that using diatomite pumice as coarse aggregates in full-lightweight pervious concrete can reduce its density and thermal conductivity by 3-4 times compared to previous concrete containing natural aggregate. Kaffetzakis and Papanicolaou [26] experimented with the bond behavior of reinforcement in full-lightweight aggregate self-compacting concrete. They reported that the maximum bond stress under normalization increased when each of the following is increased: rebar diameter, bond length, and the oven-dry density of the mix. In short, these new FLACs possessed satisfactory physical properties, but their mechanical properties were still inferior to ordinary concrete.

Several groups have attempted to improve the poor mechanical properties of FLAC through modification. Miller and Tehrani [27] mixed rubber into FLAC to prepare 36 beam specimens. The results showed that the tire-derived aggregates had reduced the mechanical strength, but did induce a partial enhancement of the ductility and toughness. Aslam et al. [28] produced high-strength specified density lightweight aggregate concrete by using blended coarse-lightweight aggregates. Test results showed that oil palm shell in oil-palm-boiler clinker concrete contributed to reductions in density and in the mechanical property indexes. Ma et al. [29] manufactured modified expanded-clay ceramsite concrete with an inorganic polymer compound and conducted failure tests at room temperature before and after exposure to high temperatures. Results showed that the polymer selected for the modification material decomposed gradually to produce volatiles as the temperature increased, which are risks for concrete spalling. However, creating channels for vapor release may mitigate spalling. Chung et al. [30] evaluated the effects of crushed and expanded waste glass aggregates on the material properties of lightweight concrete, respectively. The derived results supported the feasibility of both glass aggregates being used as alternative lightweight aggregates.

The above studies focused either on lightweight aggregate or on FLAC, specifically modified FLAC. Although the attempted modifications produced satisfactory results, the modification mechanism and the relations among strength indexes were not understood, which constrained further study and reduced the number of potential engineering applications. More importantly, no report that describes the mechanical properties and conversion relations of strength indexes for specified density shale aggregate concrete with ordinary stone or ordinary river sand has been found up to now. The work described in this paper is an attempt at addressing these unknowns.

In this investigation, FLAC graded as LC35 was deemed as the control concrete. According to the exchange method of equal volume, the shale ceramsite and pottery sand in control concrete were replaced by stone and river sand to prepare stone-LAC and sand-LAC, respectively. The investigation focused on the failure mechanism and influence of the replacement rate of stone and river sand in order to establish conversion formulas for strengths, as well as deformation, and tension-compression ratio. The information gained from this research may help expand the number of engineering applications for concrete structures produced by specified density shale aggregate.

2. Test Program

2.1. Test Raw Materials. The cement was 42.5-grade ordinary portland cement produced by a company in Jiaozuo, China. The mix water was tap water; coarse aggregate was comprised of natural crushed stone and ceramsite; ceramsite is shown in Figure 1(a). Ceramsite is a ceramic material with different particle sizes and is made from natural shale after crushing, sieving, high-temperature calcination, and screening. The main properties of coarse aggregate are displayed in Table 1. Two kinds of fine aggregate are pottery sand and ordinary river sand, pottery sand is shown in Figure 1(b), the main properties of fine aggregate are shown in Table 2, It can be seen that the accumulation density of ceramsite and pottery is lower than that of stone and river sand. The higher porosity of ceramsite and pottery make them absorb water more easily. The ceramsite was submerged in water for 12 hours before it was used. Fly ash adopted third grade fly ash. The fly ash was 25% of the total amount of cementing material. The main ingredient of the water reducing agent was a β-high condensation compound of naphthalene sulfonic acid formaldehyde, and its mixing amount is 1% of the total cementing material.

2.2. Test Method. The replacement rate of coarse aggregate in stone-LAC is defined as the loose volume of stone within the aggregate. Five kinds of replacement rate of coarse aggregate in stone-LAC are expressed as r (0%, 25%, 50%, 75%, and 100%). When $r = 0\%$, the stone-LAC is an all-lightweight aggregate concrete. The fine aggregate replacement rate of sand-LAC is the loose volume of river sand, which accounts for the loose aggregate volume. Three aggregate replacement rates for the sand-LAC fine aggregate (0%, 50%, 100%) have been selected. When $r = 0\%$, the sand-LAC is an all-lightweight aggregate concrete. Each replacement rate contained six $150 \times 150 \times 150$ mm cube specimens, which are divided into two groups based on the intended test: compressive strength and splitting tensile strength. Each replacement rate also contained three prism specimens of $150 \times 150 \times 300$ mm, which were used for the determination of axial compressive strength, elastic modulus, and Poisson's ratio. Lastly, each rate included another three prism specimens of $150 \times 150 \times 550$ mm, which were used for the determination of flexural strength. The test specimens were formed in standard sizes.

The design strength of all-lightweight aggregate concrete was LC35. According to the technical specification for lightweight aggregate concrete (JGJ51-2002), the loose volume method is used for the design and calculation of the mix ratio for all-lightweight aggregate concrete. According to previous research [31–33], the mix proportion of all-lightweight aggregate concrete was based on the proportion of the loose volume of coarse or fine aggregate. The quality coordination of stone-LAC and sand-LAC under each replacement rate are shown in Table 3. Under different mix proportions in Table 3, the slump values of specified density shale aggregate concrete were measured. As shown in Figure 2, after the fresh concrete was prepared, the slump cylinder was rinsed and placed on an wetting plate, then the representative concrete was loaded into the cylinder fully. According to the Chinese code for test method of performance on ordinary fresh concrete (GB/T50080-2016), the slump cylinder was lifted and placed beside the cone concrete, and the vertical distance from the top of the cylinder to the center of the concrete top was the slump value. The result shows that the slump values of stone-LAC and sand-LAC are at the range of 150 mm to 180 mm, it indicates that specified density shale aggregate concrete has good working performance.

The strength index of specified density shale aggregate concrete was determined according to the test method of mechanical properties of ordinary concrete (GB50081-2002), where a lateral and longitudinal strain patch was attached to the middle of a side prismatic specimen. The elastic modulus and Poisson's ratio were obtained by using the method of force controlled loading, and the strain values were collected.

3. Results and Discussion

3.1. Failure Process and Failure Mode. At the initial stage of loading, there was no obvious change in the surface of the cube. As loading increased, the specimen internal stress increased and produced a weak "crackling" sound. Continued loading led to the emergence of smaller cracks and microcracks on the surface of the test specimen. These cracks gradually expanded and passed through the specimen bulk until upon the limit load, the test specimen was ultimately destroyed. As shown in Figure 3, the failure surface of the cube test specimen was oriented about 60 degrees relative to the center of the test specimen, while the upper and lower specimen surfaces were essentially intact, and more and more defect appeared near the middle area, so the final failure mode was similar to the inverted pyramid. The damage to the specified density shale aggregate concrete was caused by the joint between the ceramsite itself and the cement mortar.

The prism began to fail when microcracks emerged on the surface, which then expanded through the specimen before it was finally destroyed by massive flaking. Destroyed prismatic Specimens are shown in Figure 4. The damage to the specified density shale aggregate concrete specimen was also caused by the joint between the ceramsite and the cement mortar. Figure 4(a) shows that the increase in the stone replacement rate is accompanied by more inclined cracks on the stone-LAC prism specimen; at 100% replacement, the inclined cracks of the stone-LAC appeared through the upper and lower specimen. On Figure 4(b), the increase of river sand replacement rate was accompanied by a reduced degree of destruction for the sand-LAC specimen, where the failure area was concentrated in the middle of the test specimen, while the upper and lower ends remained intact.

The splitting and flexural failure modes of specified density shale aggregate concrete specimens are shown in Figures 5 and 6. Due to the low strength of shale ceramsite

FIGURE 1: Shale aggregate: (a) ceramsite; (b) pottery sand.

TABLE 1: The basic property indexes of ceramsite and crushed stone.

Type	Particle size (mm)	Bulk density (kg/m^3)	Needle-like content (%)	Mud content (%)	Cylinder pressure strength (MPa)
Ceramsite	5–15	660	—	—	4.5
Crushed stone	5–15	1434	<10.0	<1.0	10.7

TABLE 2: The basic property indexes of pottery sand and river sand.

Type	Particle size (mm)	Bulk density (kg/m^3)	Mud content (%)	Fineness modulus
Pottery sand	≤5	880	<1.0	3.15
River sand	≤5	1472	<1.0	2.85

TABLE 3: Quality mix proportion of stone-lightweight and sand-lightweight aggregate concrete (kg/m^3).

Type	r	Cement	Fly ash	Ceramsite	Crushed stone	Pottery sand	River sand	Water	Water reducer
	0%	472	159	444	0	408	0	171	6.31
	25%	472	159	333	241	408	0	171	6.31
Stone-LAC	50%	472	159	222	482	408	0	171	6.31
	75%	472	159	111	723	408	0	171	6.31
	100%	472	159	0	964	408	0	171	6.31
	0%	472	159	444	0	408	0	171	6.31
Sand-LAC	50%	472	159	444	0	204	341	171	6.31
	100%	472	159	444	0	0	682	171	6.31

and pottery sand, the fracture surfaces of stone-LAC and sand-LAC existed not only in cement paste but also in a large number of lightweight aggregate. The splitting tensile test specimen was destroyed along two directions: first in the vertical direction, and while the second direction was oriented at an angle of 70–90° with respect to the horizontal plane of the fracture resistant specimen.

3.2. Influence on Mechanical Strength

3.2.1. Compressive Strength. The cube compressive strength (f_{cu}) and axial compressive strength (f_c) of specified density shale aggregate concrete were measured via a common concrete mechanics performance test method (GB50081-2002), $150 \times 150 \times 150$ mm cube specimens were used to

measured f_{cu} and $150 \times 150 \times 300$ mm prism specimens were used to measured f_c, the results are shown in Table 4. It can be seen that the cube compressive strength and the axial compressive strength of specified density shale aggregate concrete increased with the increase of the replacement rate of stone and river sand (r). This is explained by the stone and the river sand possessing higher strength than the ceramsite and the pottery. The increase of the replacement rate is accompanied by a higher proportion of stone and river sand in the light and sand light concrete, which then increases the compressive strength of the specified density shale aggregate concrete.

The increase in compressive strength for the stone-LAC at each replacement rate (Table 4) is 10.3%, 5.1%, 12.8% and 18.8%, respectively; the increase in compressive strength for the sand-LAC is 17.6% and 6.4%, respectively. It can be seen that with the increase of replacement rate, the increase of

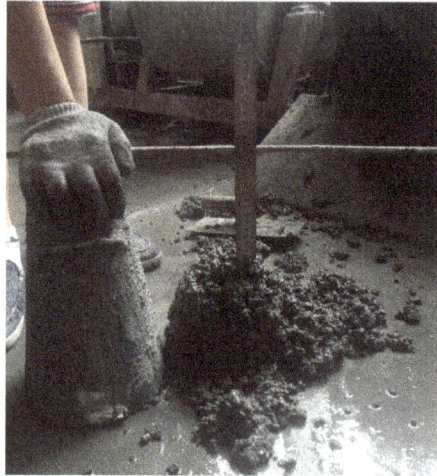

FIGURE 2: Measuring method of slump value.

FIGURE 3: Cube compressive failure mode: (a) stone-LAC; (b) sand-LAC.

FIGURE 4: Axial compression failure mode: (a) stone-LAC; (b) sand-LAC.

cube compressive strength of stone-LAC decreased first and then increased, while the increment of sand-LAC compressive strength decreased. The cube compressive strength for the stone-LAC at each replacement rate increased by 10.3%, 15.9%, 30.7%, and 55.3% compared with all-lightweight aggregate concrete, respectively. The cube

FIGURE 5: Splitting tensile failure mode: (a) stone-LAC; (b) sand-LAC.

FIGURE 6: Flexural failure mode: (a) stone-LAC; (b) sand-LAC.

TABLE 4: Compressive strength of specified density shale aggregate concrete.

Type	r (%)	Dry apparent density (kg/m^3)	f_{cu} (MPa)	f_c (MPa)	f_c/f_{cu}
	0	1330	35.8	24.7	0.69
	25	1450	39.5	33.6	0.85
Stone-LAC	50	1550	41.5	35.4	0.85
	75	1740	46.8	40.7	0.87
	100	1940	55.6	47.3	0.85
	0	1330	35.8	24.7	0.69
Sand-LAC	50	1510	42.1	35.9	0.85
	100	1550	44.8	42.4	0.95

compressive strength for the sand-LAC at each replacement rate increased by 17.6% and 25.1%, respectively, compared with all-lightweight aggregate concrete. At $r = 100\%$, the stone-LAC experienced a much greater increase in cube compressive strength compared with all-lightweight aggregate concrete.

At each replacement rate, the stone-LAC axial compressive strength increased by 36%, 5.4%, 15% and 16.2%, respectively; the sand-LAC compressive strength grew by 45.3% and 18.1% for each rate, respectively. It can be seen that as replacement rate increased, the increase of stone-LAC axial compressive strength decreased firstly and then increased, while the increase of sand-LAC axial compressive strength decreased; this behavior is similar to that of the cube compressive strength. The stone-LAC axial compressive strength under each replacement rate increased by 36%, 43.3%, 64.8% and 91.5%, respectively, compared to that of all-lightweight aggregate concrete; sand-LAC axial compressive strength increased by 45.3% and 71.7%, respectively,

compared with all-lightweight aggregate concrete under different replacement rates.

The cube compressive strength of stone-LAC and sand-LAC is linear with the replacement rate. The values of the measured strength were treated with a dimensionless method. The expressions obtained by using the principle of least square method are as follows:

$$\text{stone-LAC:} \quad \frac{f_{cu}}{f_{cu0}} = 0.52r + 0.96, \quad R^2 = 0.915,$$

$$\text{sand-LAC:} \quad \frac{f_{cu}}{f_{cu0}} = 0.25r + 1.02, \quad R^2 = 0.879, \tag{1}$$

where f_{cu0} represents the cube compressive strength of all-lightweight aggregate concrete specimens.

The axial compression strength of stone-LAC and sand-LAC is also linear with the replacement rate, and is also treated with a dimensionless method. The expressions obtained by using the principle of least square method are as follows:

$$\text{stone-LAC:} \quad \frac{f_c}{f_{c0}} = 0.76r + 1.03, \quad R^2 = 0.963,$$

$$\text{sand-LAC:} \quad \frac{f_c}{f_{c0}} = 0.42r + 1.02, \quad R^2 = 0.958,$$

(2)

where f_{c0} represents the axial compressive strength of all-lightweight aggregate concrete specimens.

According to Technical specification for lightweight aggregate concrete (JGJ 51-2002), the dry apparent densities of stone-LAC and sand-LAC cube specimens in different replacement rates were measured, which were shown in Table 4. It can be seen that the dry apparent densities of stone-LAC and sand-LAC all increase as the increasing of replacement rate. For stone-LAC, the dry apparent density of specified density shale aggregate concrete increased from 1330 to 1940, and the biggest amplitude of variation was 45.9%. For sand-LAC, the dry apparent density of specified density shale aggregate concrete increased from 1330 to 1550, and the biggest amplitude of variation was 16.5%. The relationship between cube compressive strength and dry apparent density of stone-LAC and sand-LAC are approximately linear, as shown in Figure 7. In other words, the ratio of f_{cu} and dry apparent density was close to a constant, and based on f_{cu}, the optimum replacement rate was 100%, without the consideration of self-weight.

3.2.2. Splitting Tensile Strength. The splitting tensile strength (f_{ts}) of stone-LAC and sand-LAC is shown in Table 5. It can be seen that the splitting tensile strength of specified density shale aggregate concrete increased with the replacement rate in both cases. Natural aggregate has higher strength than that of lightweight aggregate, but the splitting tensile strength of sublightweight concrete still increased as the natural aggregate was replaced by lightweight aggregate. With each increase in replacement rate, the splitting tensile strength growth rate for stone-LAC was 35.1%, 13%, 15.3% and 12.9%, respectively; the splitting tensile strength growth rate for sand-LAC was 20.1% and 20.2%, respectively. It can be seen that the growth rate of splitting tensile strength for sand-LAC is more stable than that of stone-LAC. The splitting tensile strength of stone-LAC increased by about 35.1%, 52.7%, 76% and 98.7%, respectively, more than that of all-lightweight aggregate concrete under different replacement rates; the corresponding increase for sand-LAC was 20.1% and 44.4%, respectively, when compared with all-lightweight aggregate concrete under different replacement rates.

Table 5 shows the splitting tensile strength for stone-LAC to be higher than that of sand-LAC for the same r. For $r = 50\%$ and 100%, the splitting tensile strength for stone-LAC was 27.1% and 37.3% higher, respectively, than that of sand-LAC. It can be seen that the replacement of ceramsite with crushed stone results in a greater increase of splitting tensile strength than for the replacement of pottery sand with river sand, and the increase became more and more

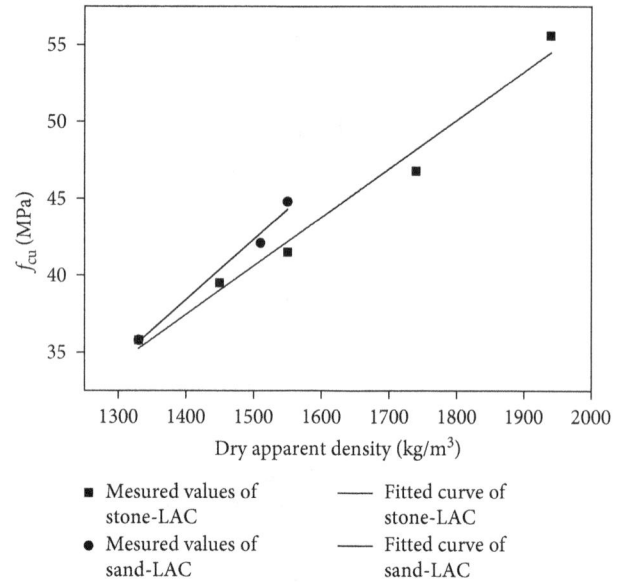

FIGURE 7: The relationship between cube compressive strength and dry apparent density.

TABLE 5: Splitting tensile strength of specified density shale aggregate concrete.

r (%)	0	25	50	75	100
Stone-LAC	3.13	4.23	4.78	5.51	6.22
Sand-LAC	3.13	—	3.76	—	4.52

large as replacement rate increases. This is explained by the fact that the coarse aggregate in specified density shale aggregate concrete has a stronger bearing function than that of the fine aggregate from river sand.

The splitting tensile strength of stone-LAC and sand-LAC is linear with the replacement rate, and was treated with a dimensionless method. The expressions obtained by the principle of least square method are as follows:

$$\text{stone-LAC:} \quad \frac{f_{ts}}{f_{ts0}} = 0.90r + 1.06, \quad R^2 = 0.973,$$

$$\text{sand-LAC:} \quad \frac{f_{ts}}{f_{ts0}} = 0.44r + 0.99, \quad R^2 = 0.995,$$

(3)

where f_{ts0} indicates the splitting tensile strength of all-lightweight aggregate concrete specimens.

3.2.3. Flexural Strength. Measured values for the flexural strength of specified density shale aggregate concrete (f_f) are shown in Table 6. It can be seen that the flexural strength of stone-LAC and sand-LAC increased with r for stone and river sand. With each increase in r, the stone-LAC flexural strength grew by 10.3%, 2.7%, 8%, and 9.9%, respectively; the sand-LAC flexural strength grew by 9.8% and 14.9%, respectively. In comparison to all-lightweight aggregate concrete, the flexural strength of stone-LAC increased by 19.7%, 23.0%, 32.8%, and 45.9%, respectively; for sand-LAC, the

TABLE 6: Flexural strength of specified density shale aggregate concrete.

r (%)	0	25	50	75	100
Stone-LAC	6.1	7.3	7.5	8.1	8.9
Sand-LAC	6.1	—	6.7	—	7.7

TABLE 7: Elastic modulus of specified density shale aggregate concrete.

r (%)	0	25	50	75	100
Stone-LAC	22.06	23.30	23.49	24.23	25.48
Sand-LAC	22.06	—	26.59	—	28.70

flexural strength increase was 9.8% and 23.0%, respectively. It can be seen that the flexural strength of specified density shale aggregate concrete saw significant improvement, but not to the same degree as was observed for splitting tensile strength.

According to Table 6, the flexural strength of stone-LAC is higher than that of sand-LAC for the same replacement rate. The flexural strength of stone-LAC at $r = 50\%$ and 100% is 11.9% and 15.6% higher, respectively, than that of sand-LAC. This behavior is similar to that observed for splitting tensile strength.

The flexural strength of stone-LAC and sand-LAC is linear with the replacement rate, and has also been treated with a dimensionless method. The expressions obtained by using the principle of the least square method are as follows:

$$\text{stone-LAC:} \quad \frac{f_f}{f_{f0}} = 0.40r + 1.04, \quad R^2 = 0.933,$$

$$\text{sand-LAC:} \quad \frac{f_f}{f_{f0}} = 0.26r + 0.99, \quad R^2 = 0.965,$$

(4)

where f_{f0} indicates the flexural strength of all-lightweight aggregate concrete specimens.

3.3. Influence on Indexes of Deformation Performance

3.3.1. Elastic Modulus. The secant modulus between $0.5f_c$ and the origin is taken as the elastic modulus (E) for specified density shale aggregate concrete. Measured values of the elastic modulus of stone-LAC and sand-LAC under different replacement rates are shown in Table 7. It can be seen that the elastic modulus of stone-LAC and sand-LAC gradually increased with r for stone and river sand. The elastic modulus of stone-LAC under each replacement rate increased by 3.26%, 6.48%, 9.84%, and 15.50%, respectively, more than occurred in all other lightweight concrete. The elastic modulus of sand-LAC under each replacement rate increased by 20.53% and 30.10%, respectively, more than that of all-lightweight concrete. It can be seen that the addition of stone and river sand resulted in a specified density shale aggregate concrete better able to resist deformation to varying degrees. Under the same replacement rate, the elastic modulus of sand-LAC was higher than that of stone-LAC. When $r = 50\%$ and 100%, the elastic modulus of sand-LAC is 13.2% and 12.6%, respectively, higher than that of stone-LAC.

The elastic modulus of specified density shale aggregate concrete has no unified empirical formula at present. However, according to the changing rule of the elastic modulus of sublightweight concrete, new empirical formulas can be derived using the elastic modulus of ordinary

concrete, and the influence parameters α and β. This procedure results in the following equations:

$$\text{stone-LAC:} \quad E = \frac{10^5}{2.2 + 34.7/f_{cu0}}\alpha,$$

$$\text{sand-LAC:} \quad E = \frac{10^5}{2.2 + 34.7/f_{cu0}}\beta.$$

(5)

The measured elastic modulus and cube compressive strength of stone-LAC and sand-LAC are taken into Equation (5) to obtain values for α and β at each replacement rate, as shown in Table 8.

The fitting formulas for the relationship between α, β, and r are shown in Equation (6):

$$\text{stone-LAC:} \quad \alpha = 0.10r + 0.70, \quad R^2 = 0.914,$$

$$\text{sand-LAC:} \quad \beta = -0.20(r-1)^2 + 0.90, \quad R^2 = 0.984.$$

(6)

3.3.2. Poisson's Ratio. The average value of Poisson's ratio at $0.2f_c$, $0.4f_c$, $0.6f_c$, and $0.8f_c$ are taken as Poisson's ratio of specified density shale aggregate concrete (γ). Poisson's ratio of stone-LAC and sand-LAC under each replacement rate is shown in Table 9. For stone-LAC, Poisson's ratio decreased and then increased with the rising replacement rate; when $r = 25\%$, Poisson's ratio was the smallest; when $r = 100\%$, Poisson's ratio was the largest. As for sand-LAC, Poisson's ratio decreased as r increased.

3.4. Conversion Relations of Strength Index

3.4.1. Compressive Strength Conversion. It can be seen from Table 4 that the values of stone-LAC and sand-LAC are closely related to the replacement rate r. Overall, as r increased, f_c/f_{cu} for stone-LAC and sand-LAC gradually increased. This relationship is illustrated in Figure 8. It can be seen that when $r < 25\%$, f_c/f_{cu} for stone-LAC increases linearly with r; when $r > 25\%$, f_c/f_{cu} remained essentially constant at 0.85. For sand-LAC, f_c/f_{cu} increases linearly with r. The above behaviors are presented in the following mathematical formula:

$$\frac{f_c}{f_{cu}} = \begin{cases} 0.64r + 0.69, & r \leq 0.25, \\ 0.85, & 0.25 < r \leq 1, \end{cases} \quad R^2 = 1.000,$$

$$\frac{f_c}{f_{cu}} = 0.28r + 0.70 \quad 0 \leq r \leq 1 \quad R^2 = 0.965.$$

(7)

3.4.2. Tensile and Compressive Ratios. The ratio between the splitting tensile strength and the cube compressive strength

TABLE 8: Measured values of coefficients α and β.

r (%)	0	25	50	75	100
α	0.70	0.74	0.74	0.77	0.81
β	0.70	—	0.84	—	0.91

TABLE 9: Poisson's ratio of specified density shale aggregate concrete.

r (%)	0	25	50	75	100
Stone-LAC	0.28	0.18	0.24	0.28	0.31
Sand-LAC	0.28	—	0.13	—	0.09

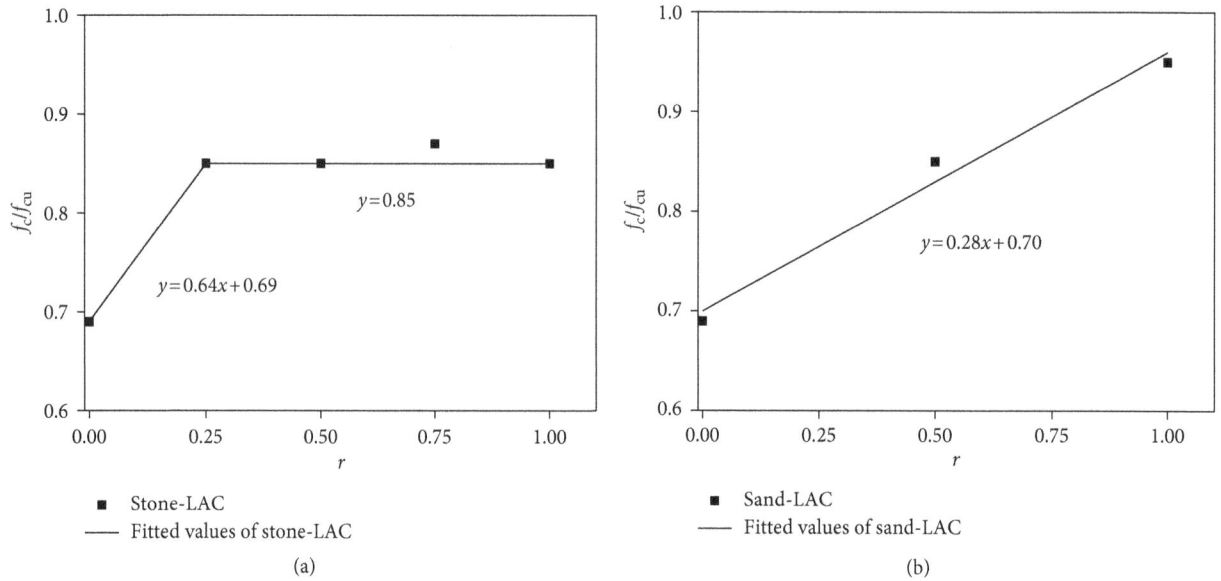

FIGURE 8: The relation curves between f_c/f_{cu} and replacement rate for specified density shale aggregate concrete: (a) stone-LAC; (b) sand-LAC.

(f_{ts}/f_{cu}) of concrete can objectively reflect the relationship between strength and brittleness of concrete. Sun [34] proposed that the tensile ratio of ordinary concrete is only related to Poisson's ratio, and established the relationship between the tensile, compressive, and Poisson's ratios, as shown in Equation (8). However, (8) does not fully apply to the specified density shale aggregate concrete. Equation (9) was derived to account for the specific changes to the tensile and compressive ratios in specified density shale aggregate concrete. Coefficients A and B indicate the impact parameters of stone-LAC and sand-LAC, respectively, and thereby link the tensile and compressive ratios with Poisson's ratio.

$$\frac{f_{ts}}{f_{cu}} = \frac{2\gamma^2}{1 + 2\gamma^2}, \tag{8}$$

$$\frac{f_{ts}}{f_{cu}} = \begin{cases} \dfrac{2\gamma^2}{1 + 2\gamma^2} A, & \text{stone-LAC}, \\[3mm] \dfrac{2\gamma^2}{1 + 2\gamma^2} B, & \text{sand-LAC}. \end{cases} \tag{9}$$

The measured values of the tension and compression's ratio of specified density shale aggregate concrete are shown in Table 10. The undetermined coefficients A and B for stone-LAC and sand-LAC under different replacement rates were calculated from the measured values of tensile, compressive, and Poisson's ratios and are shown in Table 11. A and B can now be related only to r for simplification and are correspondingly plotted in Figure 9. Fitting Equation (10) show that the fitting accuracies are all above 0.9:

$$\text{stone-LAC:} \quad A = \begin{cases} 0.64 + 4.48r, & 0 \le r \le 0.25, \\[3mm] 0.50 + \dfrac{1}{3.3r}, & 0.25 < r \le 1, \end{cases} \tag{10}$$

$$\text{sand-LAC:} \quad B = 3.96(r + 0.2)^2 + 0.60 \quad 0 \le r \le 1.$$

The replacement rate for stone-LAC and sand-LAC is replaced by Equation (10), respectively. The A and B values are calculated, then inserted into Equation (9) to obtain the calculated values of f_{ts}/f_{cu}, as shown in Table 12.

TABLE 10: Measured tensile and compressive ratios of specified density shale aggregate concrete.

r (%)	0	25	50	75	100
Stone-LAC	0.087	0.107	0.115	0.118	0.120
Sand-LAC	0.087	—	0.089	—	0.100

TABLE 11: Measured values of coefficients A and B.

r (%)	0	25	50	75	100
A	0.64	1.76	1.11	0.87	0.75
B	0.64	—	2.72	—	6.25

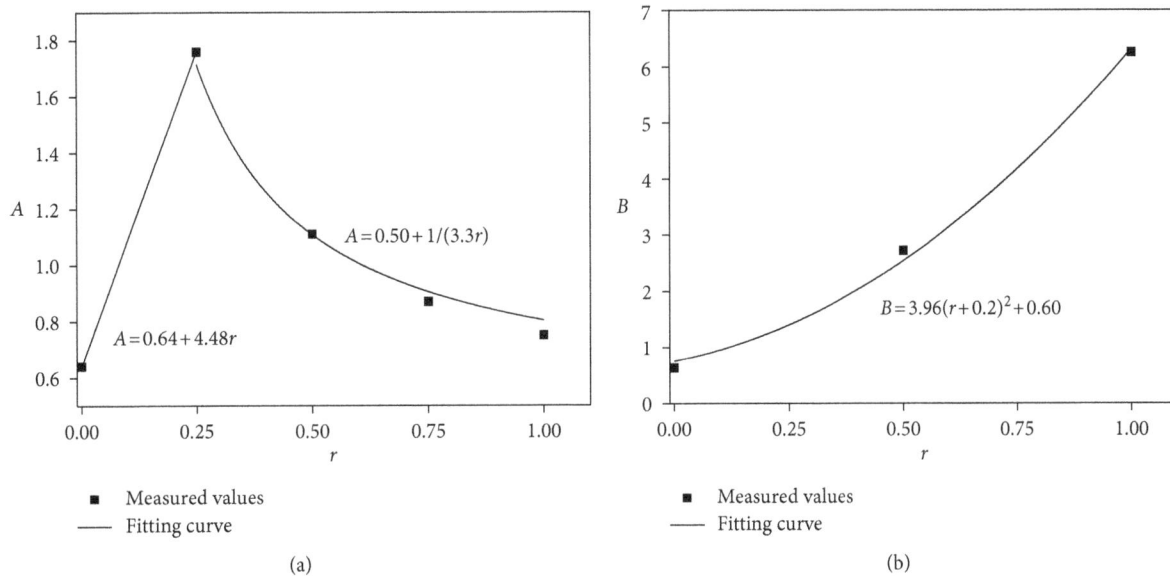

(a)

(b)

FIGURE 9: Relationship curves between A, B, and r. (a) Relationship between A and r. (b) Relationship between B and r (r represents the replacement rate of natural stone or sand; A and B are defined as influenced coefficients of f_{ts}/f_{cu} for stone-LAC, sand-LAC under different replacement rates, respectively).

TABLE 12: Comparison between calculated values and fitted values for f_{ts}/f_{cu}.

Type	r (%)	Measured value of f_{ts}/f_{cu}	Calculated value of f_{ts}/f_{cu}	Calculated value/measured value
Stone-LAC	0	0.087	0.088	1.01
	25	0.107	0.107	1.00
	50	0.115	0.114	0.99
	75	0.118	0.123	1.04
	100	0.120	0.129	1.07
Sand-LAC	0	0.087	0.816	0.94
	50	0.089	0.084	0.94
	100	0.100	0.101	1.01

The average calculated value/measured value ratios for stone-LAC and sand-LAC are 1.02 and 0.96, respectively; the variance are 0.001 and 0.002, and the coefficient of variation are 0.032 and 0.042. This statistical analysis indicates that the calculated values of specified density shale aggregate concrete are in good agreement with the measured values.

4. Conclusions

Through testing of basic mechanical properties and analysis of specified density shale aggregate concretes, the following conclusions are obtained:

(1) The compressive failure of specified density shale aggregate concrete was caused by ceramsite and

cement mortar. For flexural failure and splitting tensile failure, the lightweight aggregate was fractured at the failure interface.

(2) When the replacement rate of stone and river sand increased from 0% to 100%, the cube compressive strength of stone-LAC and sand-LAC linearly increased by 55% and by 25%, respectively, the axial compressive strength linearly increased by 91% and by 72%, respectively, splitting tensile strength increased by 99% and by 44%, respectively, and the flexural strength increased by 46% and by 26%, respectively.

(3) Increasing the replacement rate for stone-LAC and sand-LAC resulted in 16% and 30% increasing, respectively, in their elastic modulus eventually, while Poisson's ratios of sand-LAC decreased by 67%.

(4) Various parameters have been correlated with the replacement rate in this study. These parameters include elastic modulus; the relation between axial compressive strength and cube compressive strength; the relation between compression, tension, and Poisson's ratios. A series of simple calculation models were created from these correlations, and have been shown to be accurate and practical.

In this study, the mechanical failure process and fracture surfaces were analyzed on a macroscopic level. In addition, the control mix proportions for LC35 as related to conversion relations of strength indexes have been determined. However, these relations need to be verified for specified density shale aggregate concrete of other strength grades by additional tests and theoretical analysis. Future work may include techniques such as scanning electron microscopy (SEM), X-ray computed tomography (XCT), etc. to monitor and analyze the mechanical failure of specified density shale aggregate concrete on a microscopic level.

Conflicts of Interest

The authors declare that there are no conflicts of interest regarding the publication of this paper.

Acknowledgments

This work was financially supported by the Science and Technology Break-through Project of Henan Province (172102210285), the Fundamental Research Funds for the Universities of Henan Province (NSFRF170921), and the Safe Production Project of Key Technology for Major Accident Prevention and Control (Henan-0006-2016AQ). All these are gratefully appreciated.

References

[1] J. J. Li, J. G. Niu, C. J. Wan, X. Q. Liu, and Z. Y. Jin, "Comparison of flexural property between high performance polypropylene fiber reinforced lightweight aggregate concrete and steel fiber reinforced lightweight aggregate concrete," *Construction and Building Materials*, vol. 157, pp. 729–736, 2017.

[2] M. C. S. Nepomuceno, L. A. Pereira-De-Oliveira, and S. F. Pereira, "Mix design of structural lightweight self-compacting concrete incorporating coarse lightweight expanded clay aggregates," *Construction and Building Materials*, vol. 166, pp. 373–385, 2018.

[3] P. Shafigh, L. J. Chai, H. B. Mahmud, and M. A. Nomeli, "A comparison study of the fresh and hardened properties of normal weight and lightweight aggregate concretes," *Journal of Building Engineering*, vol. 15, pp. 252–260, 2018.

[4] Y. W. Zhou, X. M. Liu, F. Xing, D. W. Li, Y. C. Wang, and L. L. Sui, "Behavior and modeling of FRP-concrete-steel double-skin tubular columns made of full lightweight aggregate concrete," *Construction and Building Materials*, vol. 139, pp. 52–63, 2017.

[5] J. A. Bogas and R. Nogueira, "Tensile strength of structural expanded clay lightweight concrete subjected to different curing conditions," *KSCE Journal of Civil Engineering*, vol. 18, no. 6, pp. 1780–1791, 2014.

[6] H. Z. Cui, T. Y. Lo, S. A. Memon, and W. Xu, "Effect of lightweight aggregates on the mechanical properties and brittleness of lightweight aggregate concrete," *Construction and Building Materials*, vol. 35, pp. 149–158, 2012.

[7] A. G. Khoshkenari, P. Shafigh, M. Moghimi, and H. B. Mahmud, "The role of 0–2 mm fine recycled concrete aggregate on the compressive and splitting tensile strengths of recycled concrete aggregate concrete," *Materials and Design*, vol. 64, pp. 345–354, 2014.

[8] W. Dong, X. D. Shen, H. J. Xue, J. He, and Y. Liu, "Research on the freeze-thaw cyclic test and damage model of Aeolian sand lightweight aggregate concrete," *Construction and Building Materials*, vol. 123, pp. 792–799, 2016.

[9] X. F. Wang, C. Fang, W. Q. Kuang, D. W. Li, N. X. Han, and F. Xing, "Experimental investigation on the compressive strength and shrinkage of concrete with pre-wetted lightweight aggregates," *Construction and Building Materials*, vol. 155, pp. 867–879, 2017.

[10] M. Davraz, M. Koru, and A. E. Akdağ, "The effect of physical properties on thermal conductivity of lightweight aggregate," *Procedia Earth and Planetary Science*, vol. 15, pp. 85–92, 2015.

[11] K. Onoue, H. Tamai, and H. Suseno, "Shock-absorbing capability of lightweight concrete utilizing volcanic pumice aggregate," *Construction and Building Materials*, vol. 83, pp. 261–274, 2015.

[12] İ. B. Topçu and B. Işıkdağ, "Effect of expanded perlite aggregate on the properties of lightweight concrete," *Journal of Materials Processing Technology*, vol. 204, no. 1–3, pp. 34–38, 2008.

[13] O. Sengul, S. Azizi, F. Karaosmanoglu, and M. A. Tasdemir, "Effect of expanded perlite on the mechanical properties and thermal conductivity of lightweight concrete," *Energy and Buildings*, vol. 43, no. 2-3, pp. 671–676, 2011.

[14] Y. Z. Zhuang, C. Y. Chen, and T. Ji, "Effect of shale ceramsite type on the tensile creep of lightweight aggregate concrete," *Construction and Building Materials*, vol. 46, no. 8, pp. 13–18, 2013.

[15] Z. M. Cao, Z. G. He, and Y. Yang, "Experimental study on the influence of water cement ratio on the compressive strength of shale ceramsite concrete," *Applied Mechanics and Materials*, vol. 204–208, pp. 3895–3898, 2012.

[16] W. J. Yang, Y. D. Yang, and Y. Yang, "Study on water permeability and chloride penetrability of the shale Ceramsite Concrete," *Advanced Materials Research*, vol. 403–408, pp. 439–443, 2011.

[17] T. F. Deng and J. F. Li, "Study on preparation of thermal storage ceramic by using clay shale," *Ceramics International*, vol. 42, no. 16, pp. 18128–18135, 2016.

[18] Z. B. Bundur, M. J. Kirisits, and R. D. Ferron, "Use of pre-wetted lightweight fine expanded shale aggregates as internal nutrient reservoirs for microorganisms in bio-mineralized mortar," *Cement and Concrete Composites*, vol. 84, pp. 167–174, 2017.

[19] C. Q. Wang, X. Y. Lin, D. Wang, M. He, and S. L. Zhang, "Utilization of oil-based drilling cuttings pyrolysis residues of shale gas for the preparation of non-autoclaved aerated concrete," *Construction and Building Materials*, vol. 162, pp. 359–368, 2018.

[20] M. Rodgers, G. Hayes, and M. G. Healy, "Cyclic loading tests on sandstone and limestone shale aggregates used in unbound forest roads," *Construction and Building Materials*, vol. 23, no. 6, pp. 2421-2427, 2009.

[21] M. R. Demerchant, A. J. Valsangkar, and A. B. Schriver, "Plate load tests on geogrid-reinforced expanded shale lightweight aggregate," *Geotextiles and Geomembranes*, vol. 20, no. 3, pp. 173-190, 2002.

[22] A. Lotfy, K. M. A. Hossain, and M. Lachemi, "Lightweight self-consolidating concrete with expanded shale aggregates: modelling and optimization," *International Journal of Concrete Structures and Materials*, vol. 9, no. 2, pp. 185-206, 2015.

[23] Y. C. Gokhale, R. S. Shukla, and P. K. Jain, "Benefication of shale aggregate and production of artificial aggregate," *Bulletin of the International Association of Engineering Geology-Bulletin de l'Association Internationale de Géologie de l'Ingénieur*, vol. 30, no. 1, pp. 391-393, 1984.

[24] C. Tasdemir, O. Sengul, and M. A. Tasdemir, "A comparative study on the thermal conductivities and mechanical properties of lightweight concretes," *Energy and Buildings*, vol. 151, pp. 469-475, 2017.

[25] Y. Zaetang, A. Wongsa, V. Sata, and P. Chindaprasirt, "Use of lightweight aggregates in pervious concrete," *Construction and Building Materials*, vol. 48, no. 11, pp. 585-591, 2013.

[26] M. I. Kaffetzakis and C. G. Papanicolaou, "Bond behavior of reinforcement in lightweight aggregate self-compacting concrete," *Construction and Building Materials*, vol. 113, pp. 641-652, 2016.

[27] N. M. Miller and F. M. Tehrani, "Mechanical properties of rubberized lightweight aggregate concrete," *Construction and Building Materials*, vol. 147, pp. 264-271, 2017.

[28] M. Aslam, P. Shafigh, M. A. Nomeli, and M. Z. Jumaat, "Manufacturing of high-strength lightweight aggregate concrete using blended coarse lightweight aggregates," *Journal of Building Engineering*, vol. 13, pp. 53-62, 2017.

[29] Q. M. Ma, R. X. Guo, Y. L. Sun et al., "Behaviour of modified lightweight aggregate concrete after exposure to elevated temperatures," *Magazine of Concrete Research*, vol. 70, no. 5, pp. 217-230, 2018.

[30] S. Y. Chung, M. A. Elrahman, P. Sikora, T. Rucinska, E. Horszczaruk, and D. Stephan, "Evaluation of the effects of crushed and expanded waste glass aggregates on the material properties of lightweight concrete using image-based approaches," *Materials*, vol. 10, no. 12, 1354 pages, 2017.

[31] Y. Xu, L. H. Jiang, J. X. Xu, and Y. Li, "Mechanical properties of expanded polystyrene lightweight aggregate concrete and brick," *Construction and Building Materials*, vol. 27, no. 1, pp. 32–38, 2012.

[32] J. J. Li, Y. H. Chen, and C. J. Wan, "A mix-design method for lightweight aggregate self-compacting concrete based on packing and mortar film thickness theories," *Construction and Building Materials*, vol. 157, pp. 621–634, 2017.

[33] P. Suttaphakdee, N. Dulsang, N. Lorwanishpaisarn, P. Kasemsiri, P. Posi, and P. Chindaprasirt, "Optimizing mix proportion and properties of lightweight concrete incorporated phase change material paraffin/recycled concrete block composite," *Construction and Building Materials*, vol. 127, pp. 475–483, 2016.

[34] N. P. Sun, "Research on tension and compression's ratio of concrete," *Building Science*, vol. 30, no. 7, pp. 19–22, 2014, in Chinese.

9

Cyclic Behaviour of Expanded Polystyrene (EPS) Sandwich Reinforced Concrete Walls

Ari Wibowo ⓘ, Indradi Wijatmiko, and Christin R. Nainggolan

Department of Civil Engineering, Faculty of Engineering, Brawijaya University, Malang 65149, Indonesia

Correspondence should be addressed to Ari Wibowo; ariwibowo@ub.ac.id

Academic Editor: Pietro Russo

Precast concrete walls become increasingly utilized due to the rapid needs of inexpensive fabricated house especially as traditional construction cost continues to climb, and also, particularly at damaged area due to natural disasters when the requirement of a lot of fast-constructed and cost-efficient houses are paramount. However, the performance of precast walls under lateral load such as earthquake or strong wind is still not comprehensively understood due to various types of reinforcements and connections. Additionally, the massive and solid wall elements also enlarge the building total weight and hence increase the impact of earthquake significantly. Therefore, the precast polystyrene-reinforced concrete walls which offer light weight and easy installment became the focus of this investigation. The laboratory test on two reinforced concrete wall specimens using EPS (expanded polystyrene) panel and wire mesh reinforcement has been conducted. Quasi-static load in the form of displacement controlled cyclic tests were undertaken until reaching peak load. At each discrete loading step, lateral load-deflection behaviour, crack propagation, and collapse mechanism were measured which then were compared with theoretical analysis. The findings showed that precast polystyrene-reinforced concrete walls gave considerable seismic performance for the low-to-moderate seismic region reaching up to 1% drift at 20% drop of peak load. However, it might not be sufficient for high seismic regions, at which double-panel wall type can be more suitable.

1. Introduction

Tall buildings particularly with irregularities are prone to behave poorly and collapse when subjected to lateral loads such as earthquake excitation or strong wind. To overcome this problem, shear walls are commonly preferable to increase the lateral strength of structures significantly. However, the added massive and solid shear walls result in increasing the building weight and consequently the base shear due to earthquake excitation which might reduce the effectiveness of the shear wall use in the structures. The effort to reduce the weight of shear walls without losing lateral strength capacity is necessary.

There were many studies investigating lightweight concrete shear walls with various techniques to reduce the element weight such as using lightweight aggregates, applying porous concrete system, or inserting lightweight panel into the wall. Mousavi et al. [1] studied the effectiveness of the JK system wall, composed of EPS concrete (mortar with EPS beads as fine aggregates) and galvanized steel reinforcement, in sustaining lateral load. It was observed that JK walls had high ductility capacity, but still need further observation for the application in tall and medium buildings. Yizhou [2] investigated that the use of gangue as an aggregate in concrete shear wall provided larger energy dissipation compared with normal concrete shear wall. Furthermore, Hejin et.al. [3] focused on ash ceramsite as alternative for lightweight aggregate concrete shear wall which gave similar load-deflection behaviour and collapse mechanism to those on normal concrete ones, whereas Chai and Anderson [4] found that the performance of concrete wall panels using perforated lightweight aggregate in low-rise buildings subjected to lateral forces was generally satisfactory. Cavaleri et al. [5] investigated pumice stone in comparison with expanded clay and normal stone as aggregates in concrete shear wall which showed the benefit of the use pumice stones.

On the other side, reducing the weight of structural elements can be achieved using the sandwich system by inserting a lightweight panel inside the concrete element. This panel system is usually applied for insulation purpose as well. The lightweight wall system investigated in this paper focused on the use of the EPS panel as a filler and galvanized wire mesh for reinforcing bar as shown in Figure 1.

2. Research Methodology

The specimens were designed as structural walls composing low-rise building which were commonly found in house or school precast buildings. The squat walls are generally dominated with shear behaviour which comparably differs to tall walls commonly found in high-rise building. Concrete tall walls have been considerably well-researched and well-understood [7–10], whereas concrete squat walls have been increasingly investigated [11–14]. However, innovation studies on sandwich squat walls with the EPS panel were just initially begun. Previous experimental studies by Trombetti et al. [15] and Ricci et al. [16] showed that sandwich squat concrete walls were comparable to those of regular RC walls and able to sustain lateral load up to drift higher than 1.3%, whereas Palermo and Trombetti [17] comprehensively investigated sandwich walls experimentally and analytically with the outcomes showed that properly designed walls can accomplish high seismic performance requirement suggested by the code. However, the overall performance of sandwich RC walls with lower steel reinforcement ratios (less than minimum requirements) still needs further investigation and hence became the main focus of this study.

The laboratory tests on two specimens of sandwich reinforced concrete wall RCW4 and RCW8 have been undertaken. Figure 2 shows the typical property of the walls. All specimens had a height and width of 90 cm and 60 cm, respectively (equivalent to aspect ratio of 1.5). The RCW4 wall used a 4 cm thick EPS panel compared to the 8 cm EPS panel installed in the RCW8 wall. The specimens were reinforced with $\phi 2.5$–75 mm wire mesh on each wall side and $\phi 3.0$ mm steel wires for connecting both mesh layers. The yield and ultimate steel tensile strengths of the wire mesh were 600 MPa and 680 MPa, respectively, as shown in Figure 3. A 35 mm thick shotcrete was applied on each outer side of walls with the concrete strength of 15 MPa. The walls and the foundations were connected using $\phi 10$ mm anchor bars spaced at 75 mm.

Quasi-static cyclic load procedure was applied at the tip of the wall specimens to obtain representative hysteretic curves of lateral load versus displacement (refer Figures 4 and 5) as per ASTM E2126 code [18]. The drift-controlled order was used for the loading test comprised drift increments of 0.042% until attaining 0.167% (representing cracking point), then drift increments of 0.16% until reaching drift of 0.66% (representing yield point), and followed by inelastic stage at 0.66% drift increments. The hysteretic behaviours of walls were maintained using three loading cycles at each drift ratio.

During the testing process, at each defined discrete displacement stages, the measurement of LVDTs, dial gauges, and crack propagation were recorded. The test stopped when the peak lateral strength of the specimen reduced by 20% (lateral load failure).

3. Experimental Test Results

Hysteretic curves of lateral load-drift relationship and the crack pattern of all wall specimens are presented in Figure 6. Both specimens RCW4 and RCW8 had similar peak lateral load of about 25 kN with different behaviour characteristics. RCW4 (EPS panel thickness of 40 mm) developed more classical flexural mechanism, whilst RCW8 (EPS panel thickness of 80 mm) was more dominated with yield penetration behaviour due to the thinner concrete cover of wall foundation. As shown, the specimen RCW4 managed to complete all three cycles of quasi-static cyclic load at 1.0% drift, and then failed at the first cycle of load at drift of 1.33%, whereas specimen RCW8 produced shorter maximum drift capacity with failure at the first cycle of lateral load at 1.0% drift. A comparison of lateral strengths and drifts between the experimental results and the theoretical predictions is presented in Table 1.

The total lateral deformation consists of flexural, shear, and yield penetration components that were determined using dial gauge and LVDT and dial gauge measurements as shown in Figure 7.

The flexural displacement at the wall top at each i-segment of LVDT was determined via the following equation (refer Figure 7(a)):

$$\Delta_{\text{fl}.i} = \int_0^L \varphi(x)x\,dx = \frac{L - L_{\text{vi}}}{L_{\text{h}}}\left(\delta_{\text{f2}} - \delta_{\text{f1}}\right), \qquad (1)$$

whereas the displacement of the elastic region at upper segment was estimated analytically assuming uncracked section properties as follows:

$$\Delta_{\text{fl}.3} = \frac{\varphi L_i^2}{3} = \frac{F L_i^3}{3 E_{\text{c}} I}, \qquad (2)$$

where F = lateral load; L_i = segment length; E_{c} = concrete elastic modulus; and I = uncracked moment of inertia.

The shear deformation Δ_{sh} was predicted using diagonal LVDTs data (refer Figure 7(b)) as follows:

$$\Delta_{\text{sh}} = \frac{(\delta_{\text{s1}} - \delta_{\text{s2}})}{2}\sec\xi = \frac{(\delta_{\text{s1}} - \delta_{\text{s2}})}{2}\frac{\sqrt{L_{\text{v}}^2 + D^2}}{L_{\text{v}}}, \qquad (3)$$

where D = wall depth and δ_{si} = diagonal LVDT measurement.

The yield penetration component was measured using the vertical LVDT at the first level (refer Figure 7(c)) by assuming a rocking mechanism within the first section of wall. An upper bound of the top displacement of the column can be calculated from the product of the slip rotation θ_{slip} and the column height assuming rigid body rotation as follows:

$$\Delta_{\text{yp}} = \theta_{\text{slip}} L_{\text{column}}, \qquad (4)$$

where θ_{slip} = the slip rotation of tensile steel = $(\delta_{\text{slip}}/(d - c)) \approx \beta$; $\beta = ((\delta_{\text{f2}} - \delta_{\text{f1}})/L_{\text{h}})$ and c = the neutral axis depth at the column base interface = $(\delta_{\text{f2}}/(\delta_{\text{f1}} + \delta_{\text{f2}}))L_{\text{h}} - d_2$.

The deformation of walls comprising flexural, shear, and yield penetration components for RCW4 and RCW8 specimens is shown in Figure 8. The flexural deformation was the most dominant component of about 75% and 55%

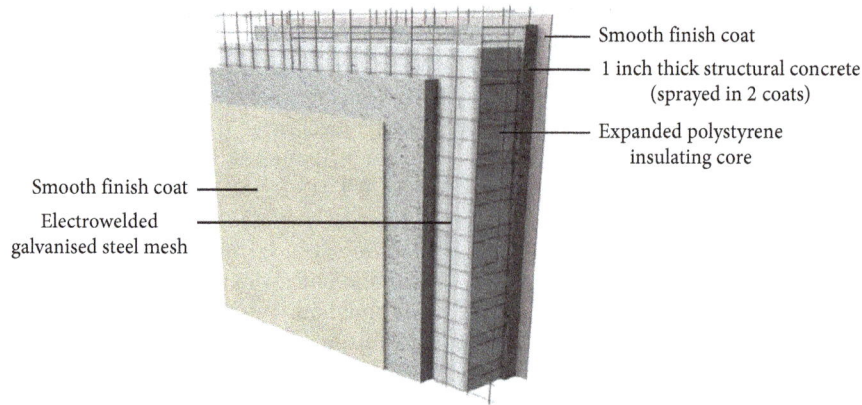

FIGURE 1: Typical EPS sandwich concrete wall panel (refer [6]).

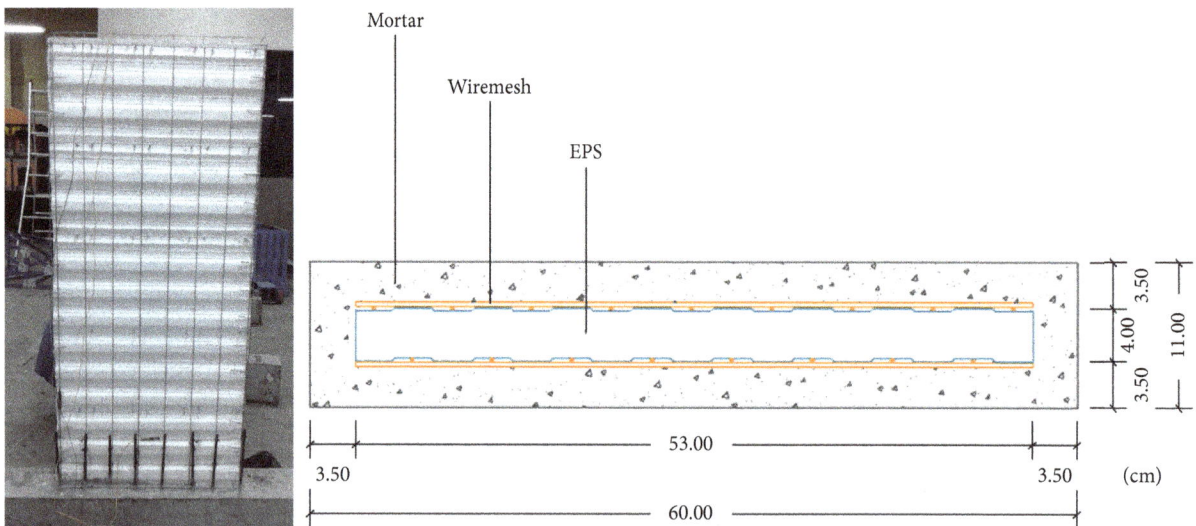

FIGURE 2: Basic properties of wall specimens.

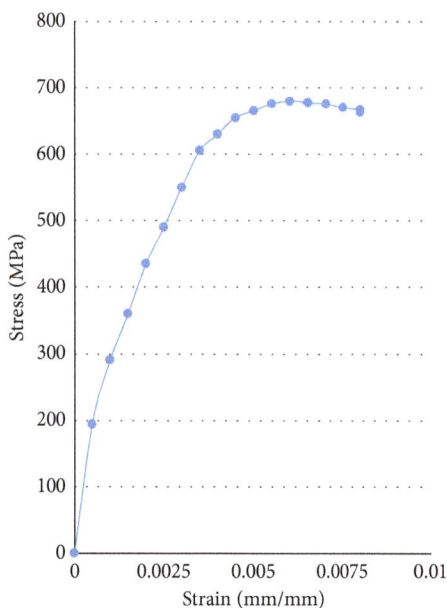

FIGURE 3: Stress-strain relationship of steel wire.

for RCW4 and RCW8 specimens, respectively, whilst, the shear deformation was the least dominant deformation component of below 5% for both specimens RCW4 and RCW8. Interestingly, the yield penetration deformation of RCW8 was about 27% compared to 21% of that of RCW4 specimen, which can be attributed to the smaller concrete cover of sloof foundation on RCW8 and hence smaller bond slip strength between steel bar and concrete at foundation.

4. Backbone Curve Models

Two simple models (backbone and simplified) were developed for design purposes or basic assessment of lateral load-displacement capacity of such walls. Both models for sandwich concrete walls are developed based on the model previously developed by authors for lightly reinforced concrete walls [19].

4.1. Model 1: Detailed. A detailed curve model is developed based on displacement-based design methodology for predicting the lateral load-drift behaviour (comprising four stages: cracking, yield, peak, and lateral load failure) as shown conceptually in Figure 9.

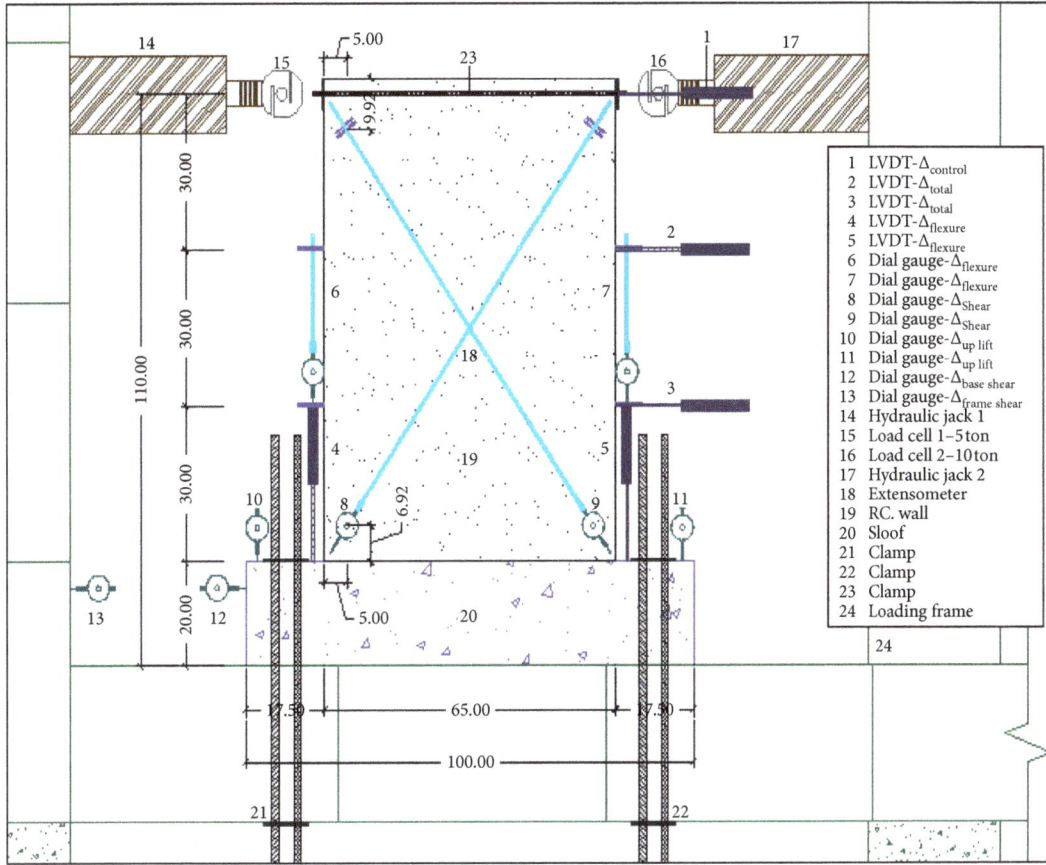

FIGURE 4: Schematic of wall loading test setup.

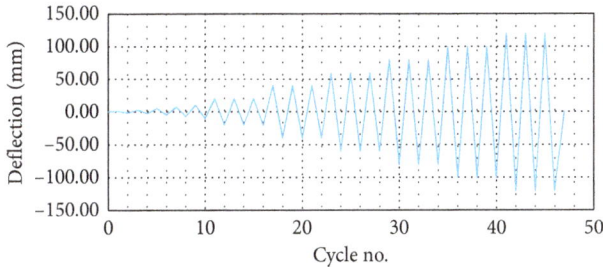

FIGURE 5: Quasi-static lateral loading history.

(a) Point A (cracking): the cracked lateral strength and drift are calculated as follows:

$$F_{cr} = \frac{M_{cr}}{L},$$

$$\gamma_{cr} = \frac{M_{cr}L}{3E_cI_g}, \tag{5}$$

where the flexural tensile strength f_t is taken as $0.6\sqrt{f_c'}$.

(b) Point B (yield): the yield drift is calculated using the effective second moment of area as follows:

$$F_y = \frac{M_y}{L},$$

$$\gamma_y = \frac{M_yL}{3E_cI_e}. \tag{6}$$

The Paulay and Priestley [8] model for effective moment of inertia is used as follows.

(i) Flexure-dominated walls:

$$I_e = \left(\frac{100}{f_y} + \frac{P_u}{f_c'A_g}\right)I_g. \tag{7}$$

(ii) Shear-dominated walls:

$$I_w = \frac{I_g}{1.2 + C},$$

$$C = \frac{30I_e}{L^2tD}, \tag{8}$$

where P_u = nominal axial load, A_g = gross cross section area of walls, and t = wall thickness.

(c) Point C (peak strength): the model was developed by investigating the curvature within the plastic hinge region using the force equilibrium equation ($N = C_c + C_s - T$) with the spalling strain ($\varepsilon_{cu} = 0.003$) used as a limit state for concrete strain. For low-rise buildings, the presence of gravity axial load is reasonably small, and hence for simplicity, the compression steel area is eliminated from the equilibrium equation. The peak flexural lateral load F_u and the drift at concrete fracture γ_u can then be obtained as follows:

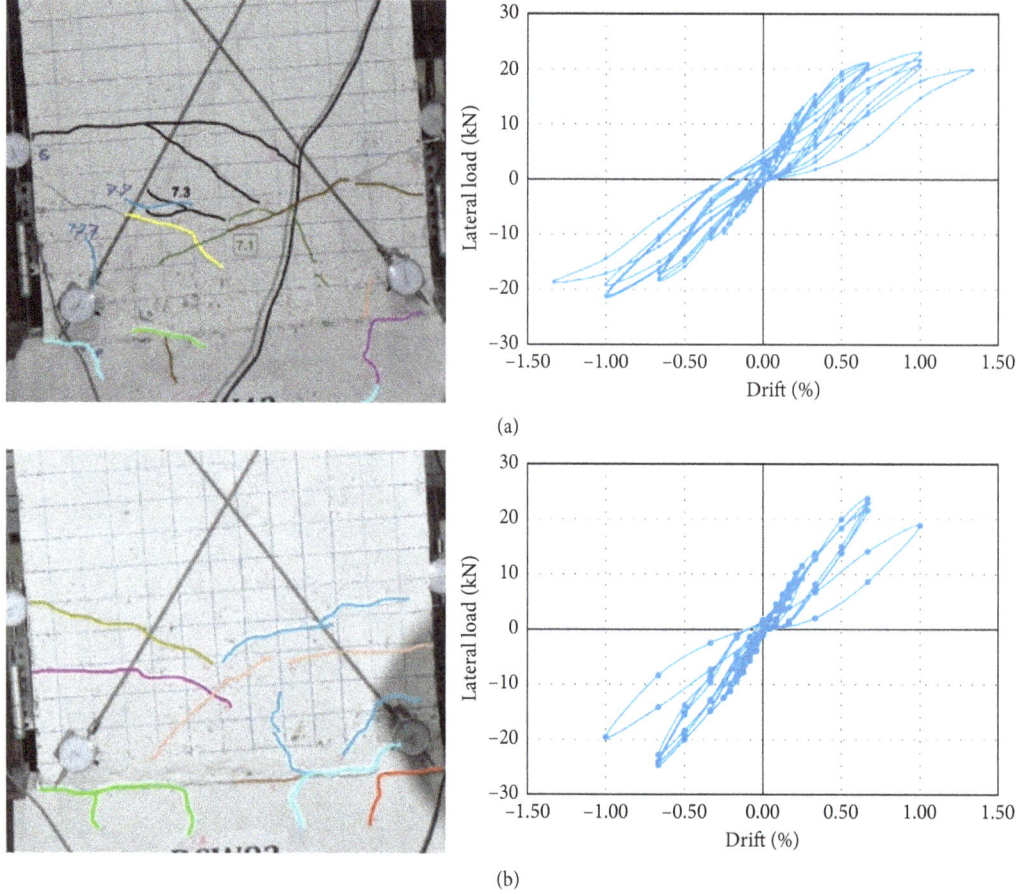

Figure 6: Crack patterns at 1% drift and hysteretic curves of all specimens: (a) RCW4; (b) RCW8.

Table 1: Strength and deformation properties of specimens RCW4 and RCW8.

		Strength (kN)			Drift (%)			
		F_{cr}	F_y	F_u	δ_{cr}	δ_y	δ_u	δ_{lf}
RCW4	Exp.	2.8	18	23.5	0.17	0.47	1.00	1.33
	Theo.	4.0	16	23	0.1	0.42	0.75	n.a
RCW8	Exp.	2.3	20	24.5	0.17	0.55	0.67	1.00
	Theo.	4.3	18	23.5	0.1	0.43	0.8	n.a

Note. The theoretical values were taken from moment-curvature analysis (flexural component only).

$$F_u = \frac{M_u}{L}, \tag{9}$$

$$\gamma_{peak} = \gamma_y + \gamma_{pl\cdot p},$$

where $\gamma_{pl\cdot p} = (\phi_{peak} - \phi_y)L_p$ in which $\phi_{peak}\,(\varepsilon_{cu}/k_u d) = (0.003/k_u d)$ and $\phi_y = (3\gamma_y/L)$, $k_u = ((N + A_{st}f_y)/0.85f'_c\gamma db)$, $\gamma = 0.85 - 0.007\,(f'_c - 28)$, $A_{st} =$ tensile steel area, and $\varepsilon_{sm} =$ steel strain-hardening strain.

The plastic hinge length L_p can be estimated using the Paulay and Priestley model [8] as follows:

$$L_p = 0.054L + 0.022\,d_b f_y. \tag{10}$$

(d) Point D (ultimate displacement): lateral load-displacement relationship of squat walls is dominated by

shear behaviour; however, for lightly reinforced squat walls, the flexure behaviour still provides large influence on lateral load-drift behaviour. The failure mechanism which is influenced by shear strength degradation is needed; hence, the lateral load failure models developed for lightly reinforced concrete columns and walls [20, 21] are modified for this model due to the similarity of lateral load-displacement behaviour between lightly reinforced concrete walls and columns.

Shear strength (V_u) of RC walls consists of concrete strength (V_c) and steel strength (V_s) components as follows:

$$V_u = V_c + V_s. \tag{11}$$

In this model, the concrete shear strength uses the formula developed based on principal tensile strength by authors [22], whilst the steel strength proposed by Wesley and Hashimoto [23] is used as follows:

$$V_c = \frac{2}{3}A_{cr}\sqrt{(f_t)^2 + \frac{f_t P}{A_{cr}}}, \tag{12}$$

$$V_s = (c_h\rho_h + c_v\rho_v)\,f_y dt, \tag{13}$$

where

$$A_{cr} = 0.85\,(n_c\rho_{st})^{0.36}\,dt, \tag{14}$$

where d is the effective depth of RC walls, which can be assumed as $0.8D$, and $C_h = 1 - c_v$, in which

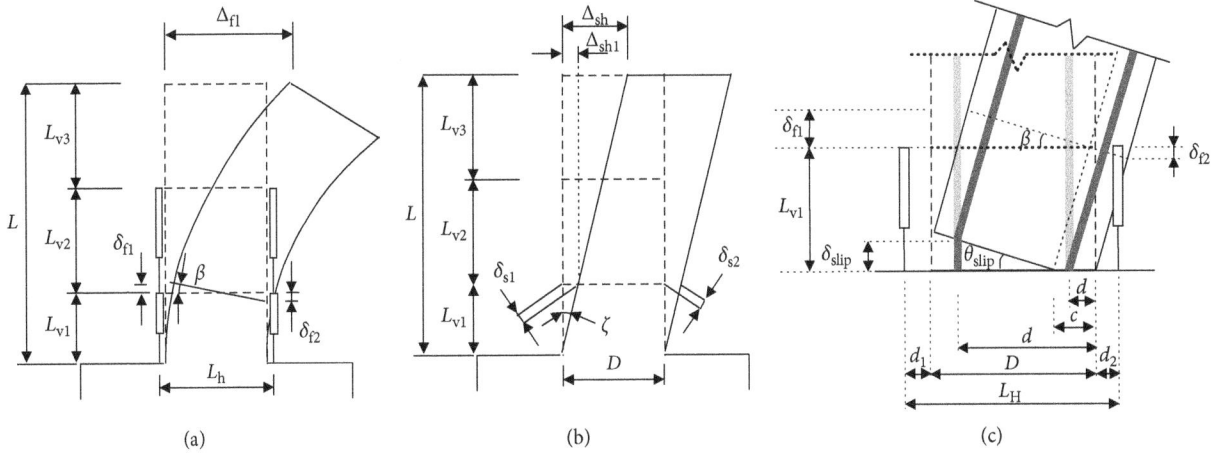

FIGURE 7: LVDT measurements for (a) flexure deformation, (b) shear deformation, and (c) yield penetration deformation.

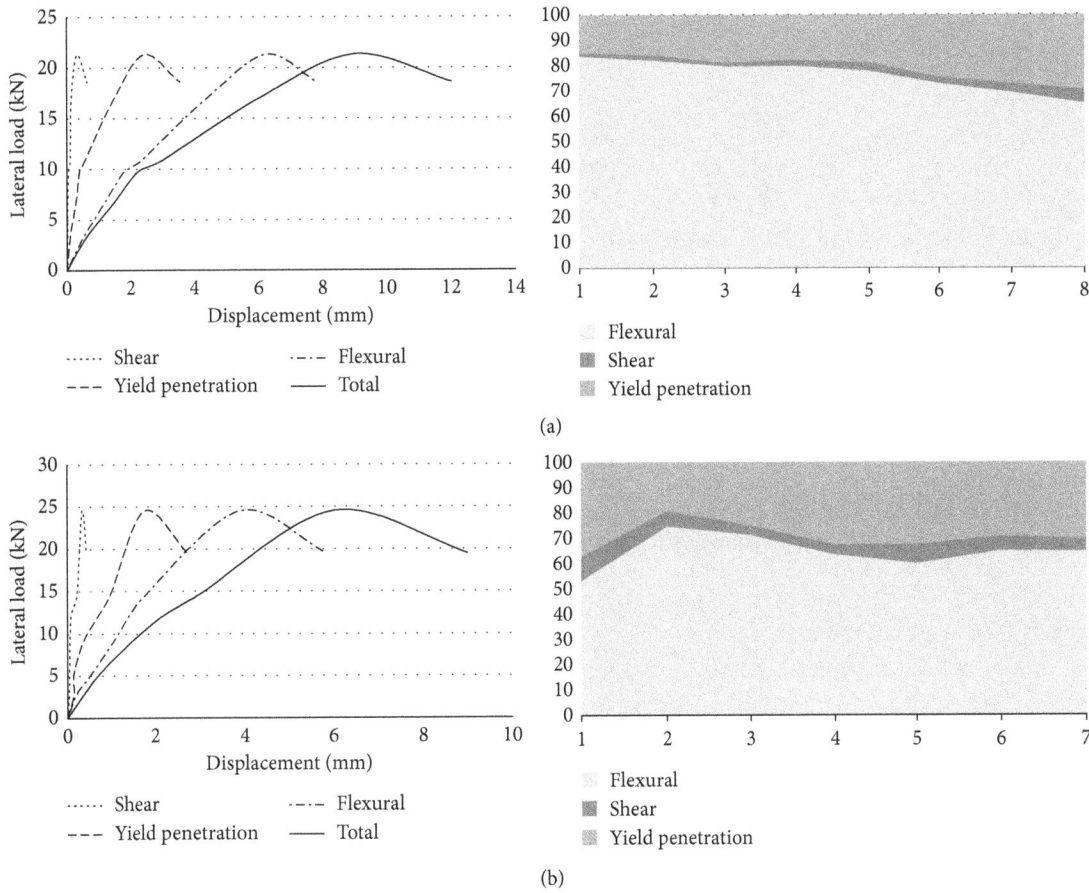

FIGURE 8: Variation of displacements due to flexure, shear, and yield penetration: axial load-deformation relationship (left) and ratio of each displacement to total displacement at each loading step (right). (a) RCW4 specimen. (b) RCW8 specimen.

$$c_v = 1, \qquad \text{for } a < 0.5$$
$$= 2(1 - a), \quad \text{for } 0.5 < a < 1 \qquad (15)$$
$$= 0, \qquad \text{for } a > 1,$$

with $n_c = E_s/E_c$, ρ_h = the transverse reinforcement ratio, ρ_v = the total longitudinal reinforcement ratio, and ρ_{st} = the tension reinforcement ratio.

As a note, for moderate and slender walls ($a > 1$, and hence $c_v = 0$), the steel strength component (equation (13)) can be rewritten as a common shear strength formula:

$$V_s = \rho_h f_y \, dt = \frac{A_v f_y d}{s}. \qquad (16)$$

The ultimate drift can be obtained as follows:

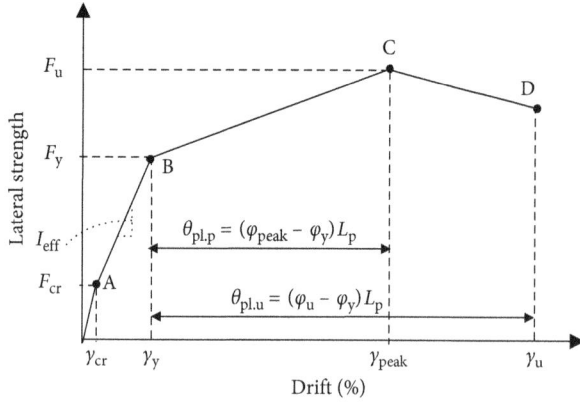

FIGURE 9: The backbone curve model of lateral load-drift capacity [20].

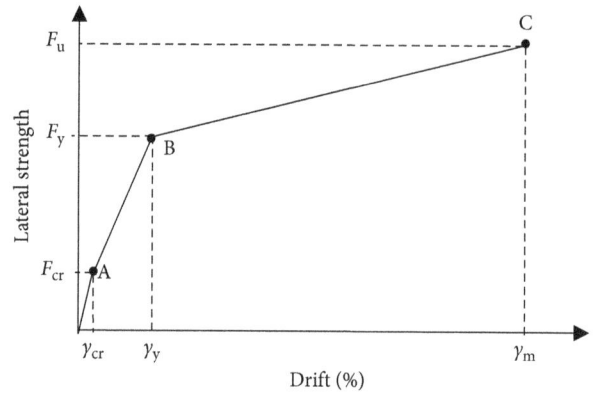

FIGURE 10: The simplified model of lateral load-drift capacity.

$$\gamma_u = \frac{\gamma_y}{k}\left[(1 - k\alpha) - 0.8\frac{F_u}{V_u}\right], \qquad (17)$$

where

$$k = \frac{0.3e^{5.7n}}{9 - a}, \qquad (18)$$

where α = the drift ductility at the start of shear strength decrease.

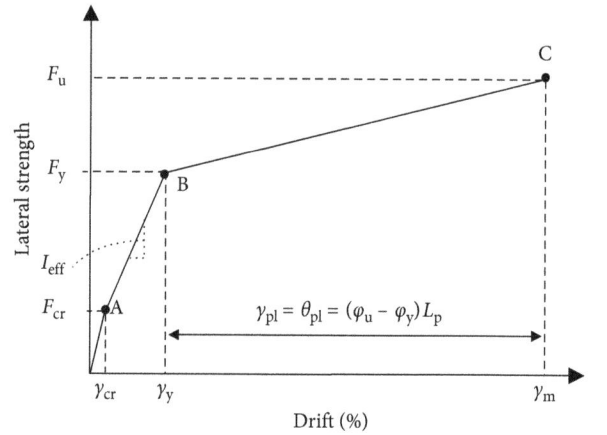

FIGURE 11: Plastic drift of simplified model.

4.2. Model 2: Simplified. The simplified model is a simple procedure for estimating lateral load-drift behaviour of lightly reinforced concrete walls. This model consists of trilinear stages with each state: cracking, yield, and ultimate as shown in Figure 10.

(a) Point A (cracking): the lateral strength at cracking point can be predicted by assuming cracking drift γ_{cr} = 0.05%.

(b) Point B (yield): the ultimate yield strength is calculated using factored ultimate strength:

$$F_y = \phi F_u, \qquad (19)$$

whereas the corresponding yield drift (γ_y) is determined using the smallest values of the following alternatives:

(i) Approximate value of γ_y = 0.2%–0.3%

(ii) Apply $I_{eff} = 0.5I_g$ (refer [24])

(c) Point C (ultimate): the ultimate drift (γ_m) can be calculated as a sum of the yield drift (γ_y) and the plastic drift (γ_{pl}) as follows (refer Figure 11):

$$\gamma_m = \gamma_y + \gamma_{pl}. \qquad (20)$$

The plastic drift can be estimated by assuming a maximum acceptable strain in the steel bar at single crack at the wall base in the order of ε_s = 5.0% and taking a more conservative approach to Priestley and Paulay [8] strain penetration length of $l_{yp} = 4400\ \varepsilon_y\ d_b \approx 15d_b$. Hence, the following models can be obtained (refer Figure 12).

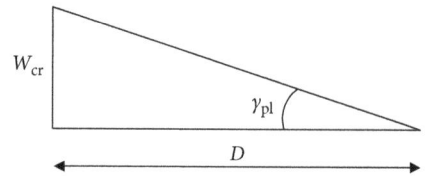

FIGURE 12: Plastic rotation at wall base.

Crack width:

$$W_{cr} = \varepsilon_s \cdot l_{yp} = 0.05 \times 15d_b = 0.75d_b, \qquad (21)$$

Plastic drift:

$$\gamma_{pl} = \theta_{pl} = \frac{w_{cr}}{D} = 0.75\left(\frac{d_b}{D}\right). \qquad (22)$$

The lateral load-drift relationship between experimental data and proposed models are considerably in good agreement as shown in Figures 13 and 14. More data are certainly required to refine the models particularly for the detailed model since it was developed using a semiempirical approach. However, interestingly, the simplified model with the pure analytical approach showed better prediction due to the dominant combinations of flexural and yield penetration behaviour.

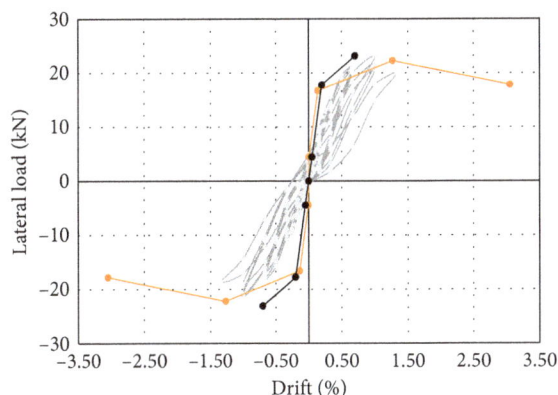

FIGURE 13: Comparison between experimental data and theoretical models for specimen RCW4.

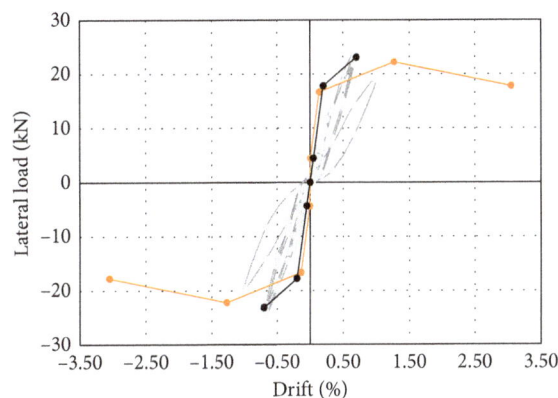

FIGURE 14: Comparison between experimental data and theoretical models for specimen RCW8.

5. Conclusion

Two specimens of light-weight sandwich concrete walls have been tested in order to investigate the lateral load-drift behaviour and collapse mechanism. Specimen RCW4 with a thinner EPS panel developed more classic flexural behaviour with drift capacity maxed at about 1.3%, whilst specimen RCW8 only managed to reached 1.0% with dominant yield penetration behaviour due to thinner concrete cover of sloof foundation. However, the tests were stopped at 20% drop of peak load instead of further failure at axial load collapse. And, hence the results can still be considered satisfactory for low-to-moderate seismic regions but might not be sufficient for high seismic regions.

Two models comprising detailed and simplified approach for predicting the load-displacement behaviour of sandwich concrete wall subjected to lateral load have been developed. The experimental data and the proposed models are in good agreement, particularly the simplified model due to the dominant behaviour of flexural and yield penetration.

Conflicts of Interest

The authors declare that they have no conflicts of interest.

References

[1] S. A. Mousavi, S. M. Zahrai, and A. Bahrami-Rad, "Quasi-static cyclic tests on super-lightweight EPS concrete shear walls," *Engineering Structures*, vol. 65, pp. 62–75, 2014.

[2] Z. Yizhou, "Mechanical properties of self-combusted gangue reinforced concrete shear walls under low-cyclic load," *Journal of Zhejiang University, Engineering Science*, vol. 34, no. 3, pp. 325–330, 2000.

[3] T. Hejin, Y. Qingrong, and S. Xinya, "Experimental study on bBehaviours of reinforced ash ceramsite concrete shear walls," *Journal of Building Structure*, vol. 15, no. 4, pp. 20–30, 1994.

[4] Y. H. Chai and J. D. Anderson, "Seismic response of perforated lightweight Aggregate concrete wall panels for low-rise modular classrooms," *Engineering Structures*, vol. 27, no. 4, pp. 593–604, 2005.

[5] L. Cavaleri, N. Miraglia, and M. Papia, "Pumice concrete for structural wall panels," *Engineering Structures*, vol. 25, no. 1, pp. 115–125, 2003.

[6] http://mpanelindonesia.com/product/wall-panel/.

[7] T. Paulay, "Earthquake-resisting shear walls-New Zealand design trends," *ACI Journal*, vol. 77, no. 18, pp. 144–152, 1980.

[8] T. Paulay and M. J. N. Priestley, *Seismic Design of Reinforced Concrete and Masonry Buildings*, Wiley Interscience Press publication, John Wiley & Sons inc., New York City, NY, USA, 1992.

[9] O. Esmaili, S. Epackachi, M. Samadzad et al., "Study of structural RC shear wall system in a 56-story RC tall building," in *Proceedings the 14th World Conference on Earthquake Engineering*, Beijing, China, October 2008.

[10] A. Carpinteri, M. Corrado, G. Lacidogna et al., "Lateral load effects on tall shear wall structures of different height," *Structural Engineering and Mechanics*, vol. 41, no. 3, pp. 313–337, 2012.

[11] P. A. Hidalgo, C. A. Ledezma, and R. M. Jordan, "Seismic bBehaviour of squat reinforced concrete shear walls," *Earthquake Spectra*, vol. 18, no. 2, pp. 287–308, 2002.

[12] T. N. Salonikios, "Shear strength and deformation patterns of R/C walls with aspect ratio 1.0 and 1.5 designed to Eurocode 8," *Engineering Structures*, vol. 24, no. 1, pp. 39–49, 2002.

[13] T. N. Salonikios, "Analytical prediction of the inelastic response of RC walls with low aspect ratio," *ASCE Journal of Structural Engineering*, vol. 133, no. 6, pp. 844–854, 2007.

[14] J. Carrillo and S. M. Alcocer, "Shear strength of reinforced concrete walls for seismic design of low-rise housing"," *ACI Structural Journal*, vol. 110, no. 3, 2013.

[15] T. Trombetti, G. Gasparini, S. Silvestri et al., "Results of pseudo-static tests with cyclic horizontal load on cast in situ sandwich squat concrete walls," in *Proceedings of Ninth Pacific Conference on Earthquake Engineering Building an Earthquake-Resilient Society*, Auckland, New Zealand, April 2011.

[16] I. Ricci, M. Palermo, G. Gasparini, S. Silvestri, and T. Trombetti, "Results of pseudo-static tests with cyclic horizontal load on cast in situ sandwich squat concrete walls," *Engineering Structures*, vol. 54, pp. 131–149, 2013.

[17] M. Palermo and T. Trombetti, "Experimentally-validated modelling of thin RC sandwich walls subjected to seismic loads," *Engineering Structures*, vol. 119, pp. 95–109, 2016.

[18] ASTM E2126, *Standard Test Methods for Cyclic (Reversed) Load Test for Shear Resistance of Walls for buildings*, ASTM International, West Conshohocken, PA, USA, 2005.

[19] A. Wibowo, J. L. Wilson, N. T. K. Lam et al., "Seismic performance of lightly reinforced structural walls for design purposes," *Magazine of Concrete Research*, vol. 65, no. 13, pp. 809–828, 2013.

[20] J. L. Wilson, A. Wibowo, N. T. K. Lam et al., "Drift behaviour of lightly reinforced concrete columns and structural walls for seismic design applications," *Australian Journal of Structural Engineering*, vol. 16, no. 1, pp. 62–74, 2015.

[21] A. Wibowo, J. L. Wilson, N. T. K. Lam et al., "Drift performance of lightly reinforced concrete columns," *Engineering Structures*, vol. 59, pp. 522–535, 2014.

[22] A. Wibowo, J. L. Wilson, N. T. K. Lam et al., "Drift capacity of lightly reinforced concrete columns," *Australian Journal of Structural Engineering*, vol. 15, no. 2, pp. 131–150, 2014.

[23] D. A. Wesley and P. S. Hashimoto, "Seismic structural fragility investigation for the zion nuclear power plant," Report NUREG/CR-2320, US Nuclear Regulatory Commission, Bethesda, MD, USA, 1981.

[24] J. W. Wallace, "Modelling issues for tall reinforced concrete core wall buildings," *The Structural Design of Tall and Special Buildings*, vol. 16, no. 5, pp. 615–632, 2007.

10

Kinetic Hydration Heat Modeling for High-Performance Concrete Containing Limestone Powder

Xiao-Yong Wang

College of Engineering, Department of Architectural Engineering, Kangwon National University, Chuncheon-si 200701, Republic of Korea

Correspondence should be addressed to Xiao-Yong Wang; wxbrave@kangwon.ac.kr

Academic Editor: Doo-Yeol Yoo

Limestone powder is increasingly used in producing high-performance concrete in the modern concrete industry. Limestone powder blended concrete has many advantages, such as increasing the early-age strength, reducing the setting time, improving the workability, and reducing the heat of hydration. This study presents a kinetic model for modeling the hydration heat of limestone blended concrete. First, an improved hydration model is proposed which considers the dilution effect and nucleation effect due to limestone powder addition. A degree of hydration is calculated using this improved hydration model. Second, hydration heat is calculated using the degree of hydration. The effects of water to binder ratio and limestone replacement ratio on hydration heat are clarified. Third, the temperature history and temperature distribution of hardening limestone blended concrete are calculated by combining hydration model with finite element method. The analysis results generally agree with experimental results of high-performance concrete with various mixing proportions.

1. Introduction

The use of limestone powder blended cement is a common practice in the modern concrete industry. The benefits from technical, economic, and ecological aspects can be achieved by using limestone blended concrete [1]. Technical benefits mainly refer to limestone powder which can increase the early-age performance of concrete. Economic benefits mean obtaining cement with a compressive strength similar to control concrete at low production costs. The ecological aspects are the reduction of greenhouse gas emission by using limestone.

Many studies in experimental or theoretical aspects have been done about early-age properties and durability of limestone blended concrete. Bonavetti et al. [2] found that the addition of limestone can increase the early-age strength of concrete. However, the late-age strength is impaired due to the dilution effect of limestone addition. Mohammadi and South [3] reported that limestone addition can reduce the bleeding of fresh concrete and increase the viscosity and cohesiveness of fresh concrete. Mohammadi and South [4]

also reported that the concrete with various limestone contents up to 12% has a similar drying shrinkage and sulfate expansion resistance with control concrete. Chen and Kwan [5] measured heat generation of concrete with different limestone stone replacement ratios and binder contents. They found that the addition of limestone can significantly reduce the heat generation of concrete. Palm et al. [6] found that high-level limestone addition can increase the carbonation depth and chloride migration coefficient of concrete. Based on life cycle assessment, Palm et al. [6] also found high-level limestone additions can reduce the CO_2 emission about 25% in comparison with average cement with the same performance.

Compared with abundant experimental studies, the theoretical models for limestone blended concrete are relatively limited. Lothenbach et al. [7] proposed a thermodynamic model for limestone blended concrete. The formation of monocarboaluminate and bulk compositions of hydrating cement is calculated by thermodynamic models. Bentz [8] proposed a hydration model which analyzed the dilution effect, nucleation effect, and chemical effect of limestone addition. Similarly, Mohamed et al. [9] also proposed a model to evaluate reaction degree of hydration for concrete with different limestone additions. However, Lothenbach et al. [7], Bentz [8], and Mohamed et al.'s [9] studies mainly focus on the degree of hydration in cement-limestone blends. The hydration heat is scarcely simulated in their studies. Poppe and Schutter [10] and Ye et al. [11] proposed models to analyze heat evolution of limestone blended self-compacting concrete. But their study mainly focuses on the isothermal condition or adiabatic temperature rise. For concrete structures in construction sites, the heat release from hydration and heat transfer to ambient environments occurs simultaneously. Poppe and Schutter [10] and Ye et al.'s [11] studies do not consider semiadiabatic temperature rise in real construction sites.

To overcome the shortcomings in current models [7–11], we proposed an integrated numerical procedure to analyze the temperature history and temperature distribution of hardening limestone blended concrete. The hydration model is combined with finite element method. The heat of hydration is calculated from the degree of hydration. The reduction of hydration heat due to limestone additions is clarified through analysis.

2. Hydration Heat Model of Limestone Powder Blended Concrete

2.1. Hydration Model of Portland Cement. Wang and Lee [12] proposed a kinetic hydration model for Portland cement which takes into account the effects of a water to binder ratio (W/B), compound compositions of cement, fineness of cement, and capillary water contents on the hydration of cement. The kinetic hydration model analyzes the involved kinetic processes of cement hydration, such as initial dormant process, phase boundary reaction-controlled process, and diffusion-controlled process. The kinetic hydration model is valid for concrete with various types of Portland cement, various mixing proportions, and various curing temperatures. The equation for the kinetic hydration model is shown as follows:

$$\frac{d\alpha}{dt} = \frac{3\rho_w C_{w\text{-free}} \left(S_w/S_0\right)}{r_0 \rho_c \left(v + w_g\right)}$$
$$\cdot \frac{1}{\left(1/k_d - r_0/D_e\right) + \left(1/k_r\right)\left(1 - \alpha\right)^{-2/3} + \left(r_0/D_e\right)\left(1 - \alpha\right)^{-1/3}}. \quad (1)$$

In (1), α is hydration degree of cement, k_d is the hydration rate coefficient in the initial dormant period, k_r is the hydration rate coefficient of phase boundary reaction-controlled process, D_e is the hydration rate coefficient in the diffusion-controlled stage, $C_{w\text{-free}}$ denotes the amount of capillary water at the exterior of hydration products, S_w denotes the effective contacting surface area between the cement particles and capillary water, and S_0 denotes the total surface area if hydration products develop unconstrained [12]. Equation (1) also considers chemical and physical aspects of cement hydration. In (1), v is the mass of chemically bound water

for one-gram hydrated cement (=0.25), w_g is the mass of physically bound water for one-gram hydrated cement (=0.15), ρ_w denotes the density of water, ρ_c denotes the density of the cement, r_0 denotes the radius of unhydrated cement particles.

The determinations of reaction coefficients k_d, D_e, and capillary water content $C_{w\text{-free}}$ are shown in (2)–(4), respectively.

$$k_d = \frac{B}{\alpha^{1.5}} + C\alpha^3, \quad (2)$$

$$D_e = D_{e0} \ln\left(\frac{1}{\alpha}\right), \quad (3)$$

$$C_{w\text{-free}} = \left(\frac{W_0 - 0.4 * \alpha * C_0}{W_0}\right)^r. \quad (4)$$

Equation (2) can be used to determine reaction coefficient k_d. In (2), B is the rate of the initial impermeable layer formation, and C is the rate of the initial impermeable layer decay.

Equation (3) can be used to determine reaction coefficient D_e. In (3), D_{e0} is the initial diffusion coefficient, and D_e decreases as cement hydration proceeds.

Equation (4) can be used to determine capillary water content $C_{w\text{-free}}$. In (4), C_0 is the cement content in mixing proportion, W_0 is the water content in the mix proportion, and r ($r = 2.6 - 4(W_0/C_0)$) is an empirical parameter that considers the accessibility of water into an inner anhydrous part through an outer hard shell of cement particles. For high-strength concrete with low W/C ratio at late ages, $C_{w\text{-free}}$ has a significant influence on the rate of hydration.

The influences of temperature on reaction coefficients can be described by using Arrhenius's law [12] as follows:

$$B = B_{20} * \exp\left(-\beta_1 \left(\frac{1}{T} - \frac{1}{293}\right)\right),$$
$$C = C_{20} * \exp\left(-\beta_2 \left(\frac{1}{T} - \frac{1}{293}\right)\right),$$
$$k_r = k_{r20} * \exp\left(-\frac{E}{R}\left(\frac{1}{T} - \frac{1}{293}\right)\right), \quad (5)$$
$$D_{e0} = D_{e20} * \exp\left(-\beta_3 \left(\frac{1}{T} - \frac{1}{293}\right)\right),$$

where β_1, β_2, E/R, and β_3 denote the activation energies of B, C, k_r, and D_{e0}, respectively. B_{20}, C_{20}, k_{r20}, and D_{e20} denote the values of reaction coefficients B, C, k_r, and D_{e0} at 293 K, respectively.

Based on the degree of hydration of concrete with various types of Portland cement and various curing temperatures, Wang [13] proposed that the reaction coefficients of hydration model, such as B_{20}, C_{20}, k_{r20}, and D_{e20}, can be determined from compound compositions of cement. The temperature sensitivity coefficients can be approximately regarded as constants for different types of cement [13]. These relationships are shown as follows:

$$B_{20} = 6 * 10^{-12} \left(C_3S\% + C_3A\%\right) + 4 * 10^{-10},$$

$$C_{20} = 0.0003 C_3 S\% + 0.0186,$$

$$k_{r20} = 8 * 10^{-8} C_3 S\% + 1 * 10^{-6},$$

$$D_{e20} = -8 * 10^{-12} C_2 S\% + 7 * 10^{-10},$$

$$\beta_1 = 1000, \tag{6}$$

$$\beta_2 = 1000,$$

$$\frac{E}{R} = 5400,$$

$$\beta_3 = 7500.$$

Summarily, the kinetic hydration model is composed of four rate determining coefficients, that is, the rate of formation of the initial impermeable layer (B), the rate of destruction of initial impermeable layer (C), the rate of phase boundary reaction-controlled process (k_r), and the rate of diffusion-controlled process (D_e). By using compound compositions of cement, the reaction coefficients of kinetic hydration model can be determined. Furthermore, the degree of hydration can be calculated by using (1). The proposed hydration model is valid for Portland cement concrete with various materials properties and curing conditions.

2.2. Effect of Limestone Addition on Cement Hydration.

Wang [13] reported that the addition of limestone presents dilution effect, nucleation effect, and chemical effect on cement hydration. Dilution effect is when cement is partially replaced by limestone, the content of cement is reduced and water to cement ratio increases correspondingly. Nucleation effect is that limestone can serve as nucleation sites of hydrating cement particles. Hydration of cement can accelerate due to nucleation effect. Chemical effect is the formation of monocarboaluminate due to limestone reaction in preference to a monosulfoaluminate.

On the other hand, because the reactivity of limestone is very weak compared with other supplementary cementitious materials, limestone can be approximately regarded as chemical inert filler [1, 11]. Hence in this study, the chemical effect of limestone is not considered.

The dilution effect of limestone powder can be considered by using (4). Wang [13] and Kishi and Saruul [14] proposed that the nucleation effect of limestone relates to the ratio of surface area of cement particles to that of limestone powder. The nucleation effect of limestone powder is significant in phase boundary reaction-controlled process and diffusion-controlled process. The nucleation effect of limestone powder can be considered as follows:

$$L_r = \frac{LS_0 * S_{LS}}{C_0 * S_C}, \tag{7}$$

$$k_{rLS} = k_r \left(1 + A_1 L_r\right), \tag{8}$$

$$D_{eLS} = D_e \left(1 + A_2 L_r\right). \tag{9}$$

In (7), L_r is the limestone nucleation effect indicator, LS_0 is the mass of limestone in concrete mixing proportions, and S_{LS} is the Blaine surface area of limestone powder.

In (8), k_{rLS} is the updated phase boundary reaction coefficient, and A_1 is enhanced coefficients of k_r. In (9), D_{eLS} is the updated diffusion coefficient D_e, and A_2 is enhanced coefficients of D_e. Based on analysis shown later (Section 3.1), the values of A_1 and A_2 are set as 0.6.

Summarily, for a cement-limestone blend, the dilution effect is considered through capillary water concentration. The nucleation effect is considered by nucleation effect indicator which considers binder proportions and surface area of binders. Furthermore, by using updated reaction coefficients, the reaction degree of cement in cement-limestone blends can be determined.

2.3. Temperature History Model of Hardening Concrete.

Hydrate heat of hydrating concrete is dependent on both cement content and degree of hydration. The relation heat from hydration of concrete can be determined as follows [15–17]:

$$\frac{dQ}{dt} = C_0 H_e \frac{d\alpha}{dt}, \tag{10}$$

where Q is hydration heat and H_e is released hydration heat from a unit mass of cement. H_e can be determined by using compound compositions and hydration heat of individual components of cement [15–17].

For hardening concrete, the temperature distribution is in a dynamic heat balance between the hydration heat generation inside the concrete and heat loss to the ambient. The heat generation comes from hydration reactions of the cement. The temperature distribution of hardening concrete is determined as follows [18]:

$$C_{\text{hc}} \frac{\partial T}{\partial t} = \text{div}\left(k\nabla T\right) + \frac{dQ}{dt}, \tag{11}$$

where C_{hc} is the heat capacity of hydrating concrete and can be calculated as the sum of the individual components of concrete, k is the thermal conductivity of concrete, T is the concrete temperature, t is time, and dQ/dt can be calculated based on the degrees of hydration of the cement (10).

For hardening concrete in construction sites, the boundary condition can be described as follows:

$$k\nabla T = \beta \left(T_s - T_a\right), \tag{12}$$

where β is the coefficient of heat convection between concrete surface and ambient; T_s is the temperature on the concrete surface; and T_a is ambient temperature.

Equation (11) is numerically solved by using finite element method. Three-dimensional (3D) eight-node isoparametric element is used to mesh the geometry model of hardening concrete. The iteration scheme of (11) is shown as follows [15–17]:

$$\left(\frac{1}{\Delta t}[C] + \theta[B]\right)\{T\}_{n+1}$$
$$= \left(\frac{1}{\Delta t}[C] - (1-\theta)[B]\right)\{T\}_n + \{P\}_n, \tag{13}$$

TABLE 1: Mixing proportions of paste for isothermal heat evolution.

	Water (g)	Cement (g)	Limestone (g)	Blaine of limestone (cm^2/g)	Limestone/(cement + limestone)	Water/cement
M100	5	10	0	—	0	0.5
M90L10-35	5	10	1.11	3500	0.1	0.5
M70L30-35	5	10	4.29	3500	0.3	0.5
M70L30-89	5	10	4.29	8900	0.3	0.5

where $[C]$ is the global mass matrix, $[B]$ is global stiffness matrix, $\{P\}$ is the global vector of temperature, and θ is integration parameter. The global stiffness matrix $[B]$, global mass matrix $[C]$, and global temperature vector $\{P\}$ can be obtained by assembling of element mass matrix, element stiffness matrix, and element temperature vector, respectively. Generally, to guarantee the stability of the numerical integration in a time domain, the value of integration parameter θ should be higher than 0.5. In this paper, according to the Galerkin method, the value of the parameter θ is used as 2/3 [12].

2.4. Summary of the Proposed Numerical Procedure. The numerical procedure consists of a kinetic hydration model and a finite element model. The kinetic hydration model considers the dilution effect and nucleation effect from limestone additions. The heat of hydration of hydrating concrete is calculated by using the degree of hydration and cement content. The calculation results of the heat of hydration are used as a source term in finite element model. By using Galerkin method, the parabolic partial differential equation about temperature distribution of hardening concrete is solved. Temperature history of semiadiabatic temperature rise is calculated considering both concrete materials properties and ambient conditions. The proposed numerical procedure is valuable for thermal cracking analysis of hardening concrete and construction plan design and materials design of concrete structures.

3. Verification of Proposed Model

3.1. Heat of Hydration. Experimental results about isothermal heat evolution shown in [14] are used to verify the proposed limestone blended hydration model. Kishi and Saruul [14] measured isothermal hydration heat of limestone blended cement paste. The mixing proportions of paste specimens are shown in Table 1. In Kishi and Saruul's study [14], to demonstrate the nucleation effect of limestone, the water to cement is the same for all specimens. Limestone is added as an additional binder, not replacing partial cement, which is different from the general applications of limestone. The limestone ranges within 10% and 30% of total binder, and the Blaine surface area of limestone ranges from 3500 cm^2/g to 8900 cm^2/g. The curing temperature is 20°C. The used cement is moderate heat Portland cement. By using Portland cement hydration model shown in Section 2.1, the hydration rate of control cement paste without limestone is calculated and shown in Figure 1(a). The y-axis of Figure 1(a) represents

the heat release for cement portions in cement-limestone blends.

Furthermore, based on experimental results about hydration heat of limestone blended cement paste, the enhanced coefficients A_1 and A_2 (shown in (8) and (9)) are set as 0.6. As shown from Figures 1(b)–1(d), the analysis results generally agree with experimental results. Figure 1(e) shows calculation results of rate of hydration heat. When cement is partially replaced by limestone, the value of the second peak of hydration heat increases, and the time corresponding to the second peak becomes much earlier. This trend agrees with experimental results about the rate of heat evolution of limestone blended cement paste [11]. When the surface area of Blaine surface of limestone increases, the nucleation effect becomes more significant, and the value of the second peak of hydration heat also increases. Hence limestone with a high Blaine surface is more effective in improving the hydration of cement.

3.2. Parameter Analysis about Hydration Heat. Parameter studies are carried out to analyze the degree of hydration and hydration heat of hardening concrete with different limestone replacement ratios and water to binder ratios. The water to binder ratio ranges from 0.3 to 0.5, and limestone content ranges from 15% to 30%. The curing temperature is assumed as 20°C.

The calculated degree of hydration is shown in Figure 2. Due to dilution effect and nucleation effect, the degree of hydration in cement-limestone blends is higher than that in control concrete. Moreover, for concrete with a lower water to binder ratio 0.3, when cement is partially replaced by limestone, the change of water to cement ratio is significant; hence the improvement of the degree of hydration is also obvious.

The relative degree of hydration means the ratio of the degree of hydration of limestone blended concrete to that of control concrete. Figure 3 shows relative degree of hydration. Given a certain limestone content, concrete with a lower water to binder ratio has a higher relative degree of hydration. This agrees with Bonavetti et al. [19] and Bentz et al.'s [20] study. They stated that limestone filler used in low water to cement ratio concrete is a rational option for saving energy.

As shown in (10), hydration heat relates to both cement content and degree of hydration. For limestone blended concrete, the degree of hydration increases which increases hydration heat, while the cement content decreases due to limestone replacing partial cement which reduces hydration heat. The total hydration heat depends on the combined

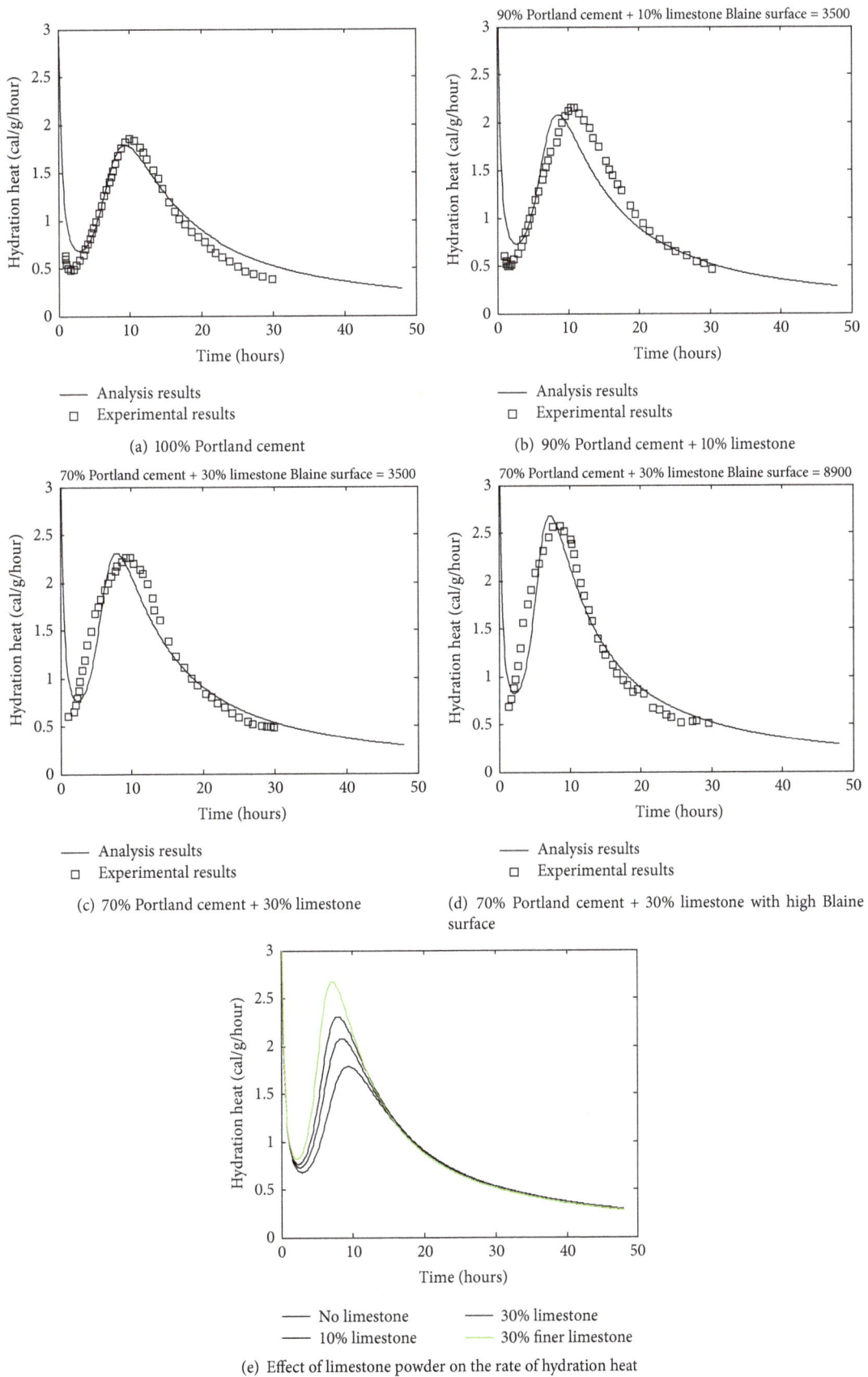

(a) 100% Portland cement

(b) 90% Portland cement + 10% limestone

(c) 70% Portland cement + 30% limestone

(d) 70% Portland cement + 30% limestone with high Blaine surface

(e) Effect of limestone powder on the rate of hydration heat

FIGURE 1: Rate of hydration heat.

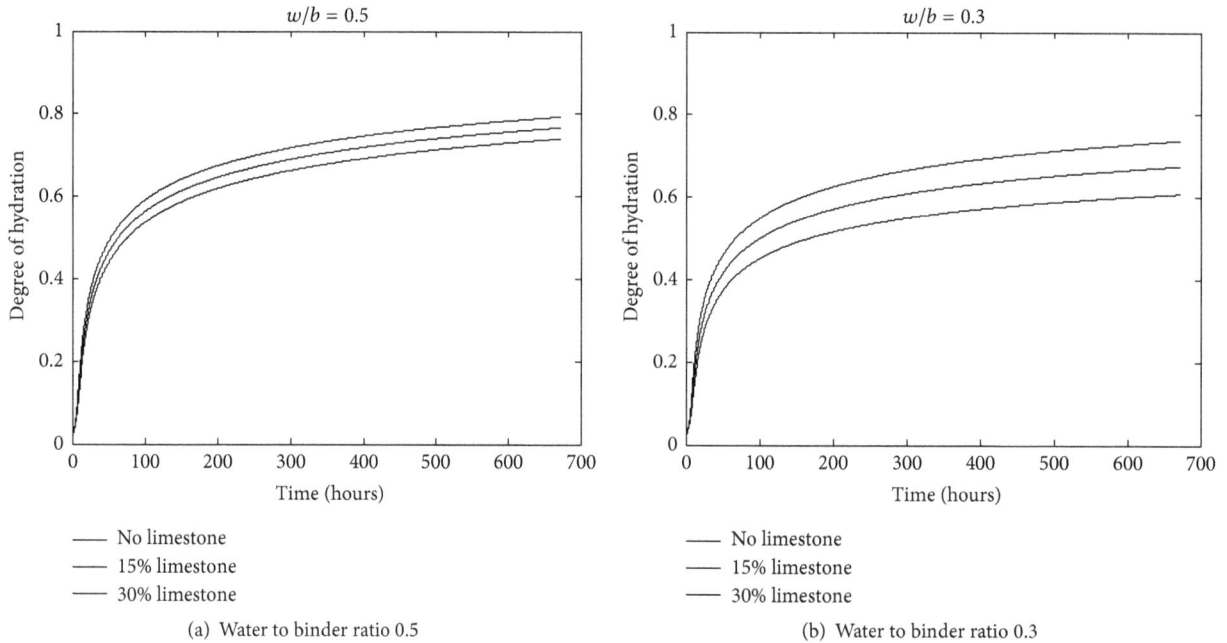

(a) Water to binder ratio 0.5

(b) Water to binder ratio 0.3

FIGURE 2: Degree of hydration.

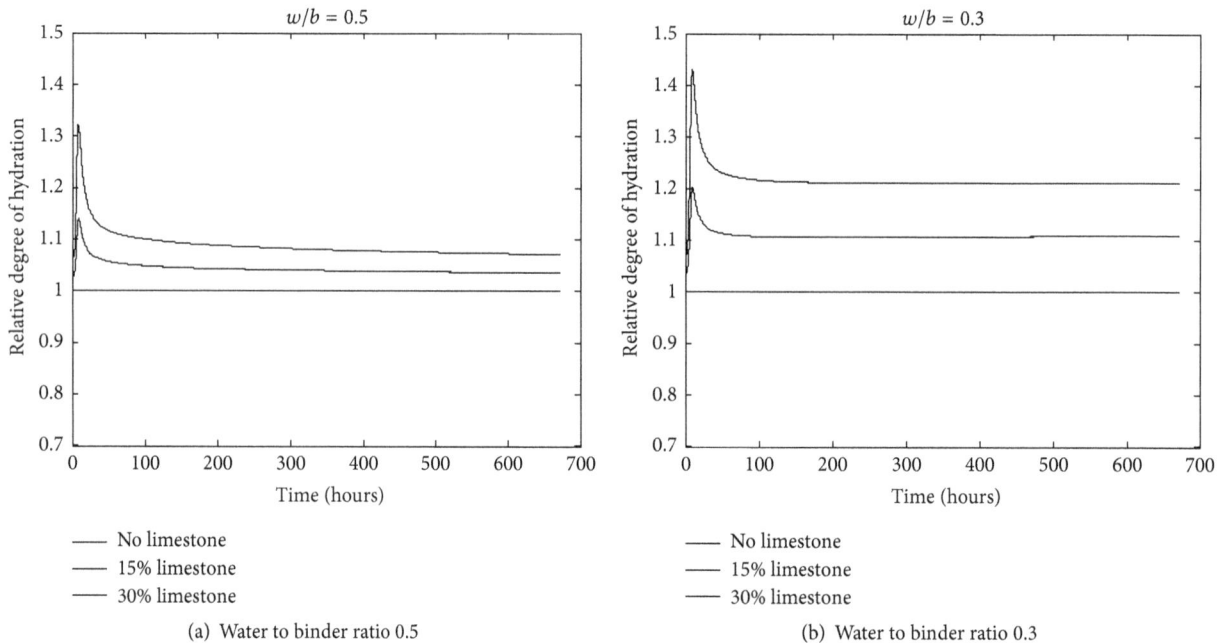

(a) Water to binder ratio 0.5

(b) Water to binder ratio 0.3

FIGURE 3: Relative degree of hydration.

action of increasing factor and decreasing factor, that is, degree of hydration and cement content. The relative heat of hydration means the ratio of heat of hydration of limestone blended concrete to that of control concrete. Figure 4 shows relative heat of hydration. Limestone blended concrete has a lower hydration heat than control concrete. Given a certain limestone content, concrete with a lower water to binder ratio has a higher relative heat of hydration. This agrees with Chen and Kwan's study [5]. Limestone addition can reduce the hydration heat of concrete and decrease the tendency of thermal cracking of hardening concrete.

3.3. Temperature History of Semiadiabatic Temperature Rise. Experimental results about semiadiabatic temperature rise shown in [14] are used to verify the proposed semiadiabatic temperature rise model. Table 2 shows the mixing proportions of concrete. The initial temperature and ambient temperature for each specimen are also shown in Table 2.

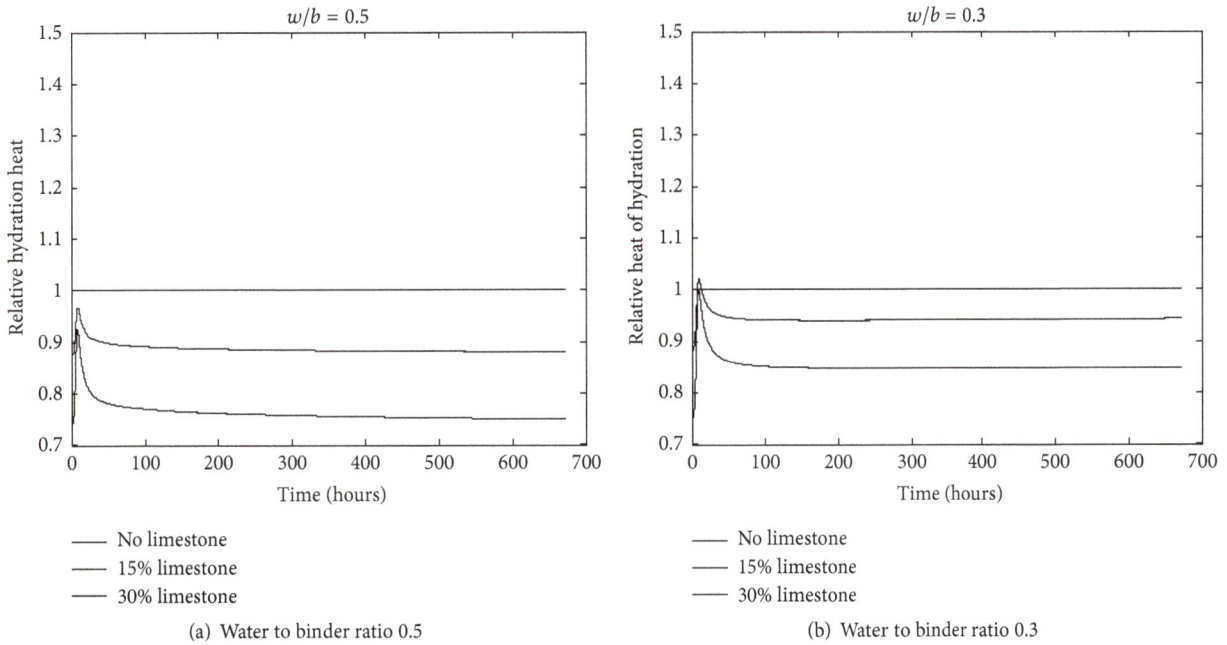

(a) Water to binder ratio 0.5

(b) Water to binder ratio 0.3

FIGURE 4: Relative heat of hydration.

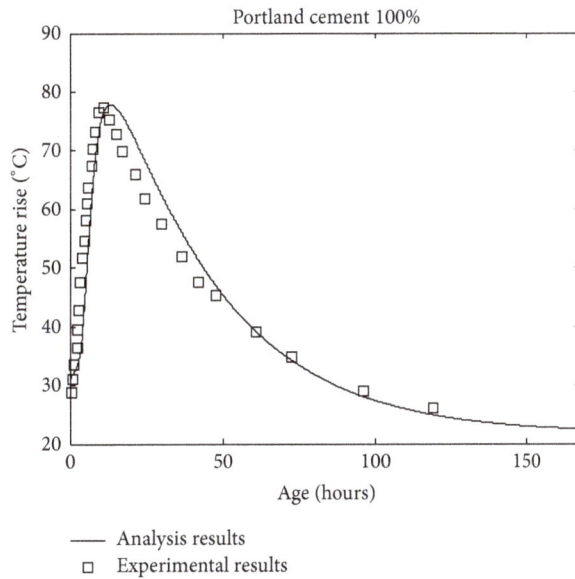

(a) Temperature history

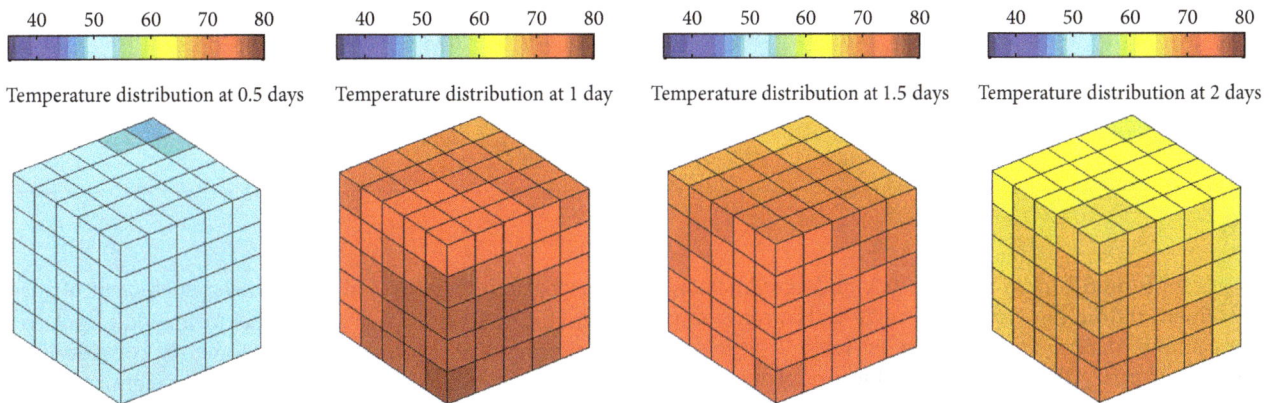

(b) Temperature distribution

FIGURE 5: OPC concrete.

Limestone 30%

— Analysis results
□ Experimental results

(a) Temperature history

Temperature distribution at 0.5 days   Temperature distribution at 1 day   Temperature distribution at 1.5 days   Temperature distribution at 2 days

(b) Temperature distribution

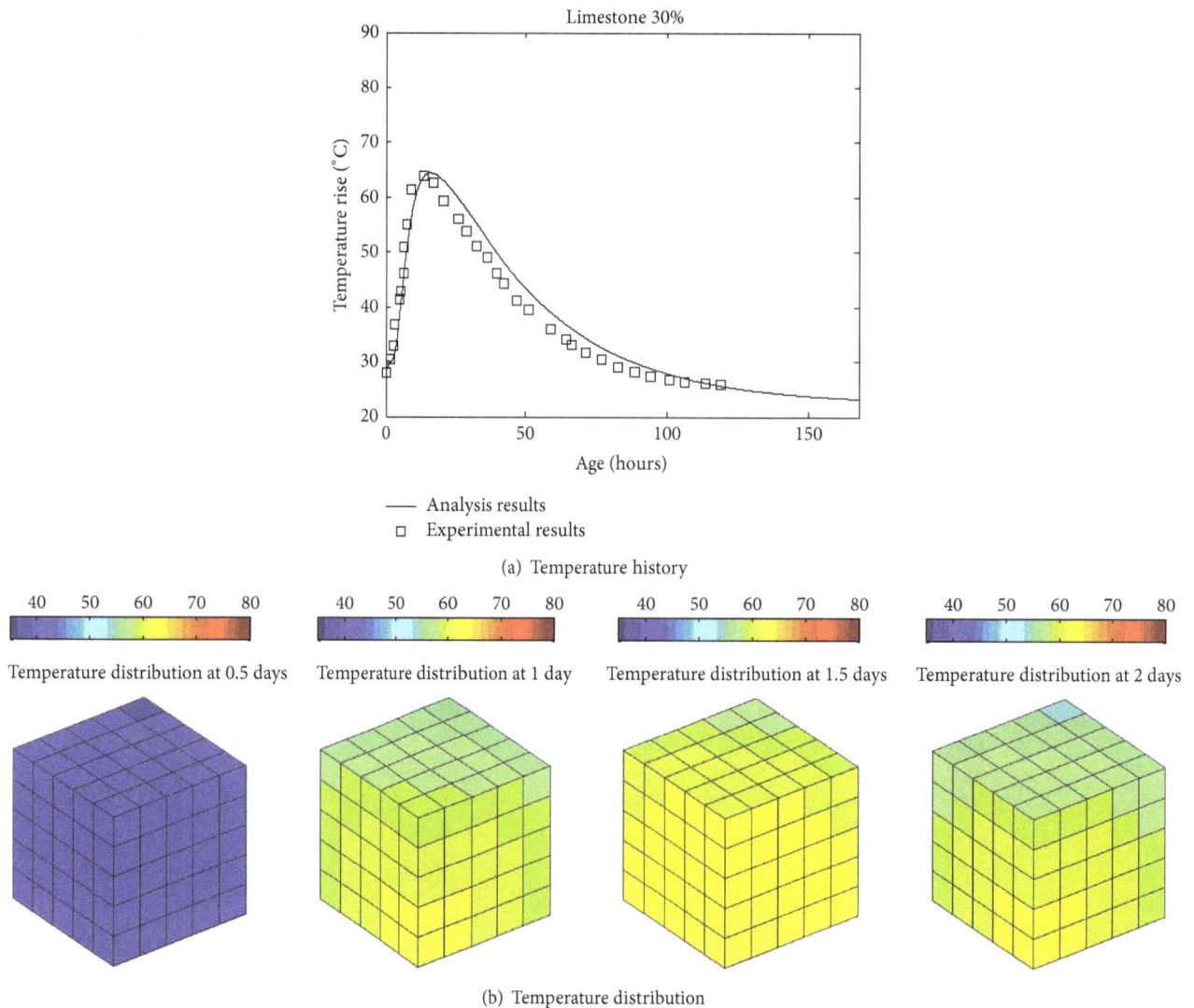

FIGURE 6: Limestone 30% concrete.

TABLE 2: Mixing proportions of concrete.

	Water/(cement + limestone)	Water (kg/m^3)	Cement (kg/m^3)	Limestone (kg/m^3)	Sand (kg/m^3)	Gravel (kg/m^3)	Superplasticizer (kg/m^3)	Initial temperature $(°C)$	Ambient temperature $(°C)$
OPC	30.3	171	565	—	820	915	1.0	30	23
Limestone 30%	32.7	177	396	146	820	915	0.7	28	23
Limestone 55%	32.6	168	226	291	820	916	0.7	28	23
Limestone 70%	32.0.	165	155	362	820	915	0.7	28	28

The water to binder ratio is about 0.3, and the limestone replacement ratio ranges from 30% to 70%. The compound compositions of cement are shown in Table 3. The used cement is ordinary Portland cement. The size of specimens is 445 mm ∗ 445 mm ∗ 445 mm. The temperature at center point is measured from mixing time to the age of five days.

Thermal conductivity and heat transfer coefficient of specimens are 41 kcal/m/day/k and 35 kcal/m²/day/k, respectively.

Because of symmetries of geometry condition and boundary condition of the specimen, a one-eighth specimen is adopted to represent the full specimen. The 8-node brick isoparametric element is used to mesh the specimen in

(a) Temperature history

(b) Temperature distribution

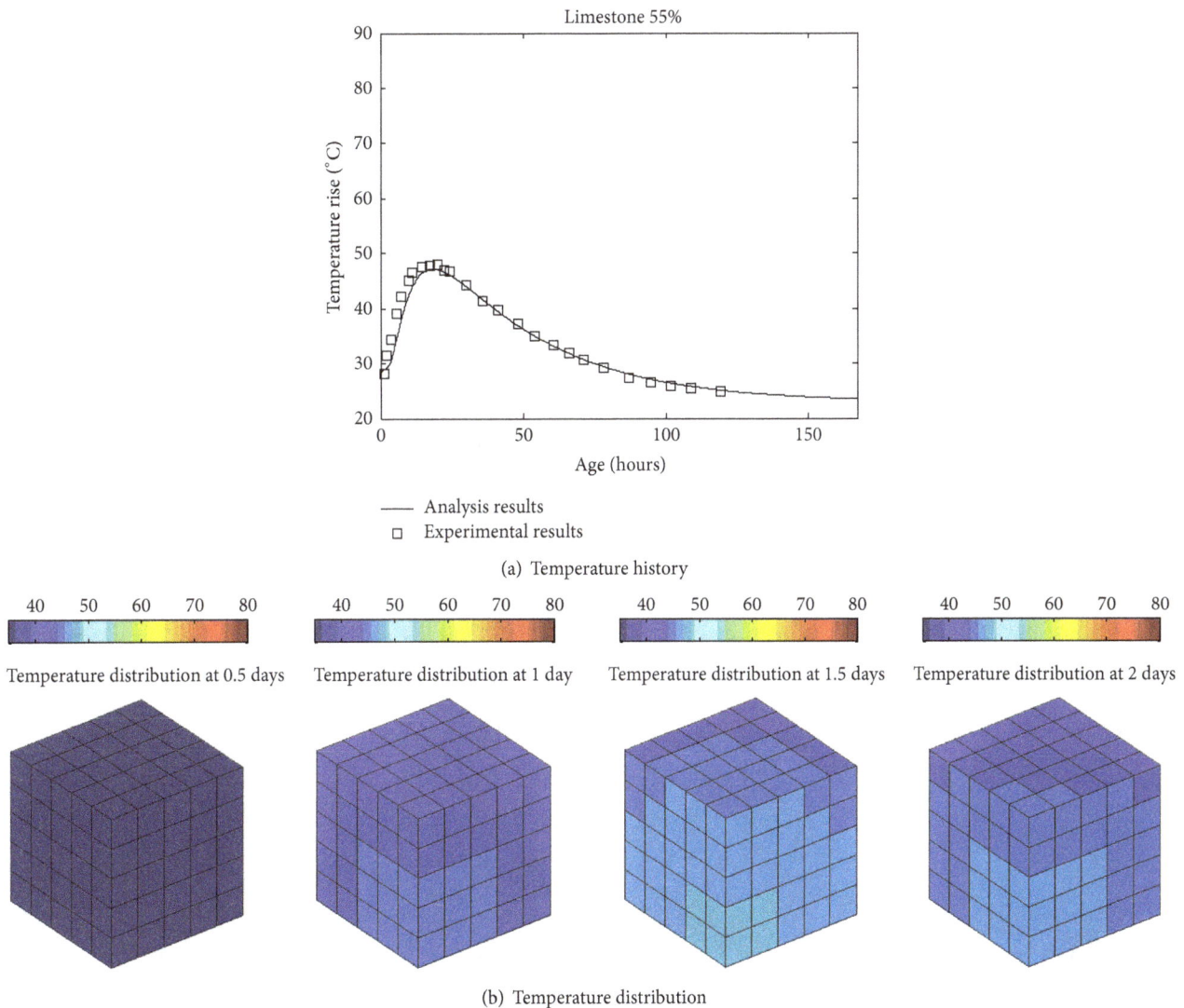

FIGURE 7: Limestone 55% concrete.

TABLE 3: Compound compositions of cement.

	C_3S %	C_2S %	C_3A %	C_4AF %	Gypsum %	Blaine (cm^2/g)
Cement	53.9	18.8	11.0	10.4	5.83	3350

three-dimensional spaces. Total 125 elements ($5 * 5 * 5 = 125$) are used. In each time step, the heat evolution rate of each element is determined from current temperature of the element. Figures 5–8 show the calculation results about temperature history and temperature distribution of hardening concrete. With the increasing of limestone content, the maximum temperature rise of concrete decreases. For control concrete, 30% limestone concrete, and 50% limestone concrete, the calculation results about temperature history generally agree with experimental results, while, for 70% limestone blended concrete, at early ages, the calculation result is slightly lower than experimental results. This may be due to heat release from the chemical reaction of limestone.

For concrete with very high content limestone, the chemical reaction of limestone becomes significant and can contribute to the temperature rise of hydrating concrete. The proposed model considers the acceleration of cement hydration due to temperature increasing. However, the reduction of solubility of concrete due to temperature increasing is not considered. The point should be taken into account in the future study.

4. Conclusions

This study proposes a numerical procedure for predicting temperature history of hardening limestone blended concrete. The numerical procedure combines a kinetic limestone blended cement hydration model with a finite element method.

First, the hydration model analyzes the dilution effect and nucleation effect due to limestone additions. For concrete with a lower water to binder ratio, the dilution effect due to limestone addition becomes obvious, and the degree of

(a) Temperature history

(b) Temperature distribution

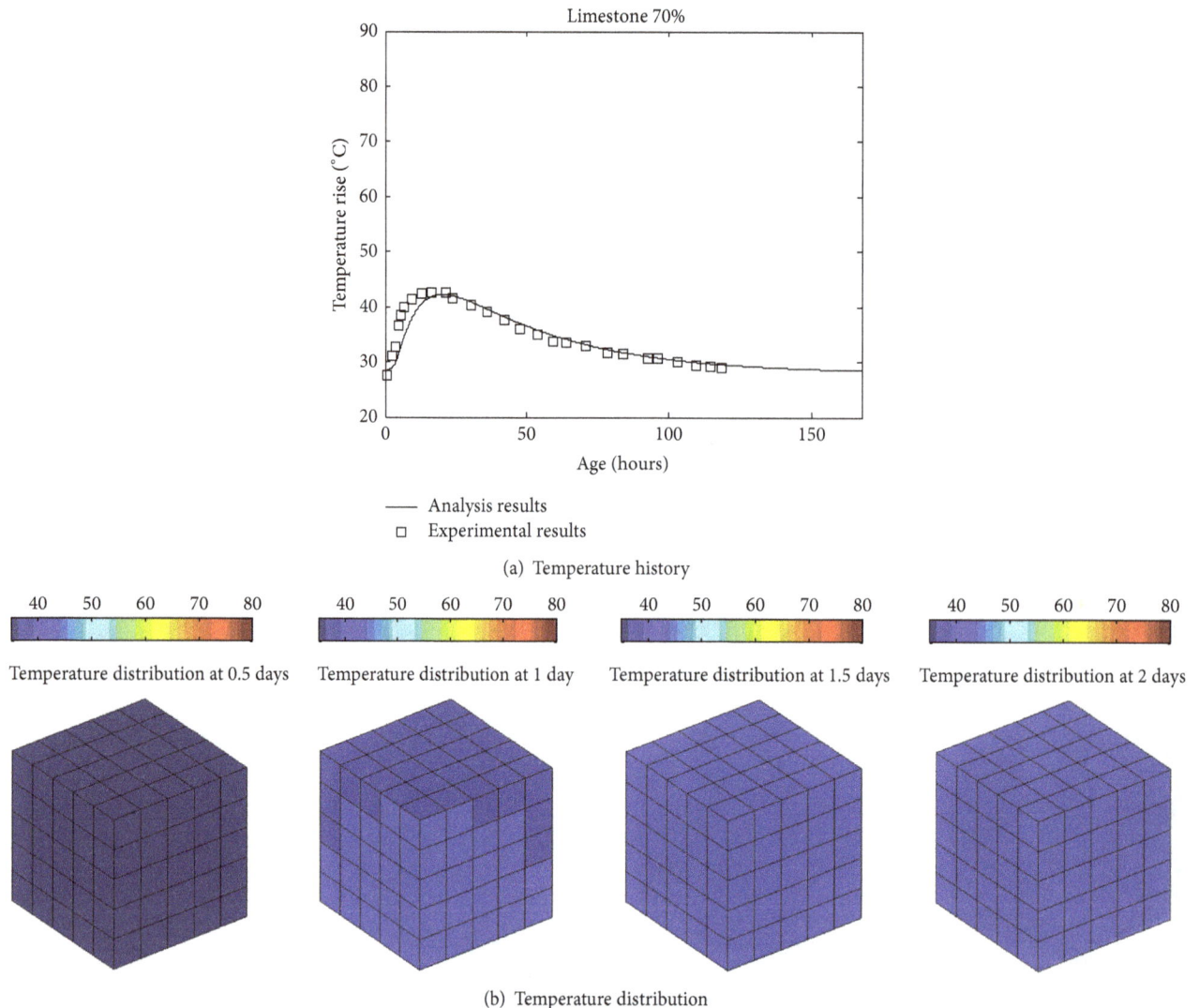

FIGURE 8: Limestone 70% concrete.

hydration is significantly improved compared with control concrete without limestone.

Second, the released heat of concrete relates to both cement content and degree of hydration. Limestone additions increase the degree of hydration but reduce cement content. The total hydration heat depends on the combined action of cement content and degree of hydration. The results of parameter analysis show that limestone additions can reduce the heat of hydration.

Third, the calculation results of hydration heat from hydration model are used as input parameters of finite element model. The combined hydration and finite element model can be used to evaluate temperature history and temperature distribution of semiadiabatic hardening concrete. The analysis results show that, with the increasing of limestone content, the maximum temperature rise of concrete decreases.

Conflicts of Interest

The author declares that they have no conflicts of interest.

Acknowledgments

This research was supported by a grant from Smart Civil Infrastructure Research Program (13SCIPA02) funded by Ministry of Land, Infrastructure and Transport (MOLIT) of Korean Government and Korea Agency for Infrastructure Technology Advancement (KAIA).

References

[1] S. H. Liu and P. Y. Yan, "Effect of limestone powder on microstructure of concrete," *Journal Wuhan University of Technology, Materials Science Edition*, vol. 25, no. 2, pp. 328–331, 2010.

[2] V. Bonavetti, H. Donza, V. Rahhal, and E. Irassar, "Influence of initial curing on the properties of concrete containing limestone blended cement," *Cement and Concrete Research*, vol. 30, no. 5, pp. 703–708, 2000.

[3] I. Mohammadi and W. South, "Decision-making on increasing limestone content of general purpose cement," *Journal of Advanced Concrete Technology*, vol. 13, no. 11, pp. 528–537, 2015.

[4] J. Mohammadi and W. South, "Effects of intergrinding 12% limestone with cement on properties of cement and mortar," *Journal of Advanced Concrete Technology*, vol. 14, no. 5, pp. 215–228, 2016.

[5] J. J. Chen and A. K. H. Kwan, "Adding limestone fines to reduce heat generation of curing concrete," *Magazine of Concrete Research*, vol. 64, no. 12, pp. 1101–1111, 2012.

[6] S. Palm, T. Proske, M. Rezvani, S. Hainer, C. Müller, and C.-A. Graubner, "Cements with a high limestone content - Mechanical properties, durability and ecological characteristics of the concrete," *Construction and Building Materials*, vol. 119, pp. 308–318, 2016.

[7] B. Lothenbach, G. Le Saout, E. Gallucci, and K. Scrivener, "Influence of limestone on the hydration of Portland cements," *Cement and Concrete Research*, vol. 38, no. 6, pp. 848–860, 2008.

[8] D. P. Bentz, "Modeling the influence of limestone filler on cement hydration using CEMHYD3D," *Cement and Concrete Composites*, vol. 28, no. 2, pp. 124–129, 2006.

[9] A. R. Mohamed, M. Elsalamawy, and M. Ragab, "Modeling the influence of limestone addition on cement hydration," *Alexandria Engineering Journal*, vol. 54, no. 1, pp. 1–5, 2015.

[10] A.-M. Poppe and G. De Schutter, "Analytical hydration model for filler rich binders in self-compacting concrete," *Journal of Advanced Concrete Technology*, vol. 4, no. 2, pp. 259–266, 2006.

[11] G. Ye, X. Liu, A. M. Poppe, G. De Schutter, and K. Van Breugel, "Numerical simulation of the hydration process and the development of microstructure of self-compacting cement paste containing limestone as filler," *Materials and Structures/Materiaux et Constructions*, vol. 40, no. 9, pp. 865–875, 2007.

[12] X. Wang and H. Lee, "Modeling the hydration of concrete incorporating fly ash or slag," *Cement and Concrete Research*, vol. 40, no. 7, pp. 984–996, 2010.

[13] X.-Y. Wang, "Modeling of hydration, compressive strength, and carbonation of portland-limestone cement (PLC) concrete," *Materials*, vol. 10, no. 2, article 115, 2017.

[14] T. Kishi and D. Saruul, "Hydration heat modeling for cement with limestone powder," *Iabse Colloquium Phuket*, vol. 81, pp. 133–138, 1999.

[15] J. D. Logan, *A First Course in Differential Equations*, Undergraduate Texts in Mathematics, Springer, New York, 5th edition, 2011.

[16] B. F. Zhu, *Thermal Stresses and Temperature Control of Mass Concrete*, Tsinghua University Press, Beijing, China, 2014.

[17] K. Maekawa, T. Ishida, and T. Kishi, *Multi-Scale Modeling of Structural Concrete*, Taylor & Francis, New York, NY, USA, 2009.

[18] W. Xiao-Yong, "Prediction of temperature distribution in hardening silica fume-blended concrete," *Computers and Concrete*, vol. 13, no. 1, pp. 97–115, 2014.

[19] V. Bonavetti, H. Donza, G. Menéndez, O. Cabrera, and E. F. Irassar, "Limestone filler cement in low w/c concrete: A rational use of energy," *Cement and Concrete Research*, vol. 33, no. 6, pp. 865–871, 2003.

[20] D. P. Bentz, E. F. Irassar, B. E. Bucher, and W. J. Weiss, "Limestone fillers conserve cement, part 2: durability issues and the effects of limestone fineness on mixtures," *Concrete International*, vol. 31, pp. 35–40, 2009.

11

Prediction of Time-Dependent Chloride Diffusion Coefficients for Slag-Blended Concrete

Ki-Bong Park,[1] **Han-Seung Lee,**[2] **and Xiao-Yong Wang**[1]

[1]*College of Engineering, Department of Architectural Engineering, Kangwon National University, Chuncheon-Si 200-701, Republic of Korea*
[2]*Department of Architectural Engineering, Hanyang University, Ansan-Si 426-791, Republic of Korea*

Correspondence should be addressed to Xiao-Yong Wang; wxbrave@kangwon.ac.kr

Academic Editor: Kazunori Fujikake

The chloride diffusion coefficient is considered to be a key factor for evaluating the service life of ground-granulated blast-furnace slag (GGBS) blended concrete. The chloride diffusion coefficient relates to both the concrete mixing proportions and curing ages. Due to the continuous hydration of the binders, the capillary porosity of the concrete decreases and the chloride diffusion coefficient also decreases over time. To date, the dependence of chloride diffusivity on the binder hydration and curing ages of slag-blended concrete has not been considered in detail. To fill this gap, this study presents a numerical procedure to predict time-dependent chloride diffusion coefficients for slag-blended concrete. First, by using a blended cement hydration model, the degree of the binder reaction for hardening concrete can be calculated. The effects of the water to binder ratios and slag replacement ratios on the degree of the binder reaction are considered. Second, by using the degree of the binder reaction, the capillary porosity of the binder paste at different curing ages can be determined. Third, by using the capillary porosity and aggregate volume, the chloride diffusion coefficients of concrete can be calculated. The proposed numerical procedure has been verified using the experimental results of concrete with different water to binder ratios, slag replacement ratios, and curing ages.

1. Introduction

Chloride ingress is considered to be one of the major factors in the deterioration mechanisms of reinforced concrete in marine environments. The resulting corrosion of steel reinforcements causes serious detrimental effects, such as concrete cover cracking, reduced reinforcement in cross sections, decreased bonding between the steel rebar and concrete, and reduced yield strength and ductility of the steel rebar in reinforced concrete structures. However, to improve the resistance of concrete against chloride ingress, slag is widely used as a mineral admixture. Generally, incorporating GGBS into blended binders can increase the total porosity but will refine the pore size. Slag-blended concrete presents a lower chloride diffusivity and higher chloride binding capacity than control concrete [1].

The chloride diffusion coefficient is a key factor in the service life evaluation of slag-blended concrete. As for the chloride diffusivity in ordinary Portland cement (OPC)

concrete or slag-blended concrete, Papadakis [2] and Demis et al. [3] evaluated the chloride diffusivity of fully hardened concrete as a function of the ultimate porosity. Alexander and Thomas [4] evaluated the chloride diffusivity of concrete after 28 days of curing as a function of the water to binder ratios. The Life-365 program [5] also uses the Alexander and Thomas equation [4] to determine concrete diffusivity. However, the Papadakis [2], Demis et al. [3], and Alexander and Thomas [4] models are not perfect. The Papadakis [2] and Demis et al. [3] models assume that cement is completely hydrated (i.e., the hydration degree is 100%) regardless of the water to cement ratio. Wang and Lee [6] reported that concretes with lower water to cement ratios had slower rates of hydration and lower ultimate degrees of hydration. The chloride diffusivity calculated according to the Alexander and Thomas model [4] does not consider other factors, such as the binder content, binder reactivity, and curing methods.

The composition of hydration products, capillary porosity, and chloride diffusivity relate closely to the degree of hydration in hardening or hardened concrete. Han [7] and Fan and Wang [8] proposed hydration-based chloride ingress models. Time-dependent chloride diffusivity was calculated by using the development of the capillary porosity of concrete over time. However, the Han [7] and Fan and Wang [8] models do not consider the reaction of mineral admixtures and are only valid for Portland cement concrete. In the literature, some models have been proposed that evaluate the chloride diffusivity of concrete containing mineral admixtures. Song et al. [9, 10] calculated the final degrees of reaction of cement and silica fume in silica fume-blended concrete. The capillary porosity, chloride diffusivity, and water permeability were determined by considering the final degrees of the binder reaction and the concrete mixing proportions. Oh and Jang [11] evaluated the chloride diffusivity of fly ash and slag-blended concrete by considering the final degrees of the binder reaction and the concrete porosity. However, Song et al. [9, 10] and Oh and Jang [11] studies are only valid for fully hardened concrete; they do not consider the dependence of chloride diffusivity on curing ages because their models do not account for the kinetic reaction processes of mineral admixtures.

As for the time dependence of chloride diffusivity, Nokken et al. [12] and Yu and Ye [13] found that chloride diffusion into concrete decreased over time. Concrete containing mineral admixtures has shown reduced chloride diffusivity compared to reference concrete. An empirical time parameter [14–16] has frequently been used to describe the age dependence of the chloride diffusion coefficient. However, this empirical time parameter cannot fully describe the evolution of the hardening concrete microstructure. The influence of various factors, such as the cement type, water to binder ratio, and slag replacement ratio, on this empirical time parameter requires further investigation [17].

To address the shortcomings of these current models, this paper presents a numerical procedure that analyzes the time dependence of the chloride diffusion coefficient. By using a blended cement hydration model, the degree of the reaction of binders and the capillary porosity of binder paste were calculated. Furthermore, the chloride diffusion coefficients at different curing ages were determined, considering the capillary porosity and aggregate volume.

2. Cement Hydration Model and Chloride Diffusion Coefficient Model

2.1. Cement Hydration Model. Wang and Lee [6, 18] proposed a hydration model for concrete containing supplementary cementitious materials, such as silica fume, fly ash, and slag. Hydration equations were proposed for cement and mineral admixtures, respectively, and the mutual interactions between cement hydration and mineral admixture reactions are considered with the capillary water content and calcium hydroxide content. The hydration model is valid for concrete with different water to binder ratios, mineral admixture replacement ratios, and curing temperatures [6, 18].

The reaction degrees of cement and mineral admixtures are used as fundamental indicators to evaluate the development of concrete properties. The degree of cement hydration

(α) is defined as the ratio of the mass of hydrated cement to the mass of cement in the mixing proportion. The value of the degree of cement hydration (α) ranges between 0 and 1. A degree of cement hydration of $\alpha = 0$ means the absence of any hydration, and a degree of cement hydration of $\alpha = 1$ means that cement has fully hydrated.

By using an integration method, the degree of cement hydration can be determined as follows:

$$\alpha = \int_0^t \left(\frac{d\alpha}{dt} \right) dt, \qquad (1)$$

where t is time and $d\alpha/dt$ is rate of cement hydration. The detailed equations for $d\alpha/dt$ are available in our former research [6, 18].

Similarly, the reaction degree of a mineral admixture (α_M) is defined as the ratio of the mass of reacted mineral admixture to the mass of the mineral admixture in the mixing proportion. The value of the degree of the mineral admixture reaction (α_M) ranges between 0 and 1. $\alpha_M = 0$ means the absence of any mineral admixture reaction, and $\alpha_M = 1$ indicates that the mineral admixture has reacted completely. The reaction degree of the mineral admixture reaction can also be determined using an integration method in the time domain as follows:

$$\alpha_M = \int_0^t \left(\frac{d\alpha_M}{dt} \right) dt, \qquad (2)$$

where $d\alpha_M/dt$ is the rate of the mineral admixture reaction. The detailed equations for $d\alpha_M/dt$ are available in our former research [6, 18].

In cement-mineral admixture blends, the production of chemically bound water relates to both cement hydration and mineral admixture reactions. The chemically bound water content can be determined as follows:

$$W_{cbm}(t) = 0.25 * C_0 * \alpha + 0.3 * M_0 * \alpha_M, \qquad (3)$$

where W_{cbm} is the chemically bound water content, C_0 is the mass of cement in the mixing proportion, and M_0 is the mass of the mineral admixture in the mixing proportions. The expression $0.25 * C_0 * \alpha$ is the mass of the chemically bound water from cement hydration, and the expression $0.3 * M_0 * \alpha_M$ is the mass of the chemically bound water from the slag reactions [6, 18].

In the cement-mineral admixture blends, both cement hydration and mineral admixture reactions contribute to the formation of gel water. The content of gel water can be calculated as follows:

$$W_{gel}(t) = 0.15 * C_0 * \alpha + 0.15 * \alpha_M * M_0, \qquad (4)$$

where W_{gel} is the mass of gel water, $0.15 * C_0 * \alpha$ is the mass of gel water produced from cement hydration, and $0.15 * \alpha_M * M_0$ is the mass of gel water produced from slag reaction.

The mass of combined water equals the sum of chemically bound water and gel water. The mass of combined water can be determined as follows:

$$W_c(t) = W_{gel}(t) + W_{cbm}(t), \qquad (5)$$

where $W_c(t)$ is the mass of combined water.

In cement-mineral admixture blends, capillary water is consumed by both cement hydration and mineral admixture reactions. The capillary water content can be calculated as follows:

$$W_{\text{cap}}(t) = W_0 - 0.4 * C_0 * \alpha - 0.45 * \alpha_M * M_0, \quad (6)$$

where W_{cap} is the mass of capillary water in the hardening concrete and W_0 is the mass of water in the mixing proportion. The expression $0.4 * C_0 * \alpha$ is the mass of capillary water consumed by cement hydration and the expression $0.45 * \alpha_M * M_0$ is the mass of the consumed capillary water from the slag reactions.

For hardened concrete, the capillary porosity equals the sum of capillary water and chemical shrinkage. The capillary porosity can be determined as follows:

$$\phi_{\text{cap}}(t) = W_{\text{cap}}(t) + 0.0625 * C_0 * \alpha + 0.1 * \alpha_M$$
$$* M_0, \quad (7)$$

where $\phi_{\text{cap}}(t)$ is the capillary porosity of hardening concrete, $0.0625 * C_0 * \alpha$ is the chemical shrinkage from cement hydration, and $0.1 * \alpha_M * M_0$ is the chemical shrinkage from slag reactions.

2.2. Chloride Diffusion Coefficient Model. Concrete is a three-phase material consisting of a cement paste matrix, aggregate, and interfacial transition zones (ITZs) between the cement paste matrix and aggregate. The interfacial transition zone has a higher capillary porosity and contains higher calcium hydroxide volume fractions compared to the bulk matrix. Interfacial zones are formed due to the particle-packing effect and the one-sided growth effect [19]. The particle-packing effect arises because the cement particles cannot pack together as well near a flat edge as in a free space. The one-sided growth effect is reactive growth from the cement side, but not from the aggregate side. The cement paste matrix is interconnected through interfacial transition zones. Garboczi [19] found that when the aggregate volume fractions are greater than 50%, the interfacial transition zone will be fully percolated. This observation agrees with the experimental results from Princigallo et al. [20]. Because most concretes have aggregate volume fractions above 50%, the interfacial transition zones in the usual concrete are percolated.

Due to the cement particle-packing effect, the ITZ shows a higher water to cement ratio. Therefore, the effective water to cement ratio in the bulk cement paste will be reduced. Nadeau [20, 21] proposed a model to consider water to cement gradients between the aggregate and bulk cement paste. The model is a function of the overall water-cement ratio, volume fraction and radius of the aggregate, specific gravity of cement, and thickness of the ITZ. The equations for determining the effective water to binder ratio of bulk cement paste in concrete are given as follows [21, 22]:

$$\overline{V_C}$$
$$= \frac{10(1 - V_a)}{(1 + (W_0/(C_0 + M_0))G_C)[a_c V_a \varepsilon \{\varepsilon^2 + 5\varepsilon + 10\} + 10(1 - V_a)]}, \quad (8)$$
$$\overline{W_C} = \frac{1 - \overline{V_C}}{G_C \overline{V_C}},$$

where $\overline{V_C}$ is the binder volume fraction in the bulk binder paste, V_a is the volume of the aggregate, $W_0/(C_0 + M_0)$ is the overall water to binder ratio of concrete, G_C is the specific gravity of the binder, a_c is a constant equal to approximately -0.5, and ε is the thickness ratio of ITZ. $\overline{W_C}$ is the effective water to binder ratio in the bulk binder paste, whereas $1 - \overline{V_C}$ indicates the water volume fraction in the bulk binder paste.

The diffusivity of the cement paste phase D_P is mainly dependent on the capillary pores in the cement paste, which can be determined as follows [8]:

$$D_P(t) = A_1 * \left(\phi_{\text{paste}}\right)^{A_2}, \quad (9)$$

$$\phi_{\text{paste}} = \frac{\phi_{\text{cap}}}{V_P}, \quad (10)$$

where A_1 and A_2 are the relation coefficients between the capillary porosity and chloride diffusivity, respectively; ϕ_{paste} is the capillary porosity in the binder paste; and V_P is volume of binder paste calculated from the effective water to binder ratio $\overline{W_C}$. In (9), the intrinsic diffusion coefficient A_1 relates to the type of binder, such as cement or slag. Exponent A_2 ($A_2 > 1$) relates to the pore size distribution or the complexity of the microstructure of the reaction products. As shown in (9) and (10), with the progress of binder hydration, the capillary porosity of concrete decreases and the chloride diffusion coefficient decreases correspondingly.

When the diffusivity of the aggregate particle inclusions is assumed to be zero, according to the composite sphere assemblage (CSA) model [11], the chloride diffusion coefficient of concrete can be determined as follows:

$$\frac{D}{D_P} = 1 + \frac{V_a}{1/(2(D_i/D_P)\varepsilon - 1) + (1 - V_a)/3}, \quad (11)$$

where D is the diffusivity of concrete and D_i is the diffusivity in the interfacial transition zone.

By using a hard core/soft shell model, Garboczi [19] evaluated the connectivity of the interfacial zones for different choices of interfacial zone thicknesses. Furthermore, by comparison with cement mortar mercury intrusion data [23], a choice of 20 μm for the interfacial zone thickness was found to give the best agreement with the mercury data. The mean radius of coarse aggregate is approximately 10 mm. Hence, in this study, a thickness ratio of ITZ of approximately $\varepsilon = 0.002$ is used.

Bentz and Garboczi [24] studied the effects of mineral admixtures on the interfacial transition zone. They found that smaller admixtures allow better packing nearer to the aggregate edge, and the reactivity of the mineral admixture controls the consumption of the calcium hydroxide. For fly ash blended mortar, the fraction ratio of CSH between ITZ and the bulk matrix is similar to plain cement mortar [22]. Similarly, based on an analysis of the chloride diffusivity of concrete with various fly ash and slag additions, Oh and Jang [11] proposed that, for Portland cement concrete, fly ash blended concrete, and slag-blended concrete, the ratios between D_i and D_P are almost the same ($D_i/D_P = 7$).

The effects of slag addition on the chloride diffusion coefficients of concrete are summarized as follows: first,

TABLE 1: Properties of binders [15].

Types	Chemical composition (mass%)							Physical properties	
	SiO_2	Al_2O_3	Fe_2O_3	CaO	MgO	SO_3	Lg. loss	Specific gravity (g/cm^3)	Blaine (cm^2/g)
Cement	21.96	5.27	3.44	63.41	2.13	1.96	0.79	3.16	3214
Slag	32.74	13.23	0.41	44.14	5.62	1.84	0.2	2.89	4340

TABLE 2: Mixing proportions of concrete [15].

W/B^1	S/B^2	Water	Cement (kg/m^3)	Slag (kg/m^3)	Sand (kg/m^3)	Gravel (kg/m^3)	Superplasticizer (% of binder)
0.37	0	168	454	0	767	952	1%
0.42	0	168	400	0	787	976	0.9%
0.47	0	168	357	0	838	960	0.85%
0.37	0.3	168	318	136	762	946	0.8%
0.42	0.3	168	280	120	783	972	0.75%
0.47	0.3	168	250	107	835	956	0.65%
0.37	0.5	168	227	227	760	943	0.75%
0.42	0.5	168	200	200	780	969	0.7%
0.47	0.5	168	178	179	832	853	0.6%

^{1}Water to binder ratio. ^{2}Slag to binder ratio.

incorporating GGBS into blended binders can increase the total porosity because the reactivity of slag is lower than the reactivity of cement. This point is considered using slag-blended cement hydration models ((1) and (7)). Second, calcium silicate hydrate (CSH) gel produced from slag reactions has finer gel pores than the gel from cement hydration. This effect is considered using an intrinsic diffusion coefficient A_1 in (9). Third, the formation of slag reaction products can fill large capillary voids and reduce the average pore size. This effect is considered using a chloride diffusion exponent A_2 in (9). In our study, the evolution of the chloride diffusion coefficient over time is directly related to the time-dependent development of capillary porosity. We do not use the empirical time parameter [14–16] to describe the age dependence of the chloride diffusion coefficient. Compared with the empirical time parameter method [14–16], the physical meaning of the model proposed in this study is much clearer.

However, the proposed model in this study for the chloride diffusion coefficient of concrete shows some limits. First, the chloride diffusivity in the ITZ is dependent on the distance from the aggregate surface [25, 26]. The chloride diffusivity in the ITZ is not analyzed in detail in this study. Second, for hardening concrete, the moisture transport from the water-rich ITZ to the drying bulk paste is not considered [25]. Water transport plays a prominent role in the hardened ITZ microstructure [27]. Therefore, the present version of the proposed model is not perfect and needs to be improved.

3. Verification of the Proposed Model

The experimental results from Song and Kwon [15] were used to verify the proposed model. Song and Kwon [15] performed a systematic experimental study of the chloride diffusivity of slag-blended concrete. They measured a chloride diffusion coefficient for slag-blended concrete with various mixing

proportions at different curing ages. Table 1 shows the chemical and physical properties of the cement and slag. Table 2 shows the mixing proportions of the concrete specimens. Concrete specimens with three different water to binder ratios of 0.47, 0.42, and 0.37 and two different slag contents of 30% and 50% were prepared. Concrete cylinder specimens were cured under moist conditions. At curing periods of 28 days (four weeks), 90 days (three months), 180 days (six months), and 270 days (nine months), the chloride diffusion coefficients were measured through an electrical accelerated method. After the electrical accelerated test, a silver nitrate solution ($AgNO_3$ with a concentration of 0.1 mol/L) was used as an indicator to measure the chloride penetration depth that was achieved, and the chloride diffusion coefficient was calculated according to the penetration depth [15]. The diffusion coefficient determined by this test is a migration coefficient, which does not include binding, among other effects.

3.1. Reaction Degree of Cement. The addition of slag presents a dilution effect on cement hydration. A dilution effect means that, in cement-slag blends, the addition of slag makes the water to cement ratio increase. As shown in Figures 1(a) and 1(b), in cement-slag blends, the reaction degree of cement is higher than in plain cement paste. The more the slag additions, the higher the reaction degree of cement. When the water to binder ratio increases from 0.37 (Figure 1(a)) to 0.47 (Figure 1(b)), the degree of cement hydration increases because the concentration of the capillary water and available deposit space of hydration products increases.

3.2. Reaction Degree of Slag. The slag reaction relates to both the water to binder ratios and slag replacement ratios. As shown in Figures 2(a) and 2(b), when the water to binder ratio increases from 0.37 to 0.47, the reaction degree of the slag increases because the concentration of capillary water and

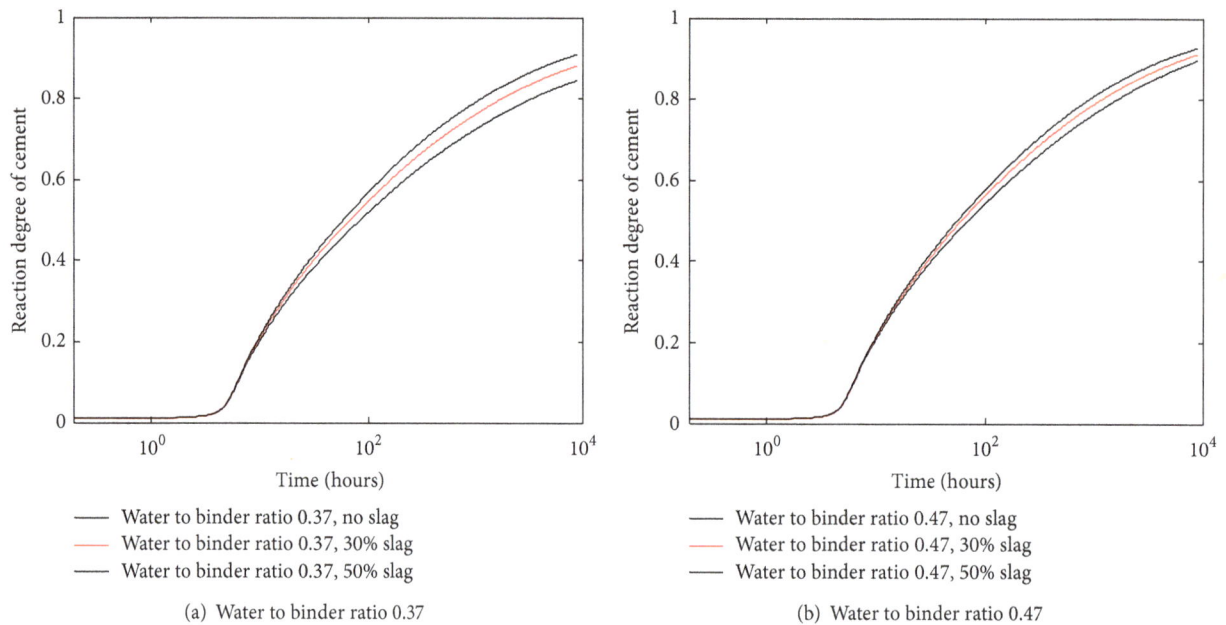

(a) Water to binder ratio 0.37

(b) Water to binder ratio 0.47

FIGURE 1: Reaction degree of cement.

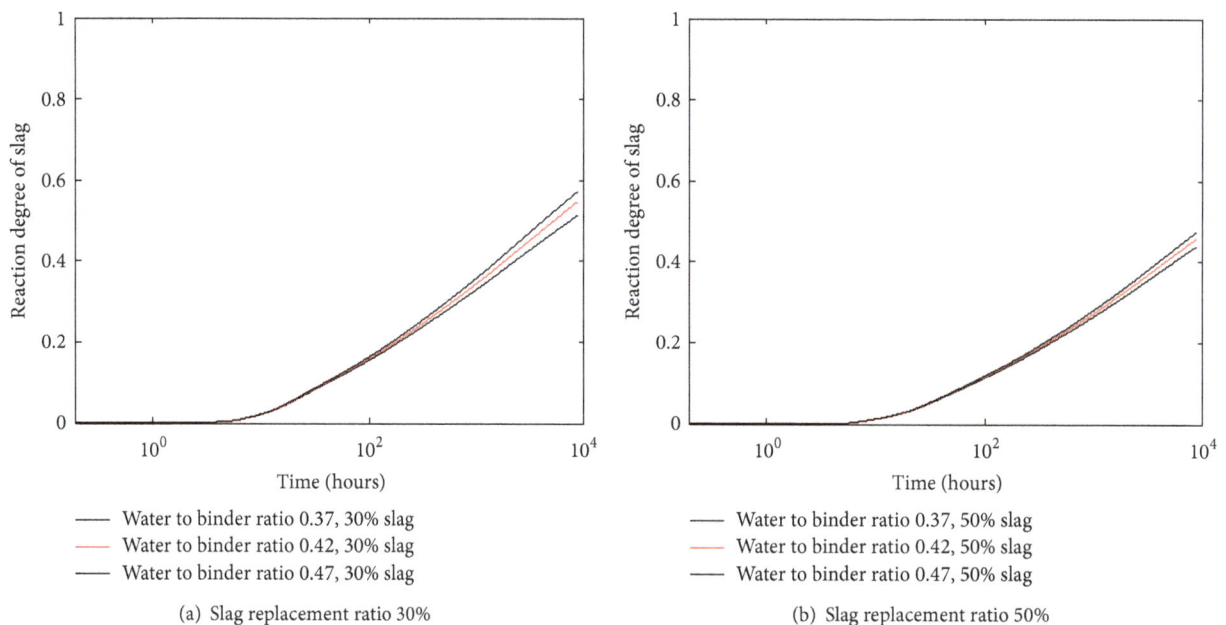

(a) Slag replacement ratio 30%

(b) Slag replacement ratio 50%

FIGURE 2: Reaction degree of slag.

available deposit space of the hydration products increases. However, when the slag replacement ratio increases from 30% (Figure 2(a)) to 50% (Figure 2(b)), the reaction degree of the slag decreases because the alkali activation effect on the slag reaction becomes weaker.

3.3. Combined Water. As shown in (3)–(5), both cement hydration and the slag reaction contribute to the production of combined water. Figure 3 shows the amount of combined water as a function of curing ages. As shown in Figures 3(a) and 3(b), when the water to binder ratio increases from 0.37 to

0.47, the amount of combined water decreases. For concrete containing 50% slag (Figure 3(b)), the amount of combined water is lower than in plain concrete (Figure 3(a)) because the reaction rate of slag (shown in Figure 2) is much slower than the reaction rate of cement (shown in Figure 1).

3.4. Capillary Porosity. In hydrating cement-slag blends, with the increase in combined water, the reaction products deposit in the capillary pore space and the capillary porosity decreases. Figure 4 shows the amount of capillary porosity in binder paste as a function of the curing ages. As shown in

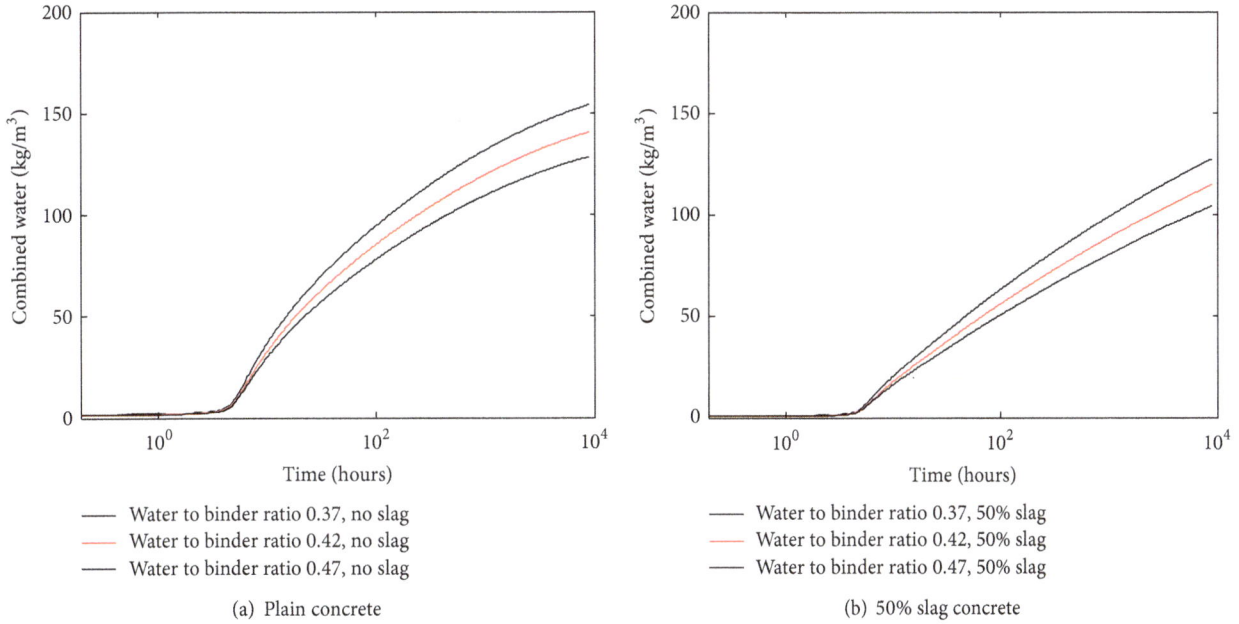

(a) Plain concrete

(b) 50% slag concrete

FIGURE 3: Combined water.

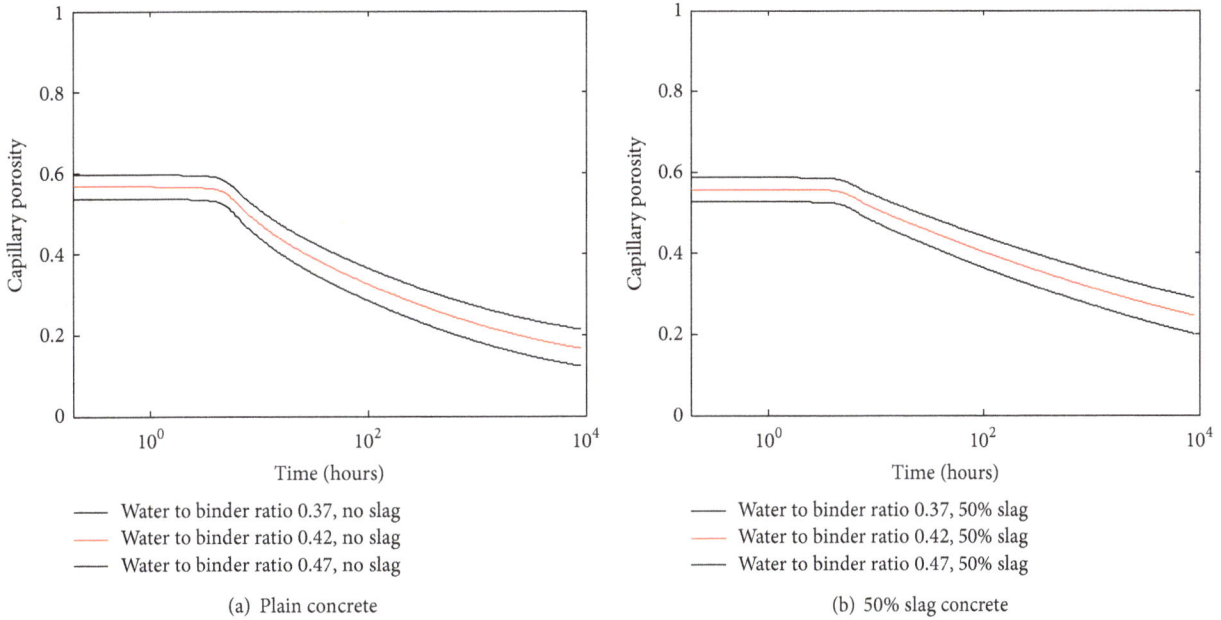

(a) Plain concrete

(b) 50% slag concrete

FIGURE 4: Capillary porosity.

Figure 4, when the water to binder ratio increases from 0.37 to 0.47, the amount of capillary porosity increases. For concrete containing 50% slag, the amount of capillary porosity is higher than in plain concrete.

3.5. *General Equation of the Chloride Diffusion Coefficient for Slag-Blended Concrete.* As mentioned in Section 2.2, the intrinsic diffusion coefficient A_1 and exponent A_2 in (9) are not dependent on the water to binder ratios and only relate to the type of binders. We assume that cement and slag contribute to both the intrinsic diffusion coefficient A_1 and the exponent A_2 relating to binder weight fractions as follows:

$$A_1 = B_1 * \frac{C_0}{C_0 + M_0} + B_2 * \frac{M_0}{C_0 + M_0}$$

$$A_2 = C_1 * \frac{C_0}{C_0 + M_0} + C_2 * \frac{M_0}{C_0 + M_0},$$

(12)

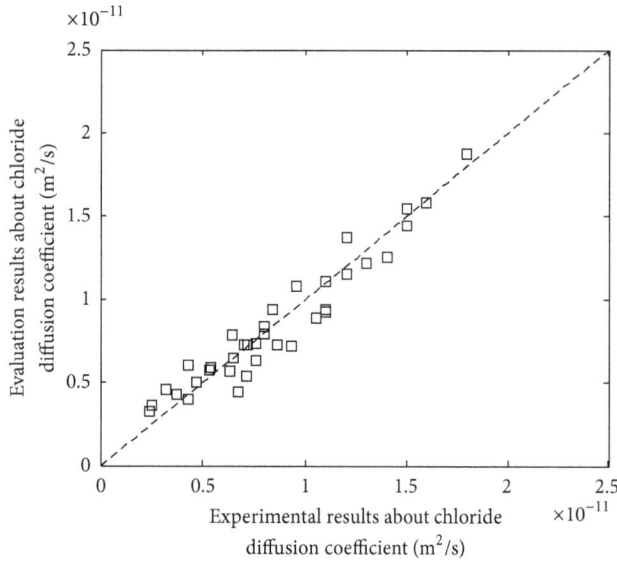

FIGURE 5: Comparisons between experimental results and analyzed results.

where B_1 and B_2 are the contributions of cement and slag, respectively, to the intrinsic diffusion coefficient A_1; C_1 and C_2 are the contributions of cement and slag, respectively, on the exponent A_2. $C_0/(C_0 + M_0)$ and $M_0/(C_0 + M_0)$ are the weight fractions of cement and slag, respectively, in cement-slag blends. As shown in (12), when the replacement ratio of slag equals zero, the chloride diffusion coefficient is only dependent on the values of B_1 and C_1, which relate to cement. When slag is used as a mineral admixture, the chloride diffusion coefficient is dependent on the values of B_1, B_2, C_1, and C_2.

Using the experimental results from chloride diffusion coefficients measured at different ages for various mixing proportions, the values of B_1, B_2, C_1, and C_2 are calibrated as $3.90e - 10$, $0.72e - 10$, 1.21, and 2.31, respectively. As shown in Figure 5, the analyzed results generally agree with the experimental results. The correlation coefficient between the analyzed results and experimental results is 0.96. The addition of slag can reduce the intrinsic chloride diffusion coefficient A_1 (because $B_1 > B_2$) because the gel produced from the slag reaction has finer gel pores than the gel produced from cement hydration. In addition, the addition of slag can increase the chloride diffusion exponent A_2 (because $C_1 < C_2$). This increase may be due to the pore size refinement effect resulting from the slag reactions. With the reduction of the intrinsic chloride diffusion coefficient A_1 and the increase of the chloride diffusion exponent A_2, the chloride diffusion coefficients decrease correspondingly.

The relationship between capillary porosity in the binder paste and the chloride diffusion coefficients is shown in Figure 6(a). As seen in Figure 6(a), given the same capillary porosity, the chloride diffusion coefficients of slag-blended

concrete are much lower than the chloride diffusion coefficients of Portland cement concrete, possibly because the formation of slag reaction products can fill large capillary voids and reduce the average pore size [6]. As shown in Figures 6(b)–6(d), the analyzed results generally agree with the experimental results. With increasing slag replacement levels or decreasing water to binder ratios, the chloride diffusion coefficients decrease. By relating the chloride diffusion coefficients to binder hydration, the proposed model can generally reflect the dependence of chloride diffusion coefficients on the curing ages, water to binder ratios, and slag content. However, because the proposed model does not consider the evolution of the Ca/Si ratio in calcium silicate hydrate (CSH) during the binder hydration process [6], the analyzed results show slight deviations from experimental results (especially for 50% slag concrete).

3.6. Comparison with Life-365 Model. In addition to our analysis results, the analyzed results from the Life-365 program [5] are shown in Figure 7. In the Life-365 program, the time-dependent chloride diffusion coefficient is calculated as follows:

$$D(t) = D_{ref} \left(\frac{t_{ref}}{t} \right)^m,$$

$$D_{ref} = 10^{(-12.06+2.4\,w/cm)}, \qquad (13)$$

$$m = 0.2 + 0.4 \left(\frac{\%\,FA}{50} + \frac{\%\,SG}{70} \right),$$

where D_{ref} is the chloride diffusion coefficient at time $t_{ref} = 28$ days, m is the diffusion decay index, w/cm is the water-cementitious material ratio, and % FA and % SG are the levels of fly ash and slag, respectively, in the mixtures. The Life-365 program assumes that fly ash or slag does not affect the early age chloride diffusion coefficient D_{ref} (chloride diffusion coefficient at 28 days). The Life-365 program considers that fly ash and slag impact the rate of reduction in diffusivity only over time. With the increase of the slag or fly ash replacement levels, the value of m increases and more reduction in chloride diffusivity occurs over time. As shown in Figure 7, the trends of the calculation results from the Life-365 program are similar to our proposed model. However, for Portland cement concrete (Figure 7(a)) and 30% slag concrete (Figure 7(b)), the chloride diffusivity calculated from the Life-365 program is lower than the experimental results, possibly because the reactivity of cement and aggregate content in our study are different from those in the Life-365 program.

4. Conclusions

This study presents a numerical procedure to predict the time-dependent chloride diffusion coefficients of slag-blended concrete. This numerical procedure starts with a slag-blended cement hydration model. The reaction degree of the binders is calculated using this hydration model. By using the reaction degree of binders, the combined water

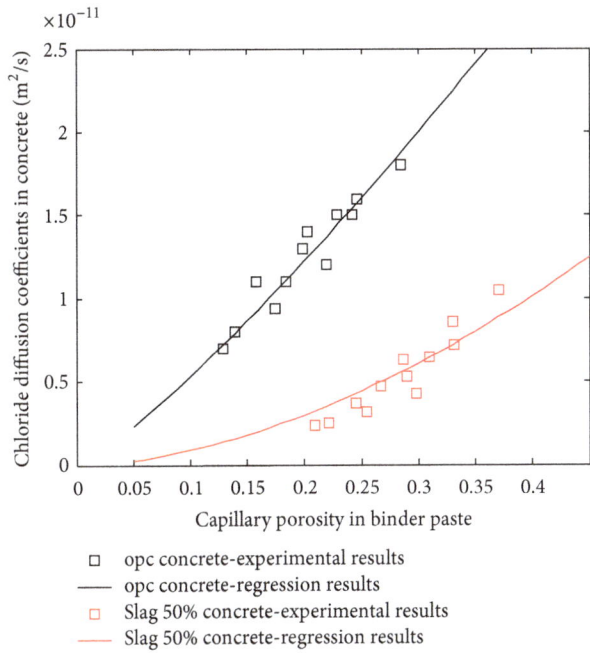

(a) Relationship between diffusion coefficients and capillary porosity

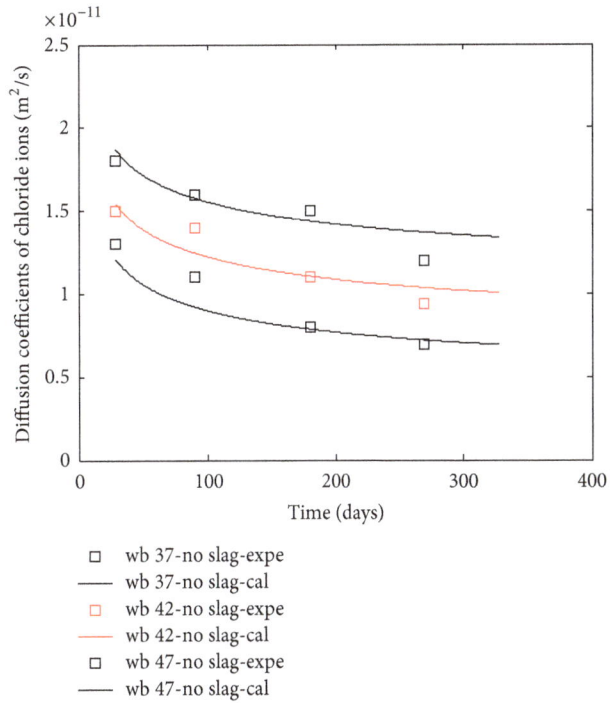

(b) Portland cement concrete

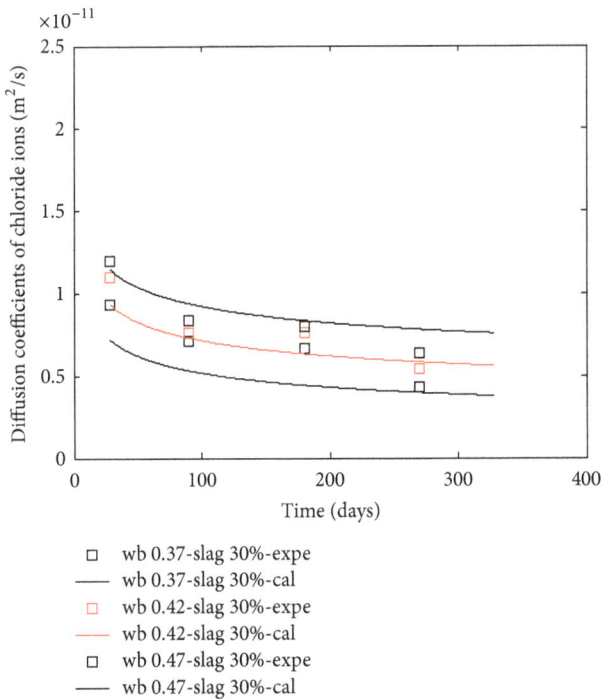

(c) 30% slag concrete

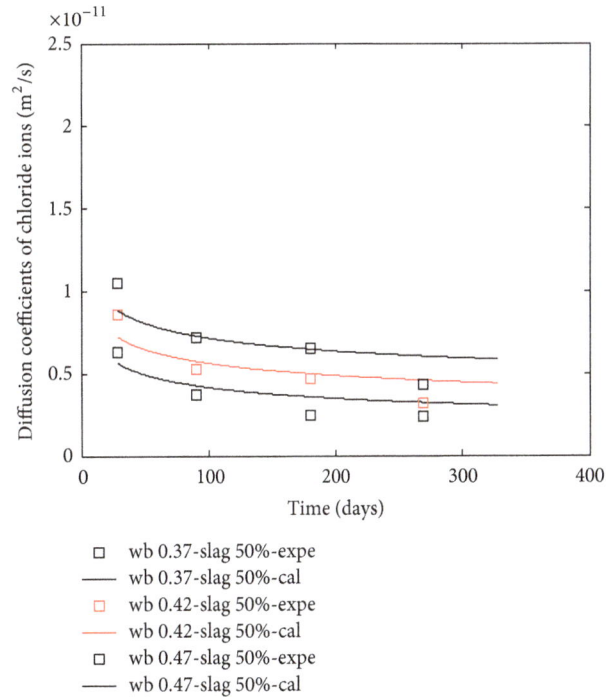

(d) 50% slag concrete

FIGURE 6: Chloride diffusion coefficients.

and capillary porosity of hardening concrete at different curing ages is determined. Furthermore, by using the capillary porosity and aggregate volume, the chloride diffusion coefficients are calculated. A general equation for the chloride diffusion coefficient for hardening slag-blended concrete is proposed. The increase of total porosity due to slag addition, pore refinement on the macroscale by slag due to the filling effect, and pore refinement on the microscale due to the

(a) Portland cement concrete

(b) 30% slag concrete

(c) 50% slag concrete

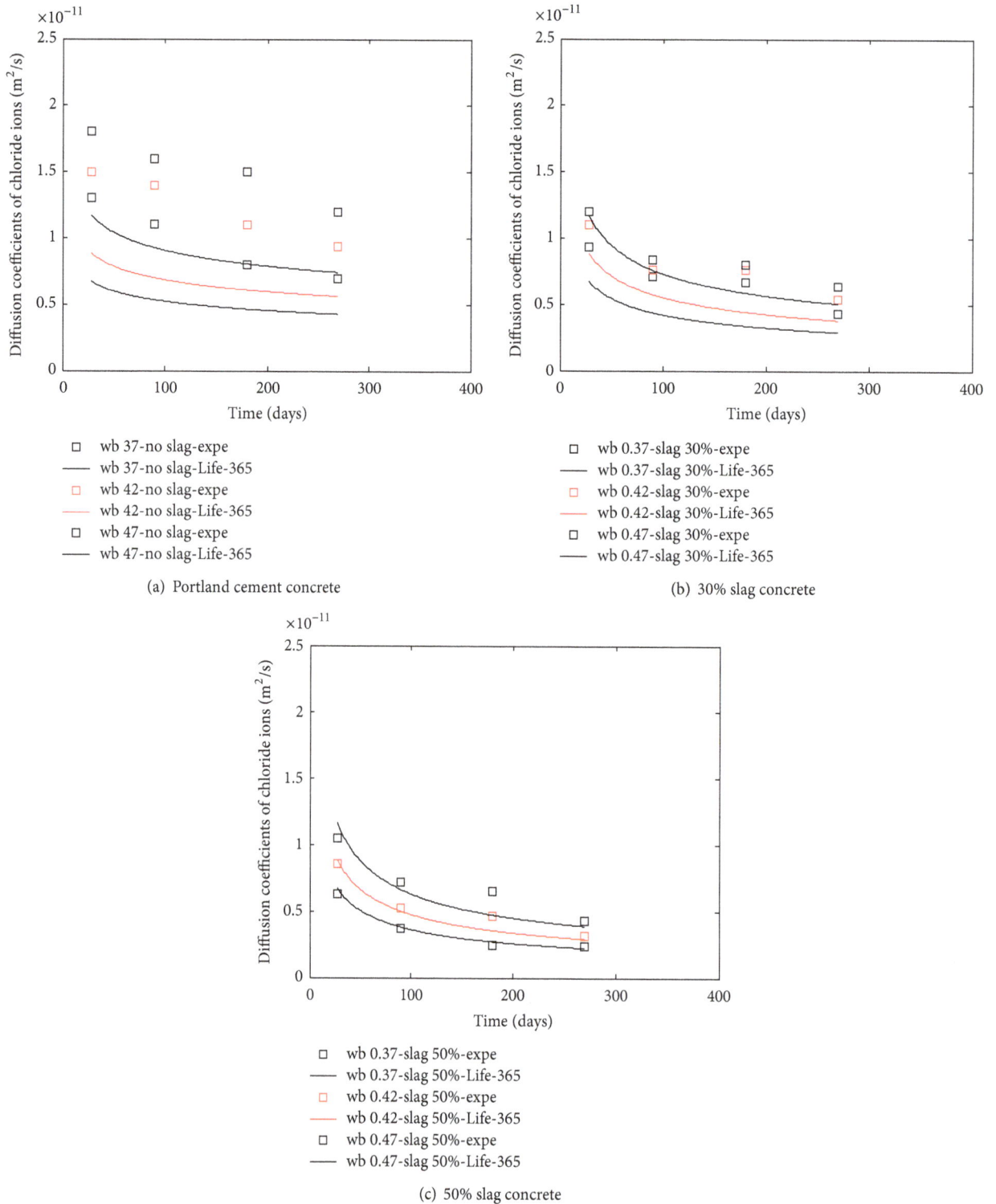

FIGURE 7: Chloride diffusion coefficients calculated from Life-365 model.

latent-hydraulic properties and pozzolanic reaction of slag are analyzed. The proposed model is valid for concrete with different water to binder ratios, slag replacement ratios, and curing ages.

Competing Interests

The authors declare that there is no conflict of interests regarding the publication of this paper.

Acknowledgments

This research was supported by Basic Science Research Program through the National Research Foundation of Korea (NRF) funded by the Ministry of Science, ICT and Future Planning (no. 2015R1A5A1037548).

References

[1] K. Maekawa, T. Ishida, and T. Kishi, *Multi-Scale Modeling of Structural Concrete*, Taylor & Francis, London, UK, 2009.

[2] V. G. Papadakis, "Effect of supplementary cementing materials on concrete resistance against carbonation and chloride ingress," *Cement and Concrete Research*, vol. 30, no. 2, pp. 291–299, 2000.

[3] S. Demis, M. P. Efstathiou, and V. G. Papadakis, "Computer-aided modeling of concrete service life," *Cement and Concrete Composites*, vol. 47, pp. 9–18, 2014.

[4] M. Alexander and M. Thomas, "Service life prediction and performance testing—current developments and practical applications," *Cement and Concrete Research*, vol. 78, pp. 155–164, 2015.

[5] http://www.life-365.org/.

[6] X.-Y. Wang and H.-S. Lee, "Modeling the hydration of concrete incorporating fly ash or slag," *Cement and Concrete Research*, vol. 40, no. 7, pp. 984–996, 2010.

[7] S.-H. Han, "Influence of diffusion coefficient on chloride ion penetration of concrete structure," *Construction and Building Materials*, vol. 21, no. 2, pp. 370–378, 2007.

[8] W.-J. Fan and X.-Y. Wang, "Prediction of chloride penetration into hardening concrete," *Advances in Materials Science and Engineering*, vol. 2015, Article ID 616980, 8 pages, 2015.

[9] H.-W. Song, J.-C. Jang, V. Saraswathy, and K.-J. Byun, "An estimation of the diffusivity of silica fume concrete," *Building and Environment*, vol. 42, no. 3, pp. 1358–1367, 2007.

[10] H.-W. Song, S.-W. Pack, S.-H. Nam, J.-C. Jang, and V. Saraswathy, "Estimation of the permeability of silica fume cement concrete," *Construction and Building Materials*, vol. 24, no. 3, pp. 315–321, 2010.

[11] B. H. Oh and S. Y. Jang, "Prediction of diffusivity of concrete based on simple analytic equations," *Cement and Concrete Research*, vol. 34, no. 3, pp. 463–480, 2004.

[12] M. Nokken, A. Boddy, R. D. Hooton, and M. D. A. Thomas, "Time dependent diffusion in concrete-three laboratory studies," *Cement and Concrete Research*, vol. 36, no. 1, pp. 200–207, 2006.

[13] Z. Yu and G. Ye, "New perspective of service life prediction of fly ash concrete," *Construction and Building Materials*, vol. 48, pp. 764–771, 2013.

[14] T. Luping and J. Gulikers, "On the mathematics of time-dependent apparent chloride diffusion coefficient in concrete," *Cement and Concrete Research*, vol. 37, no. 4, pp. 589–595, 2007.

[15] H.-W. Song and S.-J. Kwon, "Evaluation of chloride penetration in high performance concrete using neural network algorithm and micro pore structure," *Cement and Concrete Research*, vol. 39, no. 9, pp. 814–824, 2009.

[16] B. H. Oh and S. Y. Jang, "Effects of material and environmental parameters on chloride penetration profiles in concrete structures," *Cement and Concrete Research*, vol. 37, no. 1, pp. 47–53, 2007.

[17] W. Chalee and C. Jaturapitakkul, "Effects of W/B ratios and fly ash finenesses on chloride diffusion coefficient of concrete in marine environment," *Materials and Structures*, vol. 42, no. 4, pp. 505–514, 2009.

[18] X.-Y. Wang, "Properties prediction of ultra high performance concrete using blended cement hydration model," *Construction and Building Materials*, vol. 64, pp. 1–10, 2014.

[19] E. J. Garboczi, "Microstructure and transport properties of concrete," in *RILEM Report 12 Performance Criteria of Concrete Durability*, J. Kropp and H. K. Hilsdorf, Eds., pp. 198–212, E&FN Spon, London, UK, 1995.

[20] A. Princigallo, K. Van Breugel, and G. Levita, "Influence of the aggregate on the electrical conductivity of Portland cement concretes," *Cement and Concrete Research*, vol. 33, no. 11, pp. 1755–1763, 2003.

[21] J. C. Nadeau, "Water-cement ratio gradients in mortars and corresponding effective elastic properties," *Cement and Concrete Research*, vol. 32, no. 3, pp. 481–490, 2002.

[22] J. C. Nadeau, "A multiscale model for effective moduli of concrete incorporating ITZ water-cement ratio gradients, aggregate size distributions, and entrapped voids," *Cement and Concrete Research*, vol. 33, no. 1, pp. 103–113, 2003.

[23] D. N. Winslow, M. D. Cohen, D. P. Bentz, K. A. Snyder, and E. J. Garboczi, "Percolation and pore structure in mortars and concrete," *Cement and Concrete Research*, vol. 24, no. 1, pp. 25–37, 1994.

[24] D. P. Bentz and E. J. Garboczi, "Simulation studies of the effects of mineral admixtures on the cement paste-aggregate interfacial zone," *ACI Materials Journal*, vol. 88, no. 5, pp. 518–529, 1991.

[25] Y. Gao, G. De Schutter, G. Ye, Z. Tan, and K. Wu, "The ITZ microstructure, thickness and porosity in blended cementitious composite: effects of curing age, water to binder ratio and aggregate content," *Composites Part B: Engineering*, vol. 60, pp. 1–13, 2014.

[26] Y. Gao, G. De Schutter, G. Ye, H. L. Huang, Z. J. Tan, and K. Wu, "Characterization of ITZ in ternary blended cementitious composites: experiment and simulation," *Construction and Building Materials*, vol. 41, pp. 742–750, 2013.

[27] K. van Breugel, E. Koenders, Y. Guang, and P. Lura, "Modelling of transport phenomena at cement matrix—aggregate interfaces," *Interface Science*, vol. 12, no. 4, pp. 423–431, 2004.

12

Use of Preplaced Casting Method in Lightweight Aggregate Concrete

Qiang Du,[1] Qiang Sun,[1] Jing Lv,[1] and Jian Yang[2]

[1]School of Civil Engineering, Chang'an University, Xi'an, Shaanxi 710061, China
[2]School of Civil Engineering, Birmingham University, Birmingham B15 2TT, UK

Correspondence should be addressed to Qiang Du; q.du@chd.edu.cn

Academic Editor: Andres Sotelo

This study addresses the use of preplaced casting method in lightweight aggregate concrete (LC) to provide a new perspective to solve the aggregate segregation. In casting preplaced lightweight aggregate concrete (PLC), the lightweight aggregates are cast into formworks and then fresh grout is injected to fill voids. PLC and conventional lightweight aggregate concrete (CLC) with three different mixtures are compared to observe the degree of segregation. The properties of PLC and CLC are characterized by means of cubic and axial compression, splitting tension and flexural tests, static modulus of elasticity, and drying shrinkage measurements. Results show that the mechanical properties of PLC are improved with respect to that of CLC with the same mixture. The increase of shrinkage is approximately 13% for the CLC and 6% for PLC when w/c ratio ranges from 0.4 to 0.5 due to effect on interlocking. PLC shows an increased tendency in elastic modulus by approximately 2.5% of 0.5 w/c ratio, 2.7% of 0.45 w/c ratio, and 3.3% of 0.4 w/c ratio at the age of 28 days compared with CLC. In conclusion, PLC has significant reduction in the weight on the premise that it shows excellent mechanical properties.

1. Introduction

Lightweight aggregate concrete (LC) is known by its improved advantage of lightweight antiseismic performance, fire resistance, thermal-insulation, and sustainable development. With the development of high-rise buildings, long span structures, more replacing conventional aggregates with lightweight, and recycled composites are constantly focused on [1–3]. The application of LC has been also investigated and used in recent years [4–7].

Nevertheless, some limits in its engineering properties prevented its wider application. A major problem that affects fresh LC is the tendency for light aggregates with small densities to float from the mortar during the vibrating process and on the stationary state, which decreases uniformity and has greater variability in properties [8]. Therefore, the degree of heterogeneity of LC is the key difficulty to improve mechanical behavior and durability. Many researchers contributed to analysis on segregation resistance. In terms of physical properties of lightweight aggregate, Li and Ding studied

five different types of lightweight aggregate and investigated the volume quantity, vibrating time, and the characters of lightweight aggregate including shape, surface smoothness, and water absorption [9]. It is also recommended by several researchers to apply the method of mix design for LC to avoid the material segregation. Such mix methods improved the material segregation problem and stabilized the quality of LC by increasing viscosity in its fresh grout [10–12]. Barbosa proposed an image processing based technique to evaluate the segregation of LC. The tests showed that using this uniformity analysis might be efficient to classify the aggregate distribution on LC samples [13]. However, these studies mainly focused on properties inherent to LC.

The common volume fraction of coarse aggregates in conventional concrete (CC) is in the range of 35%–40% [8]. Compared with CC, LC is well recognized that it requires higher demand of cement grout [14, 15]. Thus, the high percentage of grout weakens the effect on interlocking and framework structure, which even results in segregation, laitance, and bleeding during the vibrating. As a relatively old

TABLE 1: Chemical compositions and physical properties of ordinary Portland cement.

Label	SiO_2	Al_2O_3	CaO	Fe_2O_3	MgO	SO_3	Specific gravity	Fineness (m^2/kg)
Cement	20.36	5.67	62.81	3.84	2.68	2.51	3.14	329

concrete technique, preplaced aggregate concrete (PAC) has a unique method where preplacing coarse aggregates with high volume are initially packed into the formworks after which grout is injected into the voids [16]. Such void space is effectively filled by the grout, which has a predominant effect on the properties [17, 18]. This method is of great advantage to allow aggregate particles to increase contact points and interlock with each other, which is unique to promote LC behavior and especially provide a new perspective to solve the aggregate segregation [19, 20]. The study on the performance of PAC has been a matter of great interest for limited researchers [20, 21]. The most recent study appears to be Coo and Pheeraphan [22, 23], in which they investigated the effect of sand and fly ash on PAC mechanical properties. It was found that optimized proportions of sand and fly ash replacement improve mechanical properties of PAC while no significant effects of coarse aggregate gradation were observed. For further research, they studied reinforced beam shear capacity of PAC. Najjar et al. [24] also attempted to study the damage mechanisms of PAC when exposed to chemical sulfate and physical salt attacks. It showed that PAC specimens exhibited high sulfate resistance. Obviously, previous researches have mainly concentrated on cement replacement materials of PAC grouts. However, there is little information available on the application of PAC used in lightweight concrete. PAC use can make the aggregate particles distribution of the cementitious material reach to the state of dense packing and the relative displacement of aggregate particles will not occur. Thus, the method of preplaced lightweight aggregate concrete (PLC) can be a very effective way to improve the segregation and show better homogeneity in LC.

As stimulated by above facts, this study addresses the use of preplaced casting method in the LC. Two sets of preplaced lightweight aggregate concrete (PLC) and conventional lightweight aggregate concrete (CLC) with three different mixtures are designed to observe the degree of segregation. A series of tests are conducted according to relevant standards to evaluate the properties and performance including compressive strength, unit weight, specific strength, splitting tensile strength, flexural strength, drying shrinkage, and static modulus of elasticity. This modified casting method is applied to find out whether PLC could be suitable for lightweight aggregate compared when it decreases the cement amount.

2. Experimental Program

2.1. Materials. All materials used in the experiment included ordinary Portland cement, spherical shale ceramsite used as lightweight aggregate, fine aggregate, cellulose ethers (CE),

TABLE 2: The properties of LA.

Label	LA
Bulk density (kg/m^3)	737
Particle density (kg/m^3)	1350
Cylinder compression strength (Mpa)	5.8
Water absorption (%)	8.8 (1 h)
	10.5 (24 h)

TABLE 3: The grading of LA.

Sieve size (mm)	<5 mm	5 mm	10 mm	16 mm
Residue on each sieve (%)	0.6	31.1	68.3	0

air-entraining agent (AEA), water, and a polycarboxylate-based high range water reducer (HRWR). The chemical compositions and physical properties of cement are presented in Table 1. The fine aggregate used was provided from local river with a fineness modulus of 2.7. The bulk density and absorption capacity of fine aggregate are 1350 kg/m^3 and 1.6%, respectively. The lightweight aggregate used in this research is spherical shale ceramsite with continuous grading. The properties and gradation of LA are shown in Tables 2 and 3. The water used was tap water. As water reducer, the HRWR with a solid content of approximately 45% was used to achieve the desired better workability for all concrete mixtures.

2.2. Mix Proportions. The two sets of different casting method including CLC (Set 1) and PLC (Set 2) were designed for comparative analysis and CLC was considered as control concrete. Based on the principles of designs of light aggregate concrete mix, different mix proportions were calculated by using absolute volume method according to JGJ12-2006 [25]. The dosage of lightweight aggregate per cubic meter of LC is bulk density of LA. Different strength grades were designed by changing different water to cement ratios. In the case of each set, three water to cement ratios (0.4, 0.45, and 0.5) were observed. Due to the less grout of PLC used in the experiment, the dosage of grout is decreased and calculated by staying the w/c ratio and grout flowability the same. The mix proportion and actual material ratios are shown in Table 4.

2.3. Specimens Preparation and Measurement. The method of casting PLC involves two stages: materials mixing and grouts injection into frames with preplaced lightweight aggregate. The fresh grouts were prepared using the mixture procedure through the study. First, the dry ingredients were initially mixed on a pan type concrete mixer for approximately 5 min. Then the grout mixer started to agitate with the

TABLE 4: Mix proportions (kg/m³).

Mix group	Mix code	Cement	Sand	LA	HRWR	AEA	CE	Water
	1-1	450	729	737	4.50	0.05	0.23	225
Set (1)	2-1	450	729	737	4.50	0.05	0.23	203
	3-1	450	729	737	4.50	0.05	0.23	180
	1-2	435	705	737	4.35	0.04	0.22	218
Set (2)	2-2	435	705	737	4.35	0.04	0.22	196
	3-2	435	705	737	4.35	0.04	0.22	174

TABLE 5: Testing results of compressive strength and unit weight.

Mix code	Cubic compressive strength (MPa)			Axial compressive strength (MPa)			Unit weight (kg/m³)
	7 d	28 d	56 d	7 d	28 d	56 d	28 d
1-1	31.68	35.89	39.41	25.49	28.72	31.55	1820
1-2	32.55	37.10	41.82	26.16	30.30	33.47	1780
2-1	32.06	37.01	40.65	27.29	32.99	33.57	1832
2-2	33.14	40.14	44.07	28.10	33.32	35.58	1792
3-1	35.36	42.11	46.83	28.42	34.28	37.94	1841
3-2	38.73	44.31	48.71	31.31	36.87	39.36	1809

measured water. Homogenized dry ingredients from the concrete mixer were then all loaded into the grout mixer, continually agitating until all dry particles were completely hydrated for approximately 3 min. All admixtures were then added to the mixture and mixed continuously for another 2 min. Specimen samples used for mechanical tests were all collected by injecting grout into frames with preplaced lightweight aggregates. All specimens were then transferred to the curing room and were wrapped with wet burlap when demolded after 24 hours.

Samples for grout compressive strength tested were prepared by pouring the fresh grout into 40 × 40 × 160 mm molds. Concrete specimens were cast into 100 × 100 × 100 mm molds for cubic compressive strength and splitting tensile strength. Molds of 100 × 100 × 300 mm were used for axial compressive strength. In addition, prisms of 100 × 100 × 400 mm and prisms of 100 × 100 × 515 mm were used for measuring the flexural strength and drying shrinkage. For the compressive strength, the specimens were tested at a constant rate of a load of 3.0 kN/s. Meanwhile, the compressive load was applied by using a servo-controlled hydraulic testing machine of 1000 kN capacity. The flexural strength was tested at a constant rate of loading 0.1 kN/s and the splitting tensile strength was at 0.5 kN/s.

3. Results and Discussion

3.1. Compressive Strength. The test results for compressive strength including cubic compressive strength and axial compressive strength of CLC and PLC at the age of 7 days, 28 days, and 56 days are shown in Table 5. The results presented are the average value of three specimens from each concrete mix.

Figure 1 shows the overall test results of cubic compressive strength with all mixes up to 56 days. As seen in the figure,

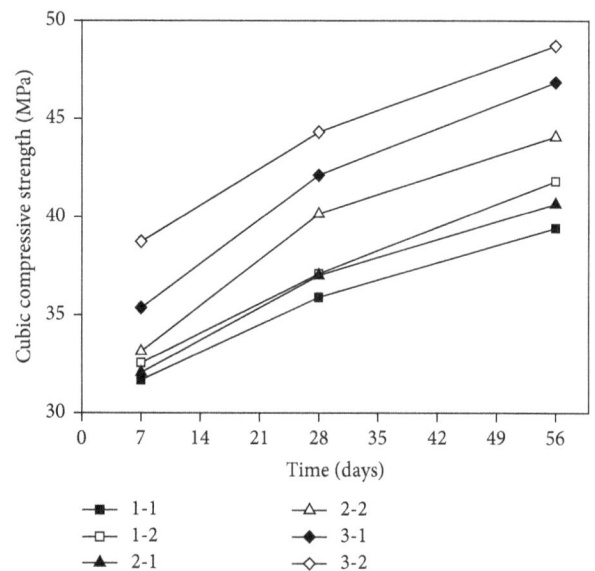

FIGURE 1: Cubic compressive strength of CLC and PLC.

tested PLC compressive strength range attained in the three different w/c ratios is from 35.89 MPa to 44.41 MPa (28 days of curing time), which is within the strength range of commonly used lightweight aggregate concrete in previous studies [11]. Compared with test results of CLC and PLC, the compressive strength losses of CLC are approximately 3.4% for 0.5 w/c ratio, 8.5% for 0.45 w/c ratio, and 5.2% for 0.4 w/c ratio. PLC mixture incorporating crushed aggregate exhibited a little higher compressive strength than CLC at the same w/c ratio. This may be attributed to the interaction that crushed lightweight aggregates provide better interlock than that of CLC [26, 27]. In Figure 3, it can be seen that the external stresses are transferred through contact

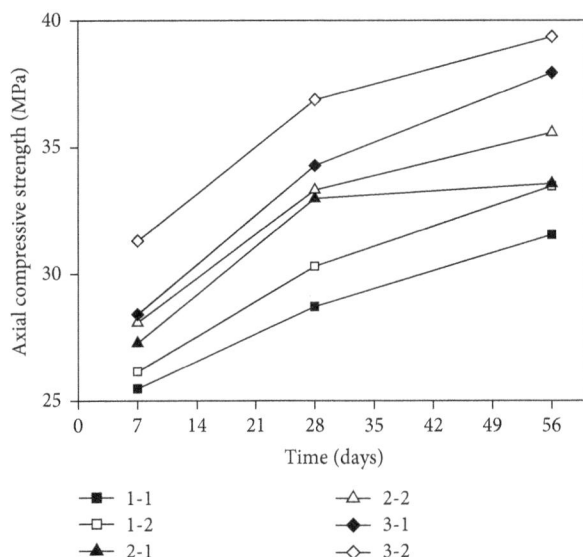

FIGURE 2: Axial compressive strength of CLC and PLC.

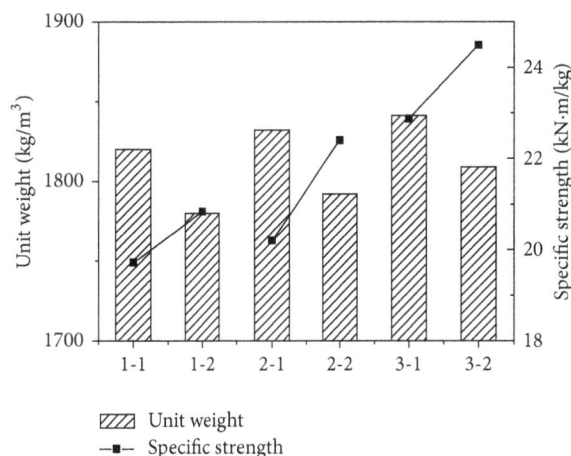

FIGURE 4: Unit weight and specific strength of CLC and PLC.

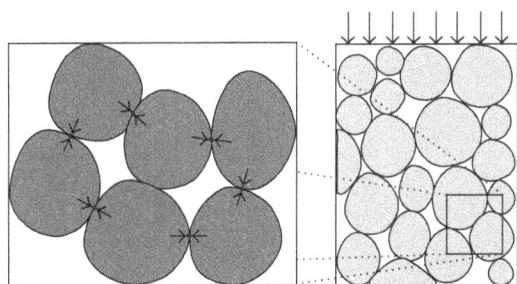

FIGURE 3: Mechanism of transmission of stresses in PLC.

points between aggregate particles due to the specific stress distribution mechanism of PLC. As expected, the interaction of water to cement (w/c) ratios shows a significant influence on compressive strength. The increase of w/c ratios from 0.4 to 0.5 resulted in a gradual decrease in compressive strength, which could be attributed to its great effect on adhesion component. Although using low w/c ratios increases the compressive strength, it will influence the grout flowability. Hence, the water reducer plays a critical role in the strength [28].

The results of axial compressive strength show the same tendency in Figure 2. It can be observed in Table 5 that increasing the w/c reduces the axial compressive strength as expected. Meanwhile, PLC mixtures exhibited higher axial compressive strength than that of the controlled conventional mixture.

3.2. Specific Strength. As seen in Table 5, the dry unit weight values of CLC are higher 2.2% for 0.5 w/c ratio, 2.2% for 0.45 w/c ratio, and 1.7% for 0.4 w/c ratio at the age of 28 days than that of PLC, which is due to less grout used in the PLC. As an important measure, specific strength is defined

as the ratio of the cubic compressive strength to the unit weight. It is indicated in Figure 4 that the specific strength of PLC is higher 5.68%, 10.89%, and 7.08%, respectively, than CLC at the similar mix design. The value of specific strength is from 20.84 kN·m/kg to 24.49 kN·m/kg, which increased as w/c ratio decreases in concrete mixture. Although the method of PLAC used in the experiment decreases the cement amount, its compressive strength has increase trend. Therefore, PLC shows significant reduction in the weight and excellent properties.

3.3. Splitting Tensile Strength. The splitting tensile strength of concrete refers to the resistant stress capacity of damage crack led by volumetric deformation because of moisture and thermal gradients [29]. The results of splitting tensile strength are listed in Table 6 at the age of 7 days, 28 days, and 56 days. It is indicated that the lower w/c ratios increase the splitting tensile strength both in PLC and CLC. However, low w/c ratios may cause a honeycombed structure of particle binding within the lightweight aggregate. Meanwhile, CLC exhibits lower tensile strength than that of PLC with similar mix design in Figure 5. For instance, the average splitting tensile strength of PLC is higher 23% for 0.5 w/c ratio, 6.7% for 0.45 w/c ratio, and 11.9% for 0.4 w/c ratio at the age of 28 days. It was due to the weaker interfacial bond of CLC between lightweight aggregate particles and the cement mixture [20]. The method of PLC used in the experiment not only decreases the cement amount, but also reduces cracks and holes between the lightweight aggregate particles and grout and the steam-fluid port in the frontal zone is also extruded, which shapes a closer union. Moreover, it was also observed that the splitting tensile strength of PLC increased along with the increase of compressive strength, which is similar to properties of CLC.

3.4. Flexural Strength. Figure 6 and Table 6 present the effects of PLC and CLC of three different w/c ratios. PLC flexural strength range in the study is from 2.97 MPa to 4.22 MPa

TABLE 6: Testing results of splitting tensile strength, flexural strength, and static modulus of elasticity.

Mix code	Splitting tensile strength (MPa)			Flexural strength (MPa)			Static modulus of elasticity (GPa)		
	7 d	28 d	56 d	7 d	28 d	56 d	7 d	28 d	56 d
1-1	1.82	2.68	3.09	1.95	2.97	3.34	16.81	20.58	21.18
1-2	2.32	3.29	3.63	2.37	3.51	3.82	17.21	21.26	21.79
2-1	2.38	3.30	3.74	2.45	3.38	3.69	17.93	21.83	22.22
2-2	2.47	3.52	3.89	2.58	3.74	4.01	18.08	22.43	23.05
3-1	2.53	3.69	4.11	2.70	3.83	4.18	18.41	23.17	23.64
3-2	2.70	4.13	4.57	2.86	4.22	4.57	19.94	23.75	25.01

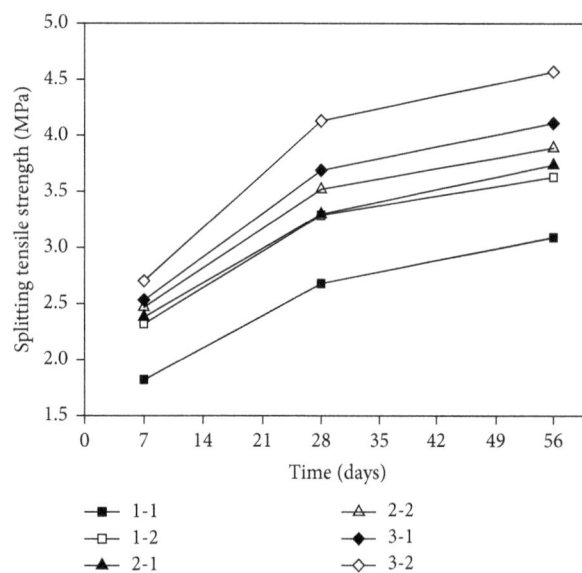

FIGURE 5: Splitting tensile strength of CLC and PLC.

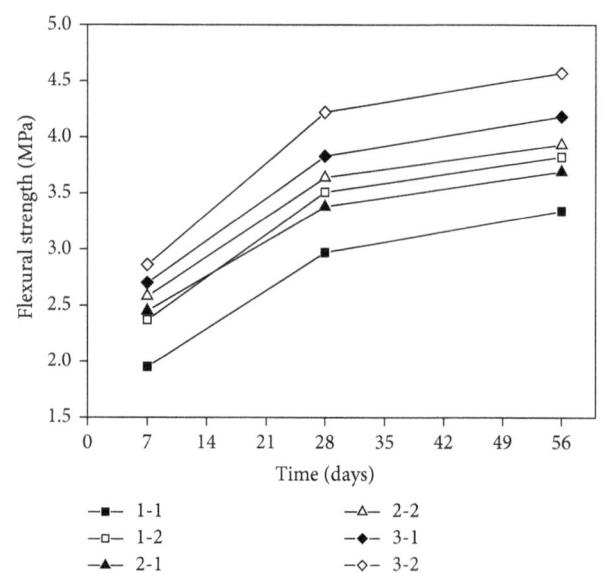

FIGURE 6: Flexural strength of CLC and PLC.

when the w/c ratio decreases from 0.5 to 0.4 (28 days of curing time). PLC shows a little higher flexural strength than the behavior of CLC. The mixing method of PLC improves its transition zone structure and makes sure the surface of each particle is covered with a homogeneous binder. It is due to incompletely utilized aggregate strength and the flexural strength relying on harden cement and interface binder; therefore the interaction of each contact point is effective connection [19, 27]. On the other hand, the data reflects that the flexural strength of PLC and CLC in each mixture has a clear tendency to increase along with the decrease of w/c ratios. Thus the significant factor that affects PAC flexural strength is the w/c ratio. The degree of increase of flexural strength at 7 days is greater than 28 and 56 days with continuous hydration process. A milder increase could be observed in flexural strength at higher age especially in the age of 56 days.

3.5. Drying Shrinkage. Shrinkage is generally defined as a volumetric change of the concrete with time owing to physically adsorbed water lost when exposed to a drying environment [30]. The results of different mixtures are plotted in Figure 7.

The shrinkage deformations were generally supposed to be zero at the final setting time and measured drying shrinkage was evaluated up to 90 days. For each mixture, there is an obvious trend that the process of hydration leads to an obvious shrinkage at early ages. This is mainly due to the fact that the amount of cement and water present in the fresh grout decreases when cement hydrates. On the other hand, it should be noted that the shrinkage of PLC as expected was found to be much lower than that of CLC. The increase of shrinkage is approximately 13% for the CLC and 6% for PLC when w/c ratio ranges from 0.4 to 0.5. This occurs because the aggregate particles in PLC are reasonably connected and close packed with the interaction of interlock, which compensates the loss of drying shrinkage.

3.6. Static Modulus of Elasticity. The modulus of elasticity is described as an ability to reflect stiffness and deformation. Table 6 summarized the modulus of elasticity of PLC and CLC at different ages. In Figure 8, PLC results in an increase in elastic modulus by approximately 2.5% of 0.5 w/c ratio, 2.7% of 0.45 w/c ratio, and 3.3% of 0.4 w/c ratio at the age of 28 days

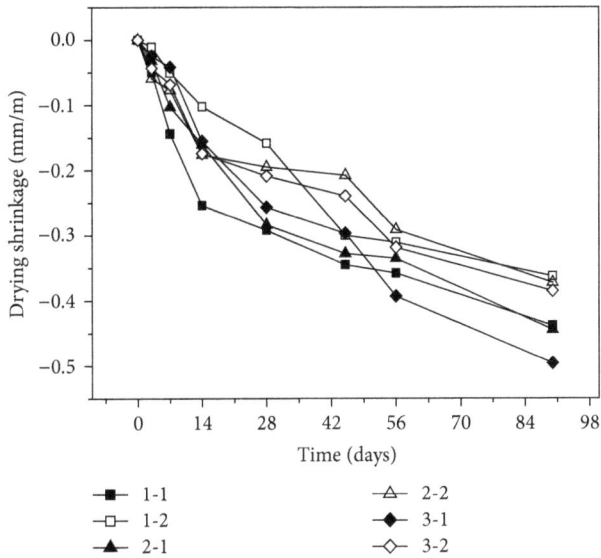

FIGURE 7: Drying shrinkage of CLC and PLC.

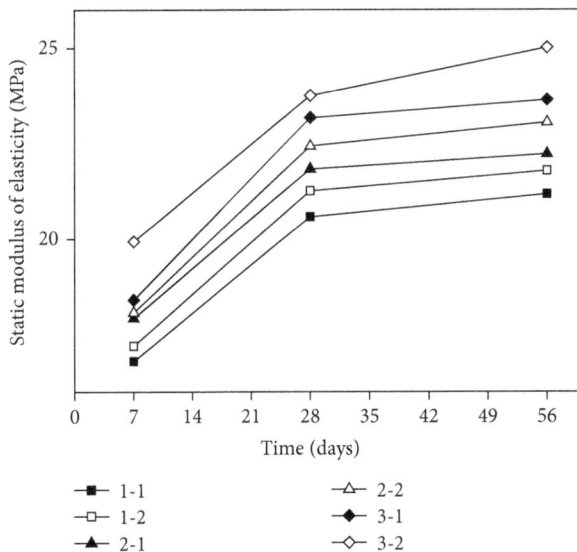

FIGURE 8: Static modulus of elasticity of CLC and PLC.

compared with CLC. It can be observed that PLC exhibits higher modulus of elasticity than that of CLC for different mixtures. This can be due to the fact that PLC forms a skeleton of lightweight aggregate particles resting on each other. Thus, external loads will transmit through connection points in the static granular piling. Generally, the modulus of elasticity has an affinity for the elastic modulus of its components and the proportion content by volume in concrete [16]. In addition, the elastic modulus of elasticity of coarse aggregates is generally higher than that of cement paste. Consequently, the higher elastic modulus of PLC can be considered as a consequence of the modulus of elasticity of the used coarse aggregate.

4. Conclusions

(1) The method of PLC used in the experiment shows higher performance in compressive strength, specific strength, splitting tensile strength, and flexural strength. Thus, PLC shows significant reduction in the weight and better properties.

(2) Compared with test results of CLC and PLC, PLC mixture incorporating crushed aggregate exhibited higher compressive strength than CLC at the same w/c ratio; this can be attributed to the fact that crushed aggregates provide better interlock than that of CLC. Thus, PLC saves the amount of grout on the premise that it shows excellent mechanical properties.

(3) PLC exhibits higher modulus of elasticity than that of CLC for different mixtures. This was attributed to the better interfacial bond between lightweight aggregate particles and the grout mixture, leading to effective connection.

(4) The increase of shrinkage is approximately 13% for the CLC and 6% for PLC when w/c ratio ranges from 0.4 to 0.5. Meanwhile, the shrinkage of PLC was found to be much lower than that of CLC due to the close pack with the interaction of interlock.

(5) Above all, the method of PLC provides a better solution for the segregation problem in the LC. PLC does not need vibration to achieve a denser structure, which can in turn save the cost in the practice. Meanwhile, PLC is considered as an effective method for the difficult section to operation and underwater concreting.

Conflicts of Interest

The authors declare that they have no conflicts of interest.

Acknowledgments

The research work was supported by China National Research Fund (51379015), Natural Science Foundation Research Project of Shaanxi Province of China (2016JM5044), and Central Universities Fund (310823172001, 310823170213, and 310823170648).

References

[1] J. N. Farahani, P. Shafigh, B. Alsubari, S. Shahnazar, and H. B. Mahmud, "Engineering properties of lightweight aggregate concrete containing binary and ternary blended cement," *Journal of Cleaner Production*, vol. 149, pp. 976–988, 2017.

[2] M. R. Hamidian, P. Shafigh, M. Z. Jumaat, U. J. Alengaram, and N. H. R. Sulong, "A new sustainable composite column using an agricultural solid waste as aggregate," *Journal of Cleaner Production*, vol. 129, pp. 282–291, 2016.

[3] P. Shafigh, M. A. Nomeli, U. J. Alengaram, H. B. Mahmud, and M. Z. Jumaat, "Engineering properties of lightweight aggregate concrete containing limestone powder and high volume fly ash," *Cleaner Production*, vol. 135, pp. 148–157, 2016.

[4] T. Y. Lo, W. C. Tang, and H. Z. Cui, "The effects of aggregate properties on lightweight concrete," *Building and Environment*, vol. 42, no. 8, pp. 3025–3029, 2007.

[5] V. Corinaldesi, A. Mazzoli, and G. Moriconi, "Mechanical behaviour and thermal conductivity of mortars containing waste rubber particles," *Materials and Design*, vol. 32, no. 3, pp. 1646–1650, 2011.

[6] B.-W. Jo, S.-K. Park, and J.-B. Park, "Properties of concrete made with alkali-activated fly ash lightweight aggregate (AFLA)," *Cement and Concrete Composites*, vol. 29, no. 2, pp. 128–135, 2007.

[7] H. K. Kim, J. H. Jeon, and H. K. Lee, "Workability, and mechanical, acoustic and thermal properties of lightweight aggregate concrete with a high volume of entrained air," *Construction and Building Materials*, vol. 29, pp. 193–200, 2012.

[8] Y. Ke, A. L. Beaucour, S. Ortola, H. Dumontet, and R. Cabrillac, "Influence of volume fraction and characteristics of lightweight aggregates on the mechanical properties of concrete," *Construction and Building Materials*, vol. 23, no. 8, pp. 2821–2828, 2009.

[9] Y. J. Li and J. T. Ding, "Experimental study on the anti-segregation performance of pumping high strength lightweight aggregate concrete," *Sichuan Building Science*, vol. 31, no. 5, pp. 103–107, 2005.

[10] A. Kiliç, C. D. Atiş, E. Yaşar, and F. Özcan, "High-strength lightweight concrete made with scoria aggregate containing mineral admixtures," *Cement and Concrete Research*, vol. 33, no. 10, pp. 1595–1599, 2003.

[11] M. N. Haque, H. Al-Khaiat, and O. Kayali, "Strength and durability of lightweight concrete," *Cement and Concrete Composites*, vol. 26, no. 4, pp. 307–314, 2004.

[12] J. A. Rossignolo and M. V. C. Agnesini, "Mechanical properties of polymer-modified lightweight aggregate concrete," *Cement and Concrete Research*, vol. 32, no. 3, pp. 329–334, 2002.

[13] F. S. Barbosa, A.-L. Beaucour, M. C. R. Farage, and S. Ortola, "Image processing applied to the analysis of segregation in lightweight aggregate concretes," *Construction and Building Materials*, vol. 25, no. 8, pp. 3375–3381, 2011.

[14] K. H. Mo, T. C. Ling, U. J. Alengaram, S. P. Yap, and C. W. Yuen, "Overview of supplementary cementitious materials usage in lightweight aggregate concrete," *Construction and Building Materials*, vol. 139, pp. 403–418, 2017.

[15] J. Alexandre Bogas, M. G. Gomes, and S. Real, "Capillary absorption of structural lightweight aggregate concrete," *Materials and Structures*, vol. 48, no. 9, pp. 2869–2883, 2015.

[16] H. S. Abdelgader and J. Górski, "Stress-strain relations and modulus of elasticity of two-stage concrete," *Journal of Materials in Civil Engineering*, vol. 15, no. 4, pp. 329–334, 2003.

[17] H. S. Abdelgader, J. Górski, J. Khatib, and A. S. El-Baden, "Two-stage concrete: effect of silica fume and superplasticizers on strength," *Concrete Plant and Precast Technology*, vol. 82, no. 3, pp. 38–47, 2016.

[18] A. S. M. Abdul Awal and I. A. Shehu, "Performance evaluation of concrete containing high volume palm oil fuel ash exposed to elevated temperature," *Construction and Building Materials*, vol. 76, pp. 214–220, 2015.

[19] A. Nowek, P. Kaszubski, H. S. Abdelgader, and J. Górski, "Effect of admixtures on fresh grout and two-stage (pre-placed aggregate) concrete," *Structural Concrete*, vol. 8, no. 1, pp. 17–23, 2007.

[20] M. F. Najjar, A. M. Soliman, and M. L. Nehdi, "Critical overview of two-stage concrete: properties and applications," *Construction and Building Materials*, vol. 62, pp. 47–58, 2014.

[21] H. S. Abdelgader and A. A. Elgalhud, "Effect of grout proportions on strength of two-stage concrete," *Structural Concrete*, vol. 9, no. 3, pp. 163–170, 2008.

[22] M. Coo and T. Pheeraphan, "Effect of sand, fly ash and limestone powder on preplaced aggregate concrete mechanical properties and reinforced beam shear capacity," *Construction and Building Materials*, vol. 120, pp. 581–592, 2016.

[23] M. Coo and T. Pheeraphan, "Effect of sand, fly ash, and coarse aggregate gradation on preplaced aggregate concrete studied through factorial design," *Construction and Building Materials*, vol. 93, pp. 812–821, 2015.

[24] M. Najjar, M. Nehdi, A. Soliman, and T. Azabi, "Damage mechanisms of two-stage concrete exposed to chemical and physical sulfate attack," *Construction and Building Materials*, vol. 137, pp. 141–152, 2017.

[25] JGJ12-2006, *Technical Specification for Lightweight Aggregate Concrete Structures*, China Standard Publishing House, Hebei, China, 2006.

[26] H. S. Abdelgader, "Effect of the quantity of sand on the compressive strength of two-stage concrete," *Magazine of Concrete Research*, vol. 48, no. 177, pp. 353–360, 1996.

[27] H. S. Abdelgader, "How to design concrete produced by a two-stage concreting method," *Cement and Concrete Research*, vol. 29, no. 3, pp. 331–337, 1999.

[28] W. Sun, H. Yan, and B. Zhan, "Analysis of mechanism on water-reducing effect of fine ground slag, high-calcium fly ash, and low-calcium fly ash," *Cement and Concrete Research*, vol. 33, no. 8, pp. 1119–1125, 2003.

[29] M. L. Nehdi and A. M. Soliman, "Early-age properties of concrete: overview of fundamental concepts and state-of-the art research," *Construction Materials*, vol. 164, pp. 55–77, 2011.

[30] A. Bentur, S.-I. Igarashi, and K. Kovler, "Prevention of autogenous shrinkage in high-strength concrete by internal curing using wet lightweight aggregates," *Cement and Concrete Research*, vol. 31, no. 11, pp. 1587–1591, 2001.

13

Degradation of Roller-Compacted Concrete Subjected to Freeze-Thaw Cycles and Immersion in Potassium Acetate Solution

Wuman Zhang ⓘ, Jingsong Zhang, Shuhang Chen, and Sheng Gong

Department of Civil and Engineering, School of Transportation Science and Engineering, Beihang University, Beijing 100191, China

Correspondence should be addressed to Wuman Zhang; wmzhang@buaa.edu.cn

Academic Editor: Nadezda Stevulova

Two sets of roller-compacted concrete (RCC) samples cured for 28 days were subjected to freeze-thaw (F-T) cycles and immersion in laboratory conditions. F-T cycles in water and water-potassium acetate solution (50% by weight) were carried out and followed by the flexural impact test. The weight loss, the dynamic elastic modulus (E_d), the mechanical properties, and the residual strain of RCC were measured. The impact energy was calculated based on the final number of the impact test. The results show that the effect of F-T cycles in KAc solution on the weight loss and E_d of RCC is slight. E_d, the compressive strength, and the flexural strength of RCC with 250 F-T cycles in KAc solution decrease by 3.8%, 23%, and 36%, respectively. The content (by weight) of K^+ at the same depth of RCC specimens increases with the increase of F-T cycles. The impact energy of RCC specimens subjected to 250 F-T cycles in KAc solution decreases by nearly 30%. Microcracks occur and increase with the increase of F-T cycles in KAc solution. The compressive strength of RCC immersed in KAc solution decreases by 18.8% and 32.8% after 6 and 12 months. More attention should be paid to using KAc in practical engineering because both the freeze-thaw cycles and the complete immersion in KAc solution damage the mechanical properties of RCC.

1. Introduction

Roller-compacted concrete (RCC) is a zero-slump concrete compacted with vibratory and rubber-tired rollers [1]. RCC has been used in the construction of dams, pavements, and airport runways because of the lower cost and the easier placement operations [2–4]. RCC requires long-term stable performance when it is applied in airport runways because reconstruction causes a great impact on the air travel industry. Although the mechanical properties of RCC have been widely recognized, its frost resistance is still the focus in this field.

Piggott [5] found that the field performance of RCC was excellent in harsh environments, including northern U.S. states and Canada. The investigation showed that RCC with a reasonable mixture composition [6], casting and curing process [7, 8] had a good frost resistance. RCC also had better salt frost resistance when it was mixed with mineral admixtures [9]. Delatte and Storey [10] found that the freeze-thaw (F-T) durability of RCC mainly depended on the amount of cement paste and the water to cement ratio, but the degree of compaction had a less effect. However, the results reported by ACI Committee 325 [11] had shown that RCC mixtures were easy to damage by F-T cycles.

For RCC used at airport pavements in cold climates, potassium acetate (KAc) is being used as a deicer because of its high performance and aggressiveness. However, recent researches showed that KAc deicers could affect concrete durability through physical deterioration of concrete and chemical reaction between the KAc deicer and the hydration products of cement [12]. It has been suggested the deterioration of the airport runway may be related to the alkali-silica reaction between the hydration products of cement and KAc [9]. Julio-Betancourt [13] found that even without alkali-silica reactive aggregates, KAc deicers can cause degradation of strength, excessive expansion, and reduce resistance to freezing and thawing. It seems that investigations look to the KAc deicer as a problem, but given the varying results, the deterioration associated with the deicer is not completely understood.

TABLE 1: Mix proportions of RCC.

Water	Cement	Fine aggregate	Coarse aggregate	SP	AG
109	315	895	1207	2	0.023

The flexural strength and impact behavior are the most important parameters for RCC used in airport pavements. However, there is little research on the impact properties of RCC after F-T cycles in KAc solution. The effect of F-T cycles in KAc solution on the mechanical properties and impact resistance of RCC also needs to be elucidated. The main objective of this research focuses on the frost resistance and impact resistance of RCC exposed to the KAc deicer.

2. Materials and Experimental Process

2.1. Materials. Ordinary Portland cement (P.I 42.5), river sand with fineness modulus 2.61, coarse aggregate with sizes of 5–25 mm, microair 202 (AG), and polycarboxylate-based superplasticizer (SP) were used in this study. The mix proportions of RCC are listed in Table 1.

2.2. Experimental Procedure

2.2.1. Vebe Time Test. The Vebe method was used to measure the workability of RCC. This test method is a variation of the simple slump test and subjects the concrete to vibration after the slump cone removal. The small vibrating table operates at a fixed amplitude and frequency, and in the test, a plastic disc is placed in contact with the upper surface of the concrete. The test is completed when the lower surface of the disc has been completely coated with cement grout. The time is the measured parameter here. The Vebe time of fresh RCC is 28 s.

2.2.2. Specimen Preparation. The fresh mixture was poured into the prism molds in three layers. The dimensions of the mold are $100 \times 100 \times 400$ mm. A vibrating hammer was fixed on a 5 kg steel plate to apply the uniform rolling load. The rolling time of each layer was 30 seconds. After 24 h, the specimens were demolded and placed in the curing room for 28 days. The temperature was $20 \pm 2°C$, and the relative humidity was 90%.

2.2.3. Strength of Specimens. Equations (1) and (2) were used to calculate the compressive strength, splitting tensile strength, and flexural strength.

$$f_c = \frac{F}{A},\tag{1}$$

where f_c is the compressive strength (MPa), F is the maximum load (N), and A is the area of the cube loading face (mm).

$$f_f = \frac{Fl}{bh^2},\tag{2}$$

where f_f is the flexural strength (MPa), F is the maximum load (N), l is the distance between the supporting rollers (mm),

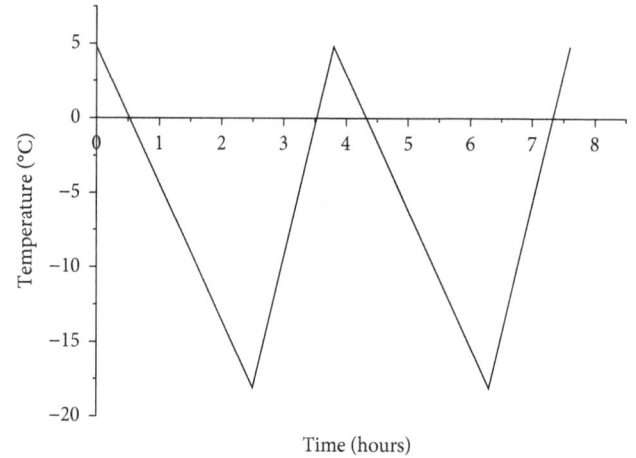

FIGURE 1: Temperature required by the standard of the specimen during the freeze-thaw cycles.

b is the width of the cross section (mm), and h is the height of the cross section (mm).

2.2.4. Freeze-Thaw Cycles and Immersion Test. Two sets of samples cured for 28 days were subjected to the following:

(a) F-T cycles, while samples were immersed in two different medias: water and 50% KAc solution (by weight).

(b) Laboratory conditions (RH 90%, $20 \pm 2°C$), while samples were immersed in 50% KAc solution (by weight) for 6 and 12 months, respectively.

The F-T cycles were carried out according to GB-T50082-2009 [14]. A thermometer embedded at the center of the specimen was used to control the temperature. The maximum temperature and the minimum temperature are 5 ± 2 and $-18 \pm 2°C$, respectively. Figure 1 shows the temperature of the specimen during the freeze-thaw cycles. The temperature curve was required by the standard. The real temperature of the sample itself can be different, depending, for example, on the properties of the sample and the accuracy of temperature sensors. Two different medias: water and 50% KAc solution (by weight) were used as the freezing medias. The total number of F-T cycles was 250.

The weight loss was calculated by the following equation:

$$\Delta W_n = \frac{W_0 - W_n}{W_0} \times 100,\tag{3}$$

where ΔW_n is the weight loss of the specimens at the nth freeze-thaw cycle (%), W_0 is the average weight of the concrete specimens before freeze-thaw cycles (kg), and W_n is the average weight of the concrete specimens at the nth freeze-thaw cycle (kg).

The DT-16-type dynamic modulus instrument was used to measure the relative dynamic modulus of elasticity. The relative dynamic modulus of elasticity was calculated by the following equation:

1. Lifting part.
2. Vertical guide pipe.
3. Release mechanism.
4. Part to prevent the
   second impact.
5. Holder.
6. Lifting pallet.
7. Lifting hand wheel.
8. Lead screw.
9. Optical signal tube.
10. Specimen.
11. Drop weight hammer.

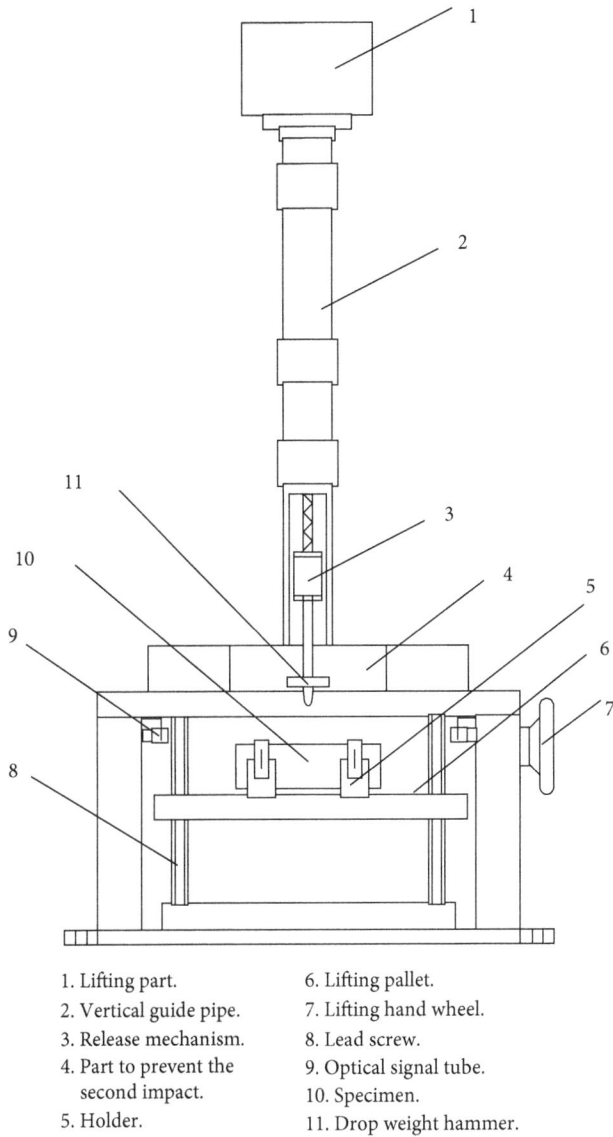

FIGURE 2: The drop hammer impact testing machine.

$$E_{\mathrm{d}} = \frac{W_i t_i^2}{W_1 t_1^2} \times 100, \tag{4}$$

where E_{d} is relative dynamic modulus of elasticity, W_1 is the initial weight of the specimen (kg), and W_i is the weight of a specimen after i times freeze-thaw cycles (kg). t_1 is the initial ultrasonic time of a specimen (s), and t_i is the ultrasonic time of a specimen after i times freeze-thaw cycles (s).

2.2.5. Flexural Impact Test.

Figure 2 shows the drop hammer impact testing machine. It consists of two stiff constraints, which restrain the specimen from moving. It is capable of dropping a mass of 1.0–10.0 kg from height of up to 2.0 m above the target specimen. 3 kg and 0.3 m were used in this study. A steel cylindrical projectile with a 40 mm diameter is the head of the drop hammer. One part is designed to prevent the second impact. For each specimen, the side surface (2 mm from the top surface) and the bottom

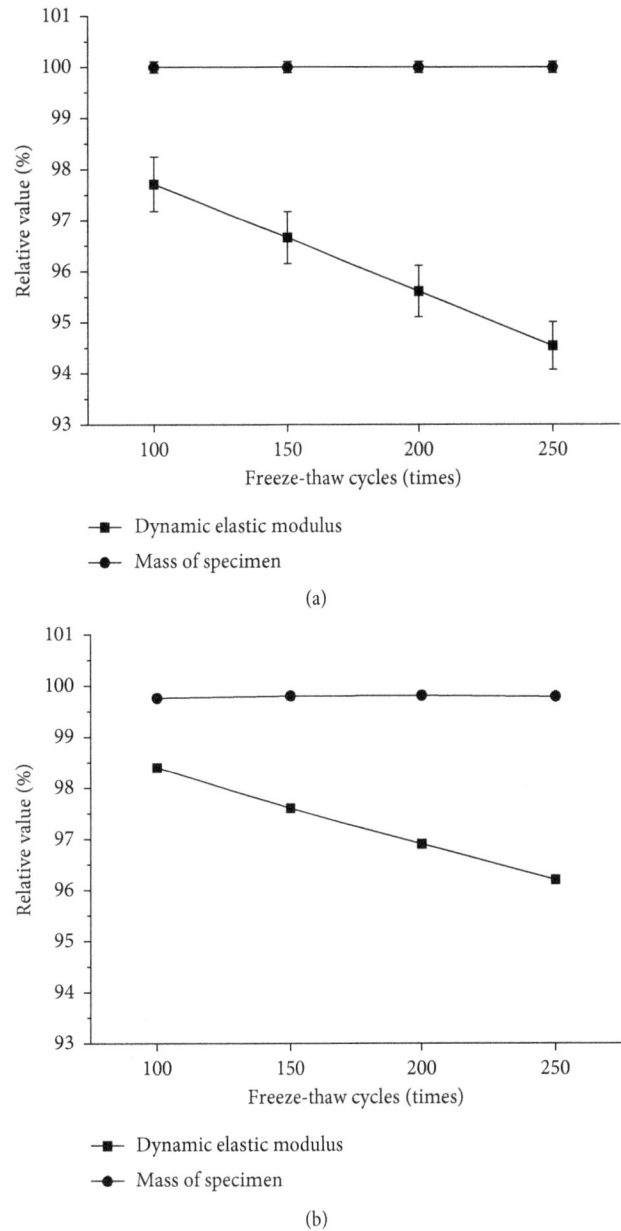

(a)

(b)

FIGURE 3: Relative weight loss and dynamic modulus of elasticity: (a) in water; (b) in KAc solution.

surface are bonded with a strain gauge, respectively. The strain was monitored by a high-speed data acquisition system. The impact energy is a constant value during the impact test. SZ120-100AA strain gauges were used to measure the strain [15, 16].

2.2.6. Microstructure and Element Content Analysis.

The field emission scanning electron microscope (SEM, JSM-7500F) with energy dispersive X-ray analysis (EDX) was used to investigate the microstructures of the specimens. The samples for SEM analysis were soaked in anhydrous ethanol to stop hydration and dried at 60°C for 4 hours. The samples were coated with 20 nm gold before testing. The EDX was used to measure the element content of

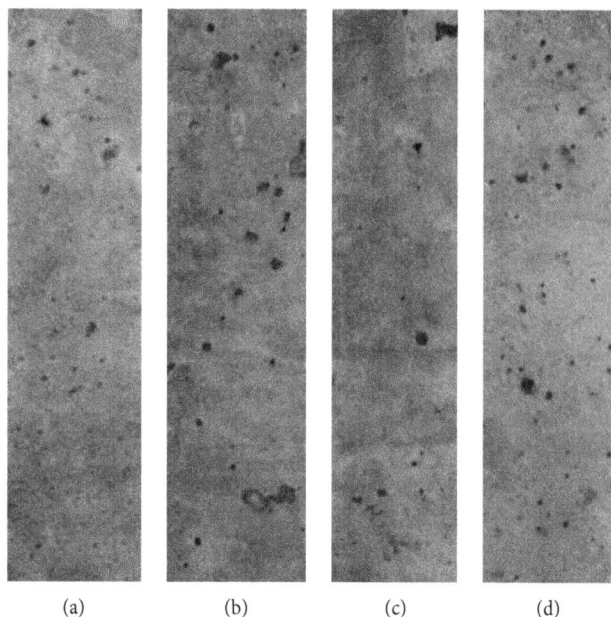

(a)   (b)   (c)   (d)

FIGURE 4: Surface of the specimens subject to the F-T cycles in KAc solution: (a) 100; (b) 150; (c) 200; (d) 250.

samples. The resolution was 129.92 eV, and the measurement time was 50 s.

3. Results and Discussion

3.1. Freeze-Thaw Cycle Test

3.1.1. Weight Loss and Dynamic Modulus of Elasticity. The relative weight loss and the relative dynamic modulus of elasticity (E_d) of RCC with F-T cycles in water and KAc solution are shown in Figure 3. It clearly indicates that there is little change in weight of RCC with F-T cycles in water. Pigeon and Marchand [7], Andersson [17], and Marchand et al. [18] also obtained the similar results. The surface of the specimens with F-T cycles in KAc solution is shown in Figure 4. Based on Figures 3(b) and 4, the KAc deicer also caused no scaling or insignificant scaling in RCC. In the study from Wang et al. [19] and also Nanni [20], it was stated that KAc minor scaling might be related to alkali carbonation of concrete surface. However, Piggott [5] reported that the overall quality and properties of concrete and internal structure of concrete surface have an effect on the surface scaling. The preparation process of RCC may be another reason that the mass loss is not remarkable.

The F-T cycles in water or in KAc solution also have an insubstantial effect on E_d of RCC. E_d of RCC with 250 F-T cycles decreases by 5.5% and 3.8%, respectively. The loss of the elastic modulus of the RCC is less than that of normal concrete in the freeze-thaw test [21, 22]. This is probably due to the layering and the vibrating during the RCC specimen preparation which results in a higher surface strength.

3.1.2. Mechanical Properties of RCC. The compressive strength and the flexural strength of RCC cured for 28 days

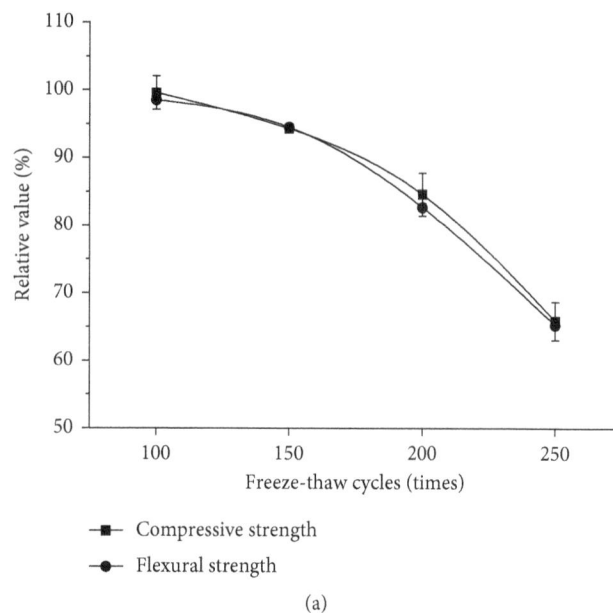

(a)

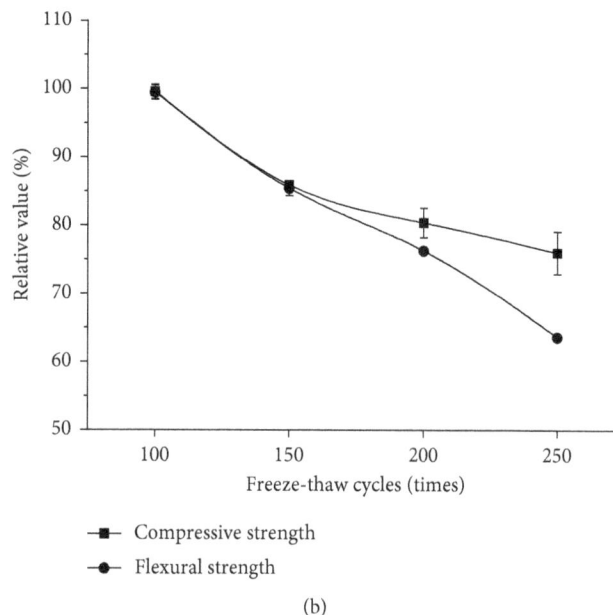

(b)

FIGURE 5: Relative compressive strength and flexural strength of RCC: (a) in water; (b) in KAc solution.

are 47.7 MPa and 9.0 MPa, respectively. Figure 5 shows the relative compressive strength and flexural strength of RCC subjected to F-T cycles in water and KAc solution. It can be seen that the F-T cycles in water or in KAc solution decrease both the compressive strength and the flexural strength of RCC. The compressive strength and the flexural strength are decreased by 33% when the RCC specimens are subjected to 250 F-T cycles in water. The results can be explained by the internal cracking caused by the expansion of water in concrete during F-T cycles.

250 F-T cycles in KAc solution decreases the compressive strength and the flexural strength of RCC by 23% and 36%, respectively. The strength loss can be explained by the traditional deterioration mechanism due to frost. In addition,

TABLE 2: The content of K^+ (by weight).

F-T cycles	100	150	200	250
Content of K^+ (%)	0.5	2.73	2.83	3.93

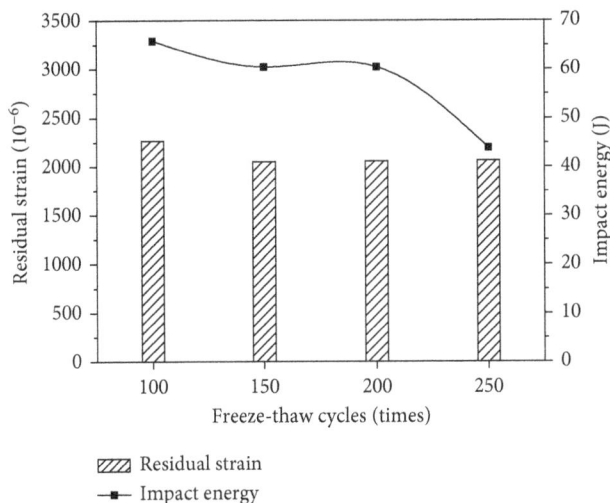

FIGURE 6: Residual strain and impact energy of RCC subjected to the F-T cycles in KAc solution.

it is postulated that KAc increases the level of saturation in concrete, possibly due to changes in surface tension and viscosity of pore water [13]. However, the decreasing trend in the compressive strength slowed. This is mainly attributed to the more penetration of KAc and the formation of an ettringite-like needle structure when the KAc deicer is used [19, 23]. The content (by weight) of K^+ at the same depth of the RCC specimens with the F-T cycles in KAc solution is shown in Table 2.

The deposition of ettringite seems to follow the appearance of cracks when F-T deterioration occurs. In the early stage, ettringite does not promote the propagation of existing cracks and cause new cracking in concrete. Cracks caused by frost damage will also give space for the crystallization of ettringite [24]. The filling and covering effect of ettringite on concrete crack could improve the compressive strength of concrete in early stage [25]. However, this filling action has slight effect on the flexural strength.

3.1.3. Impact Properties of RCC. Figure 6 shows the residual strain and impact energy of RCC subjected to the F-T cycles in KAc solution. It is clearly seen that the effect of the F-T cycles in KAc solution on the residual strain is slight. This is probably due to the damage of RCC under the impact loading is still a brittle fracture. In addition, the strain gauge bonded on the bottom surface of the specimen cannot record the strain when the specimen is broken into two sections. However, there is a decreasing trend in the impact energy of RCC, especially for the specimens subjected to 250 F-T cycles in KAc solution, and the impact energy decreases by nearly 30%. This result can be due to the decreasing action of the F-T cycles on the flexural strength of RCC.

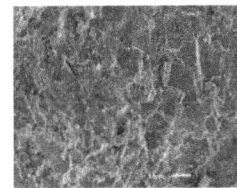

(a)

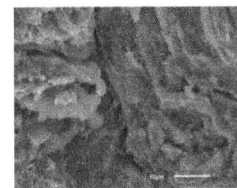

(b)

(c)

(d)

(e)

FIGURE 7: Microstructures of RCC without and with the F-T cycles in KAc solution: (a) 0; (b) 100; (c) 150; (d) 200; (e) 250.

3.1.4. Microstructures of RCC. In order to better understand the effect of the F-T cycles on the impact properties of RCC, the microstructures are observed and shown in Figure 7. Almost no microcrack occurs in the RCC matrix without the F-T cycles (Figure 7(a)). The F-T cycles will cause expansive pressure and osmotic pressure in concrete [26]. The surface spalling and internal cracking occur when the tensile stress produced by the two pressures exceeds the tensile strength of concrete. Therefore, microcracks occur and increase with the increase of the F-T cycles (Figures 7(b)–7(e)), which is consistent with the development of the strength of the specimens subjected to the F-T cycles in KAc solution.

3.2. Properties of RCC Immersed in KAc Solution. In order to determine whether the deicing fluid has a corrosive effect on the concrete, the specimens without the freeze-thaw cycles

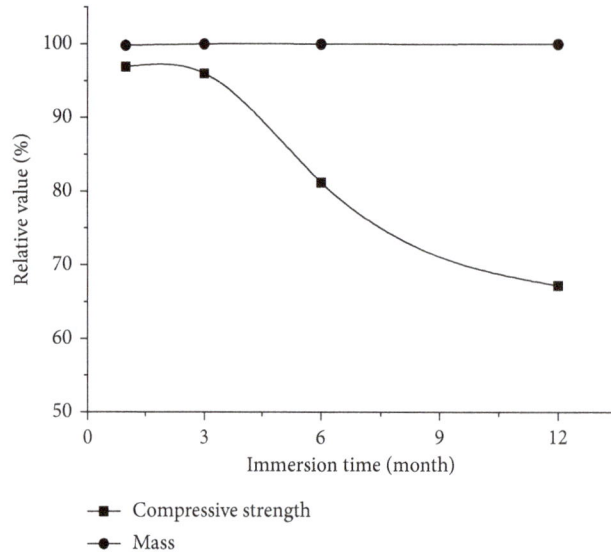

FIGURE 8: Relative mass and compressive strength of RCC immerged in KAc solution.

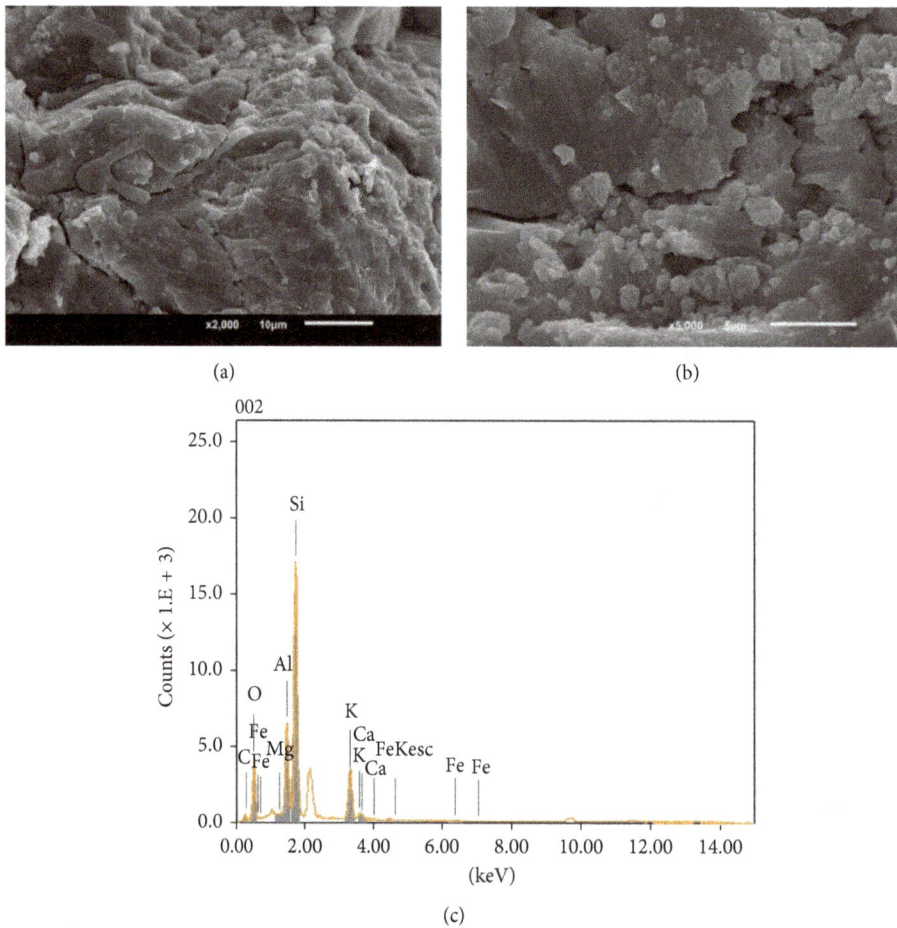

(a)

(b)

(c)

FIGURE 9: (a, b) Microstructures and (c) EDS of RCC immerged in KAc solution for 12 months.

are immersed in the KAc solution. The mass loss, the strength, and the microstructure of the specimens are measured. Figure 8 shows the mass and compressive strength of RCC immerged in KAc solution. It can be seen

that the mass is almost unchanged after the RCC specimens are immersed in KAc solution for 12 months, which indicates that the complete immersion of the KAc solution does not cause the surface spalling. The compressive strength has a smaller drop after 3 months of immersion. However, the compressive strength decreased by 18.8 and 32.8% after 6 and 12 months of complete immersion in KAc solution. The decrease of the compressive strength (32.8%) caused by 12-month immersion in KAc solution is very close to that (33%) of samples subjected to 250 F-T cycles in water. However, the decrease value is greater than that (23%) of samples subjected to 250 F-T cycles in KAc solution.

The microstructures and the energy dispersive spectrum (EDS) of RCC immerged in KAc solution for 12 months are shown in Figure 9. Comparing to the RCC matrix without the F-T cycles (Figure 7(a)), the microcracks were observed in the microstructure picture. In addition, the mass percentage of the K element in the new crystal is 5.58%, which indicates that more KAc solution has penetrated into the specimens subjected to 12 months of complete immersion in KAc solution. The osmosis pressure and the crystallization pressure of KAc solution may cause expansion and cracks in the RCC matrix, which decreases the compressive strength of specimens after a long-term contact with KAc solution.

3.3. Degradation Mechanism. The degradation mechanism of RCC subjected to the F-T cycles in KAc solution can be explained by the traditional theory due to frost. In addition to the pressure generated by osmosis and crystallization, KAc generally increases the saturation of the concrete and keeps concrete pores at or near the maximum fluid saturation, thereby increasing the risk of frost damage [13, 19].

Furthermore, KAc may induce alkali-silica reaction in the concrete-containing reactive aggregate, which causes expansion and cracks after a long-term contact. These cracks create channels for water and other solutions to penetrate into concrete and reduce freeze-thaw durability.

4. Conclusions

For the materials used and test methods applied, the following conclusions can be drawn:

(1) The effect of freeze-thaw cycles in KAc solution on the weight loss and the elastic modulus of RCC is slight. The elastic modulus of RCC with 250 freeze-thaw cycles decreases by 3.8%.

(2) 250 freeze-thaw cycles in KAc solution decrease the compressive strength and the flexural strength by 23% and 36%, respectively. The content (by weight) of K^+ at the same depth of RCC specimens increases with the increase of freeze-thaw cycles, which reduces the decreasing trend of the compressive strength caused by freeze-thaw cycles in KAc solution.

(3) The impact energy of RCC specimens subjected to 250 freeze-thaw cycles in KAc solution decreases by nearly 30%.

(4) Microcracks occur and increase with the increase of freeze-thaw cycles in KAc solution.

(5) The compressive strength of RCC without freeze-thaw cycles decreased by 18.8 and 32.8% after 6 and 12 months of complete immersion in KAc solution. The decrease of the compressive strength (32.8%) caused by 12 months immersion in KAc solution is greater than that (23%) of samples subjected to 250 F-T cycles in KAc solution.

(6) More attention should be paid to using KAc in practical engineering because both the freeze-thaw cycles and the complete immersion in KAc solution damage the mechanical properties of RCC. The decrease of mechanical properties of RCC used in airport runway will have serious effect on the flight safety. KAc solution should be replaced with a new type of harmless deicing fluid.

Conflicts of Interest

The authors declare that they have no conflicts of interest.

Acknowledgments

This research was supported by the National Natural Science Foundation of China (nos. 51378042 and 51678022) and the Fok Ying Tung Education Foundation (no. 132016).

References

[1] A. Mardani-Aghabaglou, Ö. Andiç-Çakir, and K. Ramyar, "Freeze-thaw resistance and transport properties of high-volume fly ash roller compacted concrete designed by maximum density method," *Cement and Concrete Composites*, vol. 37, no. 1, pp. 259–266, 2013.

[2] A. Yerramala and K. Ganesh Babu, "Transport properties of high volume fly ash roller compacted concrete," *Cement and Concrete Composites*, vol. 33, no. 10, pp. 1057–1062, 2011.

[3] P. W. Gao, S. X. Wu, P. H. Lin, Z. R. Wu, and M. S. Tang, "The characteristics of air void and frost resistance of RCC with fly ash and expansive agent," *Construction and Building Materials*, vol. 20, no. 8, pp. 586–590, 2006.

[4] M. Pigeon and V. M. Malhotra, "Frost-resistance of roller-compacted high-volume fly-ash concrete," *Journal of Materials in Civil Engineering*, vol. 7, no. 4, pp. 208–211, 1995.

[5] R. W. Piggott, *Roller Compacted Concrete Pavements—A Study of Long Term Performance*, Portland Cement Association, Skokie, IL, USA, R&D serial No. 2261, 1999.

[6] C. Hazaree, H. Ceylan, and K. Wang, "Influences of mixture composition on properties and freeze–thaw resistance of RCC," *Construction and Building Materials*, vol. 25, no. 1, pp. 313–319, 2011.

[7] M. Pigeon and J. Marchand, "Frost resistance of roller-compacted concrete," *Concrete International*, vol. 18, no. 7, pp. 22–26, 1996.

[8] C. Hazaree, P. Ramasamy, and W. P. David, "Roller-compacted concrete: a sustainable alternative," in *Green Building with Concrete-Sustainable Design and Construction*, G. M. Sabnis, Ed., pp. 129–180, CRC Press, Boca Raton, FL, USA, 2015.

[9] J. M. S. Silva, S. M. Cramer, M. A. Anderson, M. I. Tejedor, and J. F. Muñoz, "Concrete microstructural responses to the

interaction of natural microfines and potassium acetate based deicer," *Cement and Concrete Research*, vol. 55, pp. 69–78, 2014.

[10] N. Delatte and C. Storey, "Effects of density and mixture proportions on freeze–thaw durability of roller-compacted concrete pavement," *Transportation Research Record: Journal of the Transportation Research Board*, vol. 1914, pp. 45–52, 2005.

[11] ACI Committee, *State-of-the-Art Report on Roller-Compacted Concrete Pavements*, Vol. 32, ACI, Farmington, MI, USA, 1995.

[12] S. Ghajar-Khosravi, *Potassium Acetate Deicer and Concrete Durability*, University of Toronto, Toronto, ON, Canada, 2011.

[13] G. A. Julio-Betancourt, *Effect of De-Icer and Anti-Icer Chemicals on the Durability, Microstructure, and Properties of Cement-Based Materials*, University of Toronto, Toronto, ON, Canada, 2009.

[14] Ministry of Housing and Urban-Rural Development of People's Republic of China, *GB-T50082-2009, Standard for Test Methods of Long-Term Performance and Durability of Ordinary Concrete*, China Architecture & Building Press, Beijing, China, 2009, in Chinese.

[15] W. Zhang, S. Chen, N. Zhang, and Y. Zhou, "Low-velocity flexural impact response of steel fiber reinforced concrete subjected to freeze-thaw cycles in NaCl solution," *Construction and Building Materials*, vol. 101, pp. 522–526, 2015.

[16] W. Zhang, S. Chen, and Y. Liu, "Effect of weight and drop height of hammer on the flexural impact performance of fiber-reinforced concrete," *Construction and Building Materials*, vol. 140, pp. 31–35, 2017.

[17] R. Andersson, "Pavements of roller-compacted concrete-physical properties," *Nordic Concrete Research*, vol. 5, no. 11, pp. 7–17, 1986.

[18] J. Marchand, M. Pigeon, H. Isabelle, and J. Boisvert, "Freeze-thaw durability and deicer salt scaling resistance of roller compacted concrete pavements," in *Paul Klieger Symposium on Performance of Concrete*, pp. 217–236, SP 122-13, American Concrete Institute, Farmington Hills, MI, USA, 1990.

[19] K. Wang, D. E. Nelsen, and W. Nixon, "Damaging effects of deicing chemicals on concrete materials," *Cement and Concrete Composites*, vol. 28, no. 2, pp. 173–188, 2006.

[20] A. Nanni, "Curing of roller compacted concrete-strength development," *Journal of Transportation Engineering*, vol. 114, no. 6, pp. 684–694, 1988.

[21] Z. Wang, Q. Zeng, Y. Wu, L. Wang, Y. Yao, and K. Li, "Relative humidity and deterioration of concrete under freeze–thaw load," *Construction and Building Materials*, vol. 62, pp. 18–27, 2014.

[22] H. Cai and X. Liu, "Freeze-thaw durability of concrete: ice formation process in pores," *Cement and Concrete Research*, vol. 28, no. 9, pp. 1281–1287, 1998.

[23] C. Giebson, K. Seyfarth, and J. Stark, "Influence of acetate and formate-based deicers on ASR in airfield concrete pavements," *Cement and Concrete Research*, vol. 40, no. 4, pp. 537–545, 2010.

[24] Sulfate Task Group of Portland Cement Association, *Ettringite Formation and the Performance of Concrete*, 2001, http://www.cement.org/docs/default-source/fc_concrete_technology/is417-ettringite-formation-and-the-performance-of-concrete.pdf?sfvrsn=412.

[25] S. Ozaki and N. Sugata, "Long-erm durability of reinforced concrete submerged in the sea: concrete under severe conditions 2: environment and loading", in *Proceedings of the Second International Conference on Concrete Under Severe Conditions*, O. E. Gjørv, K. Sakai, and N. Banthia, Eds., pp. 448–457, CRC Press, Tromsø, Norway, June 1998.

[26] T. C. Powers, *Freezing Effects in Concrete*, Vol. 47, ACI Special Publication, Berkeley, CA, USA, 1975.

14

Analytical Model for Deflections of Bonded Posttensioned Concrete Slabs

Min Sook Kim,[1] **Joowon Kang,**[2] **and Young Hak Lee**[1]

[1]*Department of Architectural Engineering, Kyung Hee University, 1732 Deogyeong-daero, Yongin, Republic of Korea*
[2]*School of Architecture, Yeungnam University, 280 Daehak-ro, Gyeongsan, Republic of Korea*

Correspondence should be addressed to Young Hak Lee; leeyh@khu.ac.kr

Academic Editor: Francesco Caputo

This paper presents a finite element analysis approach to evaluate the flexural behavior of posttensioned two-way slabs depending on the tendon layout. A finite element model was established based on layered and degenerated shell elements. Nonlinearities of the materials are considered using the stress-strain relationships for concrete, reinforcing steel, and prestressing tendons. Flexural testing of the posttensioned two-way slabs was conducted to validate the developed analytical process. Comparing the analytical results with the experimental results in terms of deflections, it showed generally good agreements. Also a parametric study was performed to investigate the effects of different types of tendon layout.

1. Introduction

Posttensioned concrete slabs have many advantages, such as rapid construction, reduction of overall member depth, and reduced materials. In addition, posttensioned concrete slabs with proper posttensioning show little deflection and few cracks under service loads. Although posttensioned concrete slabs have many advantages, their performance is still not fully understood, and the behaviors of two-way slab systems are more difficult to determine than those of one-way slabs. To evaluate posttensioned concrete slabs, several experimental studies have been performed. Burns and Hemakom [1] observed the strength and behavior of posttensioned flat plates. They applied the banded tendon layout in column strips in the x-direction and distributed single tendons in the y-direction on the slabs. Through this study they found that the banded and distributed tendon layout improved the flexural and shear capacities. Kosut et al. [2] experimentally evaluated the behavior of posttensioned flat plates with distributed and banded tendon arrangements in each direction. They found that banded tendons on the column strip can resist punching shear failure, and distributed tendons can improve flexural capacity. Roschke and Inoue [3] tested prestressed concrete flat slabs to investigate

strain distribution in regions adjacent to the transverse posttensioning bands.

To analyze complex posttensioned concrete slabs efficiently, some researchers have proposed finite element approaches. Van Greunen and Scordelis [4] researched a numerical procedure for the materials and a geometric nonlinear analysis for prestressed concrete slabs. Wu et al. [5] proposed a tendon model based on the finite element method that can represent the interaction between tendons and concrete. They verified the accuracy of their proposed equation against existing experimental data. El-Mezaini and Çitipitioğlu [6] developed quadratic and cubic finite elements with movable nodes to predict the behavior of different bond conditions for the tendons. Kang and Huang [7] proposed nonlinear finite element models to evaluate the behavior of unbonded posttensioned slab-column connections. The spring elements and contact formation were applied to the model to consider the interface between the concrete and prestressing tendon. Kang et al. [8] compared the structural performance of the bonded and unbonded posttensioned concrete members through experiment and analysis. Ghallab [9] suggested using simple equations to predict the prestressing tendons at ultimate stage of continuous concrete beams. The simple equations were verified by comparing with existing experimental data. Although much analytical research has been performed to evaluate the behavior of bonded and unbonded posttensioned concrete members, a relatively limited number of studies have been reported for the prediction of flexural behavior considering tendon layouts.

Two-way slab systems offer several possible arrangements for the tendon layout [10]: banded, distributed, or a mixed layout. Posttensioned slabs are used for long spans and heavy live loads, so flexural strength and ductility are important. Flexural strength usually governs the behavior of the interior panel in two-way slabs. In other words, the distribution of tendons can affect the flexural behavior and ductility of the interior panel of the two-way slabs. Though many researchers have focused on the development of finite element models, little information is available on the flexural behavior of posttensioned two-way slabs with different tendon layouts. In this study, examined was the flexural behavior of posttensioned two-way slabs depending on the tendon layout.

The objective of this paper is to present an efficient numerical analysis approach for the materials and a geometric nonlinear analysis for the posttensioned two-way slabs. In this study, developed was a nonlinear finite element model that can simulate the behavior of posttensioned two-way slabs. The reinforced concrete was modeled as combination of concrete, steel, and prestressing tendons. The test results were compared with those from the finite element model as well.

2. Finite Element Model

2.1. Layered Element Formulation. A finite element model formulated using layered and degenerate shell elements can be used in a three-dimensional global analysis of structures. Eight-node isoparametric degenerated shell elements were formulated following the procedure of Hinton and Owen [11]. It was assumed that plane cross-sections remain both plane and normal during bending. Layered elements were applied to account for the behaviors of the reinforced concrete members, which exhibited different properties in the thickness direction because of the placement of the reinforcements. Each element is divided into layers, and each layer has one integration point on its midsurface. Each layer was composed of different materials; concrete, steel reinforcement, and prestressing tendon are defined separately. The strains and stresses are calculated at midpoint of each layer. The strain-displacement matrix and the constitutive matrix are calculated at the midpoint of each layer. Stress resultants are evaluated by integrating the corresponding stress. Normal forces and bending moments can be obtained by

$$N_{x(y)} = \int_{-h/2}^{h/2} \sigma_{x(y)} dz = \frac{h}{2} \sum_{i=1}^{n} \sigma_{x(y)}^{i} \Delta\zeta^{i},$$

$$M_{x(y)(xy)} = -\int_{-\frac{h}{2}}^{\frac{h}{2}} \sigma_{x(y)(xy)} z \, dz \qquad (1)$$

$$= -\frac{h^2}{4} \sum_{i=1}^{n} \sigma_{x(y)(\tau_{xy})}^{l} \zeta^{i} \Delta\zeta^{i},$$

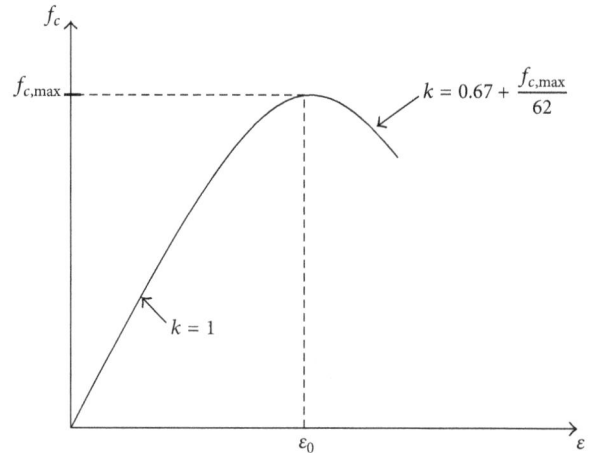

FIGURE 1: Stress-strain relationship in compression by Collins and Porasz (1989).

where N_x is normal force, M_x is bending moment, σ is normal stress, h is layer depth, and n is number of layers.

The reinforcement layers were used to model the in-plane reinforcement. Transverse reinforcement can be specified as a property of a concrete layer. The prestressing tendons at a single depth were grouped with the same prestressing force into one layer [12]. The concrete, steel reinforcement, and prestressing tendons were assumed to be perfectly bonded. The perfect bond is applicable to the analysis of reinforced concrete and posttensioned concrete with bonded tendons. The same degrees of freedom were assigned to concrete and reinforcement nodes occupying a single location.

2.2. Constitutive Models. In this paper, the failure of concrete two-way slabs is considered to be tension cracking or plastic yielding of reinforcement. Uncracked concrete was assumed to be a linear elastic material. After cracking, the concrete was treated as an orthotropic material. The total material matrix consists of concrete, steel reinforcement, and prestressing tendons.

Figure 1 shows the stress-strain relationship for concrete in compression [13]. The compressive stress of concrete can be calculated by

$$f_c = \frac{f_{c2,\max}\left(n\varepsilon/\varepsilon_c'\right)}{(n-1) + \left(\varepsilon/\varepsilon_c'\right)^{nk}},$$

$$f_{c,\max} = \frac{f_c'}{0.8 - 0.34\left(\varepsilon_c/\varepsilon_c'\right)} \le f_c', \qquad (2)$$

$$n = 0.8 + \frac{f_{c2,\max}}{17},$$

where f_c is the concrete stress; $f_{c2,\max}$ is the compressive stress of cracked concrete; ε_c is the concrete strain; ε_c' is the concrete strain corresponding to peak compressive stress; and f_c' is the compressive cylinder strength of concrete.

After cracking, the stiffness of the reinforced concrete decreases, but it does not drop to zero because the intact

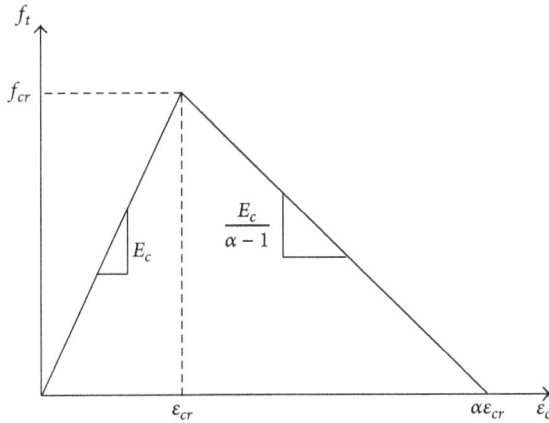

FIGURE 2: Average stress-strain relationship proposed by Lin and Scordelis (1975).

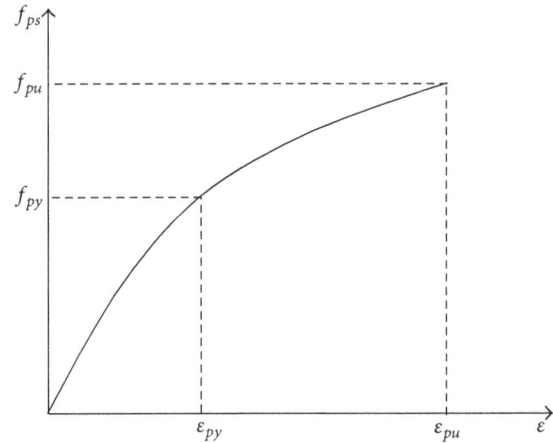

FIGURE 4: Constitutive model of prestressing tendon proposed by Menegotto and Pinto (1973).

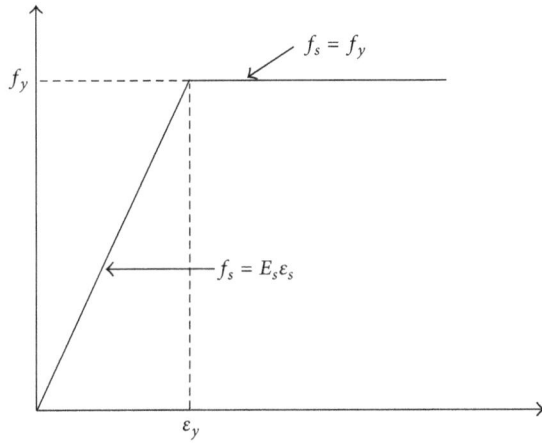

FIGURE 3: Stress-strain relationship of steel reinforcement.

concrete between adjacent cracks can still carry some tensile stress due to the bond between the reinforcement and surrounding concrete. In this paper, tension stiffening was modeled as a constitutive model [14], as shown in Figure 2.

The steel reinforcement was considered as steel layers with uniaxial behavior. A bilinear model was adopted for the elastoplastic stress-strain relationship, as shown in Figure 3. The stress-strain relationship was characterized by Young's modulus E_s and the uniaxial yield stress f_y.

For the prestressing tendon, Menegotto and Pinto's model [15] was adopted (Figure 4). The mathematical expression is given as follows:

$$\sigma_p = E_p \varepsilon_P \left[Q + \frac{1-Q}{\left[1 + \left(E_p \varepsilon_p / K f_{py} \right)^N \right]^{1/N}} \right],$$

$$Q = \frac{f_{pu} - K f_{py}}{E_p \varepsilon_{pu} - K f_{py}},$$

where σ_p is the stress of a prestressing tendon; E_p is Young's modulus of the prestressing tendon; ε_p is the strain of the

prestressing tendon; f_{py} is the yield stress; f_{pu} is the ultimate yield stress; ε_{cu} is the ultimate strain; and N, K, and Q are empirical parameters whose values were recommended by Naaman [16] as 6.06, 1.0325, and 0.00625, respectively.

2.3. Analysis Procedures. Direct method is applied as solution algorithm [11]. In each iterative step, the full load is applied to the structures. The obtained unknowns are the full displacements. In the first iteration, the materials have linearly elastic behavior and the initial displacements are zero. After that, the model can calculate the new stiffness matrix considering the appropriate material constitutive models. Full load is reapplied to the model, then the stiffness matrix is updated, and displacement can be found. The steps of evaluation and update of stiffness matrix are repeated until the satisfied convergence condition.

The displacement criterion was selected as the convergence condition. The displacement convergence is as follows:

$$\sqrt{\frac{\Sigma \left(D_a - D_p \right)^2}{\Sigma D_a^2}} \times 100\% < T, \tag{4}$$

where D_a is current step displacement, D_p is previous step displacement, and T is tolerance.

Large tolerance value can lead to inaccurate results, and the tolerance was set to 0.8.

3. Experimental Program

To validate the suggested finite element model, flexural testing was performed. The designed compressive strength of the concrete used for the fabrication of the specimens was 35 MPa. The average compressive strength measured at 28 days was 36.7 MPa. Deformed steel bars with a diameter of 13 mm and 10 mm were used for longitudinal reinforcements and stirrups, respectively. Their tensile strength and modulus of elasticity were 400 MPa and 200 GPa, respectively. Seven wire steel-strand tendons with 12.7 mm of diameter were used. Their nominal ultimate tensile strength was 1860 MPa.

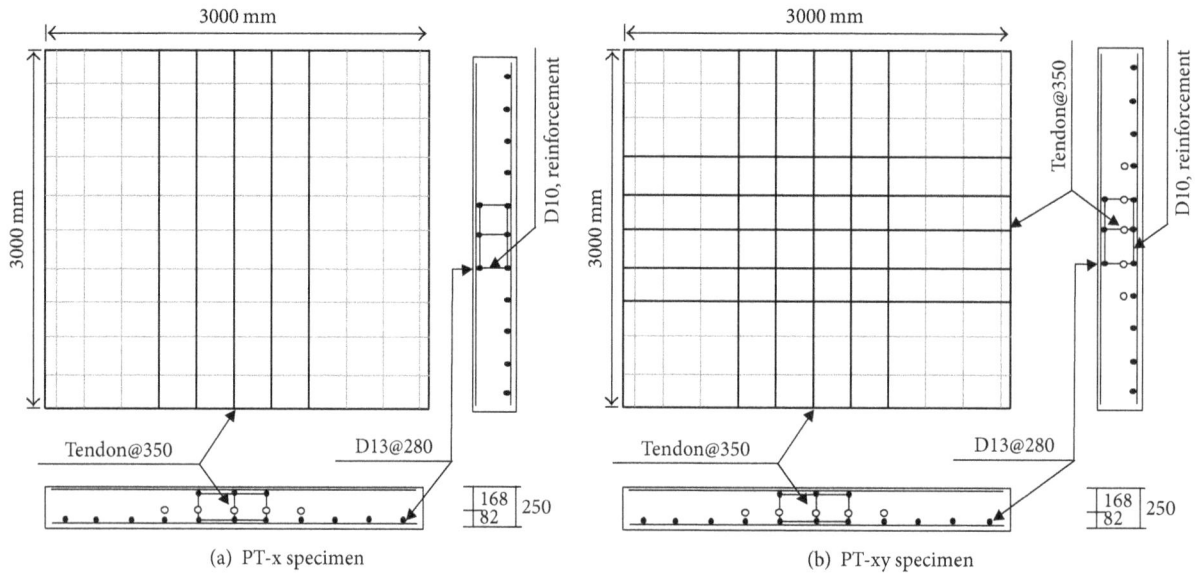

FIGURE 5: Details of the specimens.

TABLE 1: Details of specimens.

Specimen	$A_{s,x}$ (mm^2)	$A_{s,y}$ (mm^2)	f_{pe} (MPa)	f_{pe}/f_{pu}	$\rho_{p,x}$ (%)	$\rho_{p,y}$ (%)	$\rho_{s,x}$ (%)	$\rho_{s,y}$ (%)	d_p (mm)	f_c' (MPa)
PT-x	493.5	—	1488	0.8	0.198	—	0.185	0.185	168	36.7
PT-xy	493.5	493.5	1488	0.8	0.198	0.198	0.185	0.185	168	

Two posttensioned two-way slabs were manufactured for the test. One specimen had tendons distributed in only the x-direction (PT-x), and the other had tendons distributed in both the x- and y-directions (PT-xy). Details of the specimens are presented in Figure 5 and summarized in Table 1. The size of the test specimens was 3000 mm × 3000 mm with 250 mm thickness. Both specimens were posttensioned with a constant eccentricity of 43 mm. A prestressing force of 1488 MPa was applied corresponding to approximately 80% of the tensile strength of the tendon.

Load was applied to each specimen using a hydraulic jack with maximum capacity of 5000 kN. The test specimens were simply supported along the four sides. The force generated by the hydraulic jack was transmitted to a loading plate placed at the middle of the specimen. The distance from support to loading point was 1.25 m, giving a shear span to depth ratio of 6.

4. Numerical Modeling and Discussion

In order to evaluate the accuracy of the analytical model, the theoretical value is compared with test results. Comparison is made in terms of load-deflections curves. The analytical model size was set to match the posttensioned specimens, and the model contained 100 elements. It was divided into nine layers in the direction of different thickness. The thicknesses of the first and last layers were determined by considering the concrete cover. The steel reinforcements were placed at the 2nd and 8th layers in the direction of the thickness. The prestressing tendons were placed at the 6th layer in the

direction of the thickness. In this paper, Young's modulus of concrete (E_c) was determined according to ACI 318 [17]. Poisson's coefficient (ν) is 0.15, and the equation to compute the modulus of rupture of concrete is $0.32\sqrt{f_c'}$ (Table 2). To simulate the experimental support conditions, four sides of slabs are simply supported and the load is applied to the center of the slab.

The model values were compared with load-displacement curves obtained from the posttensioned two-way slabs under flexural loads. The comparisons between the test results and the analytical results are shown in Figures 6 and 7. As shown in the Figures, both test and analytical results indicated that the load-displacement curves exhibit three stages: elastic, cracking, and plastic. Overall, the analytical model predicted the deflection of the posttensioned specimens in a relatively accurate manner. However, in all cases the models show a slightly stiffer response at the cracking stage because prestressing loss and slip were not considered in the finite element analysis. Both the testing and analytical results showed that the one-way and two-way prestressing tendon layout did not significantly affect the maximum load capacity and deflection.

The validated finite element model was used to investigate the effects of changes in span length, height, and concrete strength. In total, nine posttensioned slabs were analyzed as shown in Table 3. To perform the parametric analysis, the same geometry and material properties were used same as in the verification of the proposed model, along with the one-way tendon layout. The same load to each model was

TABLE 2: Material properties used in the finite element analysis.

Concrete					Steel and prestressing tendon		
f_c' (MPa)	f_t (MPa)	E_c (MPa)	ε_{cr}	ν_c	f_y (MPa)	E_s (GPa)	ν_s
36.7	3.8	28472.9	0.002	0.15	360	210	0.25

TABLE 3: Analytical dimensions and results of parametric study.

Model	Span (mm)	Height (mm)	Compressive strength of concrete (MPa)	Displacement (mm)
PT1	3000	250	35	41.77
PT2	4500	250	35	48.54
PT3	6000	250	35	56.13
PT4	7500	250	35	67.20
PT5	3000	180	35	43.62
PT6	3000	350	35	40.14
PT7	3000	450	35	37.32
PT8	3000	250	50	40.33
PT9	3000	250	65	38.16

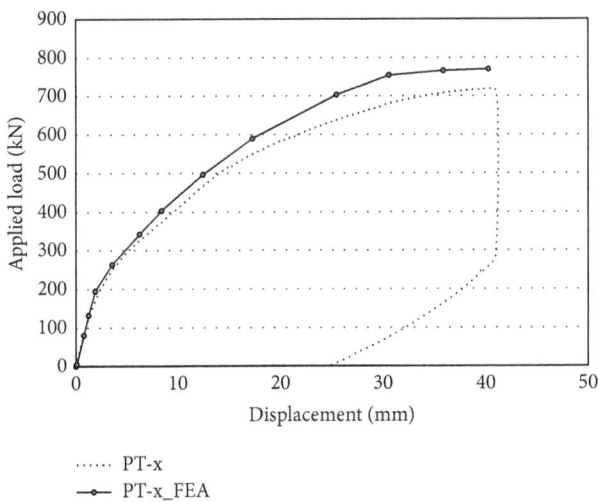

FIGURE 6: Load-displacement relations of PT-x specimen.

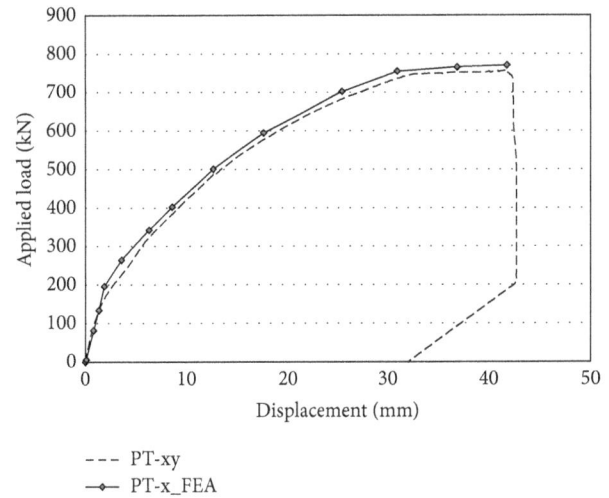

FIGURE 7: Load-displacement relations of PT-xy specimen.

applied and compared the results in terms of deflections. Figure 8–10 show the displacement according to span length, slab height, and the compressive strength of the concrete. The displacement increased with the span length and decreased as the height and concrete strength increased. The results of the parametric analysis indicate that the deflection of posttensioned two-way slab is more affected by the span length than by the other variables.

5. Conclusions

In this paper, the flexural behavior of posttensioned concrete two-way slab was analytically investigated. A finite element analysis model was proposed to predict the flexural behavior of specimens depending on the tendon layout and conducted flexural tests to evaluate the validity and applicability of the

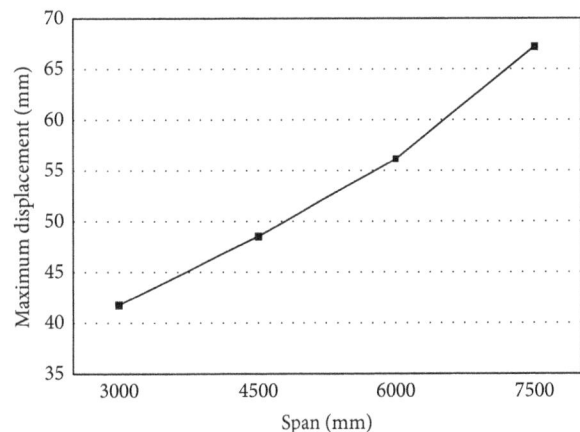

FIGURE 8: Effect of span length on displacement.

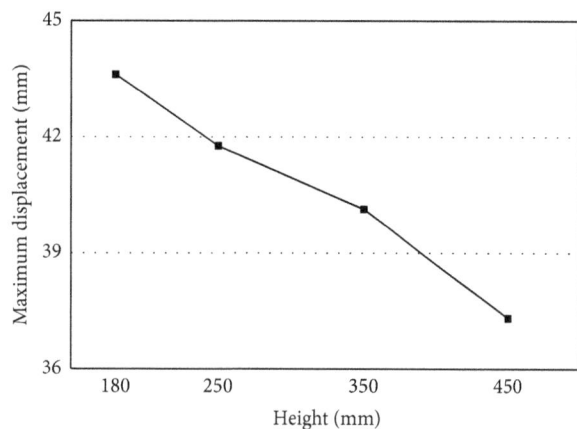

FIGURE 9: Effect of height of slabs on displacement.

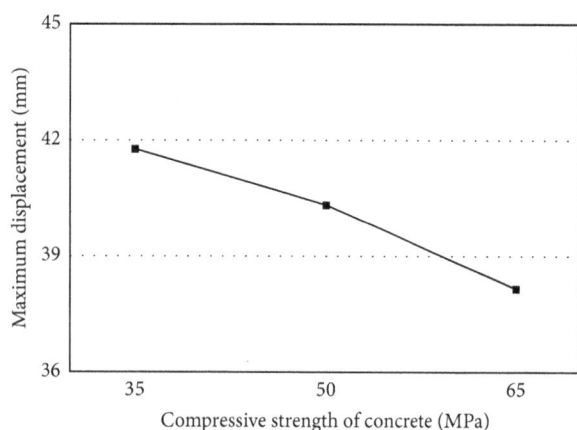

FIGURE 10: Effect of compressive strength of concrete on displacement.

the rationality of the posttensioning two-way slabs model.

(3) To investigate the effects of span length, slab height, and concrete strength on posttensioned two-way slabs, a parametric study was conducted. The displacement increased with the span length and decreased as the concrete strength and member height increased. The span length, height, and concrete strength all contribute to the flexural strength.

Competing Interests

The authors declare that they have no competing interests.

Acknowledgments

This work was supported by a National Research Foundation of Korea (NRF) grant funded by the Korean Government (MSIP) (NRF-2013R1A2A2A01067754).

References

[1] N. H. Burns and R. Hemakom, "Test of post-tensioned flat plate with banded tendons," *Journal of Structural Engineering*, vol. 111, no. 9, pp. 1899–1915, 1985.

[2] G. M. Kosut, N. H. Burns, and C. V. Winter, "Test of four-panel post-tensioned flat plate," *Journal of Structural Engineering*, vol. 111, no. 9, pp. 1916–1929, 1985.

[3] P. N. Roschke and M. Inoue, "Effects of banded post-tensioning in prestressed concrete flat slab," *Journal of Structural Engineering*, vol. 117, no. 2, pp. 563–583, 1991.

[4] J. Van Greunen and A. C. Scordelis, "Nonlinear analysis of prestressed concrete slabs," *Journal of Structural Engineering*, vol. 109, no. 7, pp. 1742–1760, 1983.

[5] X.-H. Wu, S. Otani, and H. Shiohara, "Tendon model for nonlinear analysis of prestressed concrete structures," *Journal of Structural Engineering*, vol. 127, no. 4, pp. 398–405, 2001.

[6] N. El-Mezaini and E. Çitipitioğlu, "Finite element analysis of prestressed and reinforced concrete structures," *Journal of Structural Engineering*, vol. 117, no. 10, pp. 2851–2864, 1991.

[7] T. H. K. Kang and Y. Huang, "Nonlinear finite element analyses of unbonded post-tensioned slab-column connections," *PTI Journal*, vol. 8, no. 1, pp. 4–19, 2012.

[8] T. H.-K. Kang, Y. Huang, M. Shin, J. D. Lee, and A. S. Cho, "Experimental and numerical assessment of bonded and unbonded post-tensioned concrete members," *ACI Structural Journal*, vol. 112, no. 6, pp. 735–748, 2015.

[9] A. Ghallab, "Calculating ultimate tendon stress in externally prestressed continuous concrete beams using simplified formulas," *Engineering Structures*, vol. 46, pp. 417–430, 2013.

[10] B. O. Alami, "Layout of post-tensioning and passive reinforcement in floor slabs," PTI Technical Notes 8, 1999.

[11] E. Hinton and D. R. J. Owen, *Finite Element Software for Plates and Shells*, Pineridge Press, Swansea, UK, 1984.

[12] J. Chern, C. You, and Z. P. Bazant, "Deformation of progressively cracking partially prestressed concrete beams," *PCI Journal*, vol. 37, no. 1, pp. 74–85, 1992.

[13] M. P. Collins and A. Porasz, "Shear design for high strength concrete," in *Proceeding of the Workshop on Design Aspects of High Strength Concrete*, pp. 75–83, 1989.

proposed model. The following specific conclusions were drawn from this study:

(1) A finite element analysis model has been developed to evaluate the flexural behavior of posttensioned two-way slabs considering tendon layout. The proposed finite element model, which considers the nonlinear behavior of concrete and reinforcement and neglects the bond slip and loss of prestressing force, gives relatively good predictions for the load-deflection curves.

(2) The increase in the maximum load capacity was unaffected by tendon layouts. Regardless of the direction of the tendon, the load-displacement curves indicated similar responses. The analytical response is slightly stiffer than the test results at the cracking stages, possibly because of bond slip and the loss of prestressing force. Consideration of bond slip and the loss of prestressing force could improve the model accuracy. However, the difference in deflection between the analytical and test results is relatively small. The proposed finite element analysis demonstrates

[14] C.-S. Lin and A. C. Scordelis, "Nonlinear analysis of rc shells of general form," *Journal of the Structural Division*, vol. 101, no. 3, pp. 523–538, 1975.

[15] M. Menegotto and P. E. Pinto, "Method of analysis for cyclically loaded R. C. Plane frames, including changes in geometry and non-elastic behavior of elements under combined normal force and bending," in *Proceedings of the IABSE Symposium on Resistance and Ultimate Deformability of Structures Acted on by Well-Defined Repeated Loads*, pp. 15–22, Lisbon, Portugal, 1973.

[16] A. E. Naaman, "A new methodology for the analysis of beams prestressed with external or unbonded tendons," External Prestressing in Bridges ACI SP120-16, American Concrete Institute, Detroit, Mich, USA, 1990.

[17] American Concrete Institute (ACI), "Building code requirement for reinforced concrete and commentary," ACI 318-14, American Concrete Institute (ACI), Farmington Hills, Mich, USA, 2014.

15

Experimental Investigation of the Effects of Concrete Alkalinity on Tensile Properties of Preheated Structural GFRP Rebar

Hwasung Roh,[1] Cheolwoo Park,[2] and Do Young Moon[3]

[1]Department of Civil Engineering, Chonbuk National University, Jeonju 561-756, Republic of Korea
[2]Department of Civil Engineering, Kangwon National University, Samcheok 245-711, Republic of Korea
[3]Department of Civil Engineering, Kyungsung University, Busan 608-736, Republic of Korea

Correspondence should be addressed to Do Young Moon; dymoon@ks.ac.kr

Academic Editor: Jun Liu

The combined effects of preexposure to high temperature and alkalinity on the tensile performance of structural GFRP reinforcing bars are experimentally investigated. A total of 105 GFRP bar specimens are preexposed to high temperature between 120°C and 200°C and then immersed into pH of 12.6 alkaline solution for 100, 300, and 660 days. From the test results, the elastic modulus obtained at 300 immersion days is almost the same as those of 660 immersion days. For all alkali immersion days considered in the test, the preheated specimens provide slightly lower elastic modulus than the unpreheated specimens, showing only 8% maximum difference. The tensile strength decreases for all testing cases as the increase of the alkaline immersing time, regardless of the prehearing levels. The tensile strength of the preheated specimens is about 90% of the unpreheated specimen for 300 alkali immersion days. However, after 300 alkali immersion days the tensile strengths are almost identical to each other. Such results indicate that the tensile strength and elastic modulus of the structural GFRP reinforcing bars are closely related to alkali immersion days, not much related to the preheating levels. The specimens show a typical tensile failure around the preheated location.

1. Introduction

Numerous studies have highlighted the tensile properties and performance of GFRP bars under high temperature conditions considering fire accident. Their experimental tests on the GFRP bars show that the temperature above the glass transition temperature causes deterioration in the tensile properties of the bars due to the weakening of the resin and the resin-fiber interface [1–8]. In the experiment done by Wang and Kodur [4], the tensile strength retention of the GFRP rebar is only 58% in the case of 200°C exposed temperature level. The test is conducted using the tensile loading machine which is equipped with electric furnace controlling specified high temperatures. Such type of test is called "hot tension test." Another testing type is "postheating tension test" [9–12]. In postheating tension tests the GFRP bars are firstly exposed to temperatures between 100°C and 400°C for 0.5 to 3 hours. After heating, the bars are cooled down to room temperature level and their tensile strength is evaluated. From the studies adopting the postheating tension,

the tensile strength is reduced proportionally to the exposure temperature level. Kumahara et al. [1] compared the results obtained from hot and postheating tension tests, showing that the GFRP reinforcing bars are almost recovered to their original tensile strength until 150°C of exposed temperature level. However, for the exposed temperatures level between 150°C and 250°C, the strength is recovered by about 80% of their original tensile strength. Over 400°C of exposed temperature level, the recovering capacity is almost lost. It indicates that the recovery of tensile properties of GFRP rebar closely depends on the exposed temperature level and cooling process.

Another issue on the GFRP reinforcing bars is a concrete alkaline effect. From the previous studies considering "accelerated aging test" for this issue [13–16], the bars embedded in mortar or immersed in alkaline solutions have a significant deterioration in tensile strength. Cracks and damage at the fiber-resin interface and also at the glass fiber resulting from alkaline fluid infiltration are observed by Scanning Electron Microscopy (SEM) analysis. All the previous studies

TABLE 1: Outline of experimental program.

	Preheated	Immerged in 40°C alkaline solution			
		For 0 days	For 100 days	For 300 days	For 660 days
Ref	35ea at room temp.	10ea	10ea	10ea	5ea
PH_120	35ea at 120°C	10ea	10ea	10ea	5ea
PH_200	35ea at 200°C	10ea	10ea	10ea	5ea
Sum	105ea	105ea			

mentioned above investigate the performance of GFRP reinforcing bars when they are exposed to either high temperature or concrete alkalinity, without considering the combined situation.

Few or limited studies have been conducted on the combined effects of these two conditions. One such study is conducted by Abbasi and Hogg [2]. In their study, the GFRP bars are firstly immersed for different durations in alkaline solutions at 60°C. Afterwards, the bars are placed in a heating chamber, exposed to temperatures between 80°C and 120°C, and subsequently subjected to tensile forces until failure occurs at specified temperatures. The results show that the GFRP rebar preexposed to alkaline solution has lower retention than the rebar not immersed in alkaline solution, even if both are exposed to identical temperatures. Such reduction is particularly significant at the highest considered temperature of 120°C.

In the present study, unlike the conditioning sequence adopted by Abbasi and Hogg [2], the GFRP specimens are firstly exposed to high temperature and then immersed in alkaline solution. Such scenario is to consider a postfire condition of RC structures consisting of GFRP reinforcing bars. Accelerated aging tests are performed on the preheated or thermally damaged GFRP rebars in order to investigate the effect of concrete alkaline solution on the reduction of the tensile strength and stiffness of the GFRP bars. In the test, the aging days of the GFRP rebars in alkali solution are 100 days, 300 days, and 660 days, which are normal immersion periods considered in the accelerated aging test. Also, two levels of preheating temperature conditions (120°C, 200°C) are considered as similar to the temperature ranges adopted in the previous "postheating tension test" [9–12].

2. Experimental Test Program

2.1. GFRP Bar Specimens.
The diameter of the GFRP bars is 9.5 mm and their length is 1.2 m. The bars used in the present tests are manufactured in South Korea, where they are used for strengthening concrete structures. The bars are made of E-glass/vinyl-ester and the fiber volume ratio is 65%. A total of 105 specimens shown in Table 1 are prepared to be mounted in the tensile testing machine.

2.2. Testing Conditions and Setup.
The GFRP bar specimens are preheated at two temperature levels: 120°C and 200°C. Specimens preheated at these two temperature levels are hereafter referred to as "PH_120" and "PH_200," respectively. Details of the preheating procedure are described in the following subsection. The glass transition temperature of the

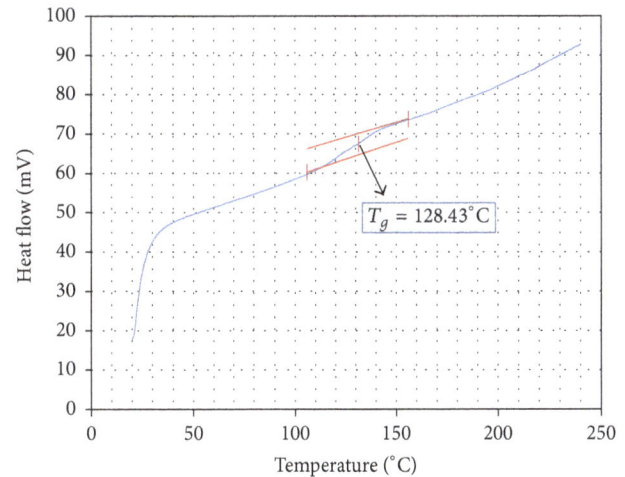

FIGURE 1: Differential Scanning Calorimetry (DSC) analysis result.

resin contained in the GFRP reinforcing bars is identified through Differential Scanning Calorimetry (DSC) analysis and it is found to be about 128°C as shown in Figure 1. The preheating temperature is set to be slightly below this value (120°C) for the 35 GFRP bar specimens and far above it (200°C) for the other 35 specimens. Unpreheated specimens are also prepared and tested for the comparison purposes, and these specimens are named "Ref" specimens. The total number of the Ref specimens is also 35. Strong alkaline solution having a pH of 12.6 is considered to represent the alkalinity of concrete. Herein, the present study assumes that the damaged parts of concrete due to fire event are replaced with new concrete; thus the GFRP bars are also assumed to be placed in the same concrete environment before and after fire event. The temperature of the alkaline solution is maintained with 40°C. Details of this treatment are described in Section 2.4. As summarized in Table 1, the specimens are subjected to tensile tests just after being immersed for different time durations such as "100 days," "300 days," and "660 days." The condition of zero days (i.e., "0 days" in Table 1) means that the specimens are subjected to tensile tests after exposure to 120°C or 200°C without immersing in alkaline solution. All preheated rebar specimens are cooled down for 3 hours before conducting the tensile test.

2.3. Preexposure to High Temperature: Preheating.
Heating tape which is able to raise the temperature up to 250°C is used to preheat the specimens. As shown in Figure 2, the heating tape is wrapped around the center (±30 mm from the center) of the specimens and connected to a temperature

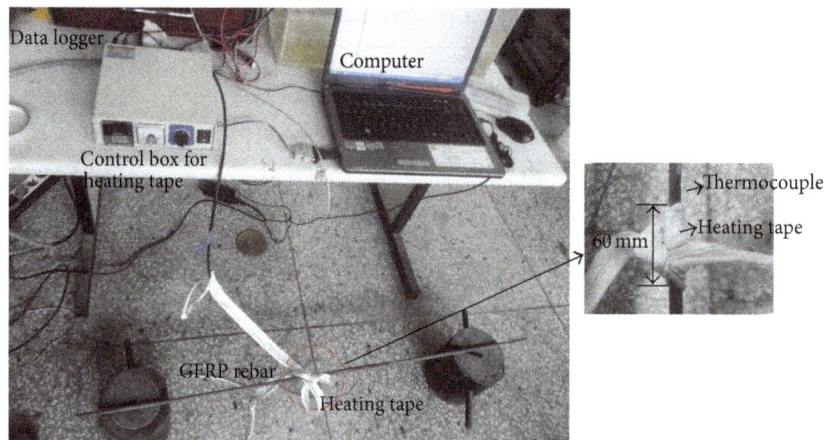

FIGURE 2: Preheating of GFRP reinforcement bars.

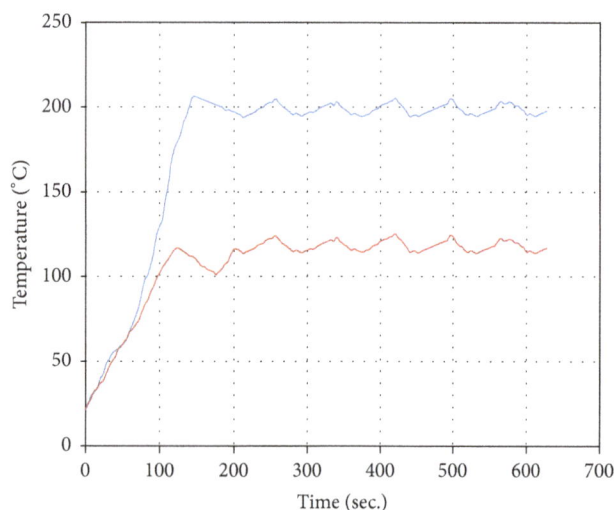

FIGURE 3: Temperature time histories for preheated specimens.

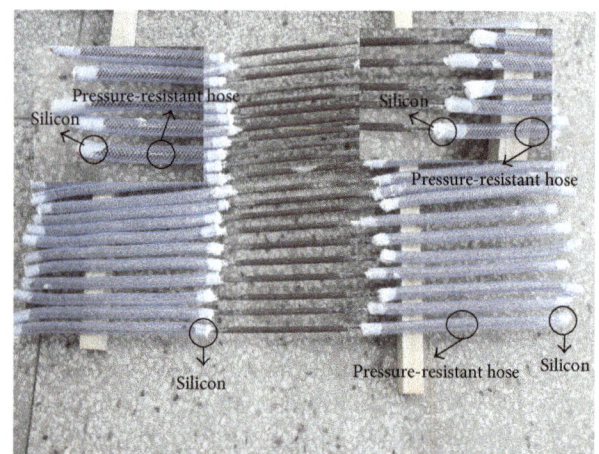

FIGURE 4: Protection of specimens anchoring parts.

control box to specify the temperature levels. A thermocouple is taped to the surface of the bars and connected to a data logger and computer in order to monitor the temperature of the specimens. The specified maximum temperature is set to 120°C and 200°C, and the heating is applied. After reaching the specified temperature levels the heating tape is removed. This heating procedure is repeated to ensure the same amount and extent of thermal conditioning. Figure 3 shows one example of the temperature history measured on the surface of the specimens. The exposure time to high temperatures is selected based on the reference of Katz et al. [17] which shows that after 10 minutes of exposure time under 120°C~200°C the adhesive strength between concrete and GFRP rebars with the same diameter considered in the present study is significantly decreased due to the slip behavior.

2.4. Immersion in 40°C Alkaline Solution. PH_120 and PH_200 bar specimens, whose length is 1.2 m, are fully immersed in a large plastic cistern containing strong alkaline solution for the predetermined time periods indicated in

Table 1. During immersing into the alkaline solution, both ends parts of the specimens are protected to prevent damage since the parts are anchored to the tensile testing machine. For this, as shown in Figure 4, a length of 400 mm at each end of the bar specimens is inserted in pressure-resistant hoses, and then alkali-resistant silicone is inserted inside the hoses for a length of 30 mm to prevent infiltration of the alkaline solution. According to previous studies [14, 15, 18], the temperature of the alkaline solution is usually set to about 40°C or 60°C. In the present study, the alkaline solution, which is 1 Mole NaOH solution, is maintained to be 40°C. While the GFRP bar specimens are immersed into the NaOH alkali solution, the plastic cistern is shielded with cover in order to avoid its evaporation. Moreover, a circulator is used to ensure a uniform temperature distribution of the NaOH solution inside the cistern. The alkalinity of the solution is measured every morning using a pH-meter and maintained to be the designed pH. Figure 5 shows the GFRP bar specimens immersed into the NaOH solution.

2.5. Tensile Tests and Testing Equipment. The bar specimens are anchored to the tensile test machine according to the

TABLE 2: Tensile strength and elastic modulus.

Specimen ID	Mean (MPa)				SD (MPa)				CoV (%)			
	0 days	100 days	300 days	660 days	0 days	100 days	300 days	660 days	0 days	100 days	300 days	660 days
					Tensile strength							
Ref	644.7	574.8	183.0	114.9	24.3	47.7	8.7	29.3	3.8	8.3	4.7	25.5
PH_120	643.7	582.0	162.8	125.6	38.4	60.2	29.9	23.9	6.0	10.3	18.4	31.9
PH_200	612.8	503.7	159.6	120.7	27.6	47.7	18.0	17.1	4.5	9.5	11.3	14.1
					Elastic modulus							
Ref	53.4	52.0	17.6	16.4	6.8	3.9	1.7	2.7	12.7	7.4	9.8	16.5
PH_120	52.0	51.0	16.2	16.0	3.0	3.2	1.7	2.1	5.7	6.3	10.3	13.1
PH_200	50.2	49.6	16.4	15.2	3.5	3.8	1.3	2.2	6.9	8.0	7.6	14.2

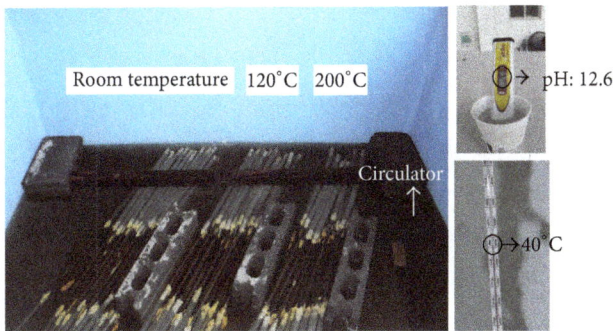

FIGURE 5: Immersion of GFRP rebar specimens into alkaline solution.

standards specified in ASTM D 3916 [19]. The aluminium plate is used to accommodate the specimens as shown in Figure 6. The specimens are inserted into the aluminium plate and anchored with bolts. Universal Testing Machine (UTM) providing 500 kN maximum loading capacity is used for the tensile tests. The tests are performed under displacement control and the applied loading rate is 5 mm/min. A Linear Variable Displacement Transducer (LVDT) is attached to the bar specimens to measure the axial elongation during loading.

2.6. Calculation of Tensile Strength and Elastic Modulus. The tensile strength (f_u) and elastic modulus (E_L) are calculated using (1) and (2), respectively. The tensile strength is calculated by dividing the ultimate load (P_u) by the cross section area (A) and the elastic modulus is calculated based on the test methods for fiber reinforced polymers described in ACI 440.3R-04 specifications [20]. The parameters P_1 and P_2 are the axial loads corresponding to 25% and 50% of the ultimate load (P_u), respectively. The parameters ε_1 and ε_2 denote the corresponding strains.

$$f_u = \frac{P_u}{A} \tag{1}$$

$$E_L = \frac{P_1 - P_2}{(\varepsilon_1 - \varepsilon_2) A}. \tag{2}$$

2.7. Preparation of Optical Microscopic Analysis. Optical microscopy images are acquired using the Leica DM-750M microscope to examine the local damage to the specimens due to preheating and alkaline solution. An additional set of 100 mm long GFRP rebar specimens are prepared for this analysis; these specimens are preheated and immersed in alkaline solution for 30 days. The preheating levels are the exactly same levels considered in the present tensile test. The center of the specimens is cut into short pieces (10 mm long) by using precision cutter and the short pieces are cold-mounted in a 30-mm diameter mounting cup by pouring acrylic resin. In order to acquire clear images, the mounted short pieces are subjected to 8-stage polishing process.

3. Experimental Test Results

For the discussion of the test results, the average tensile strength and elastic modulus are considered, which are summarized in Table 2, including their standard deviations and covariances.

3.1. Alkaline Solution Effects without Exposure to High Temperature (Ref Specimens). The results of accelerated aging test for the "Ref" specimen are compared with those reported in previous works [13–16]. As shown in Figure 7, the tensile strength of the GFPR bar specimens is very similar until 100 immersion days, compared to test results of Chen et al. [15]. Also, it can be recognized that the tensile strength of GRFP bars greatly depends on the resin type, the alkali solution temperature, and the conditioning method in the alkali solution (bare bar or mortar-wrapped bar). The polyester resin is found to have a lower alkali resistance than the vinyl-ester and thermoplastic resins. As expected, a more rapid reduction in the tensile strength is observed in the case of immersion in 60°C alkaline solution than in the case of immersion in 40°C alkaline solution. Comparing the present test results with those of Robert et al. [16], in which the rebar is made of the same resin type and immersed into alkaline solution maintained with identical temperature level, the tensile strength of the bare bars is similar to that of the mortar-wrapped bars until 120 immersion days. However, the mortar-wrapped bar retains about 90% of its original tensile strength until 240 immersion days; on the contrary, the tensile strength of the bare bars is drastically decreased to 30% of the original strength until 300 immersion days.

(a)

(b)

(c)

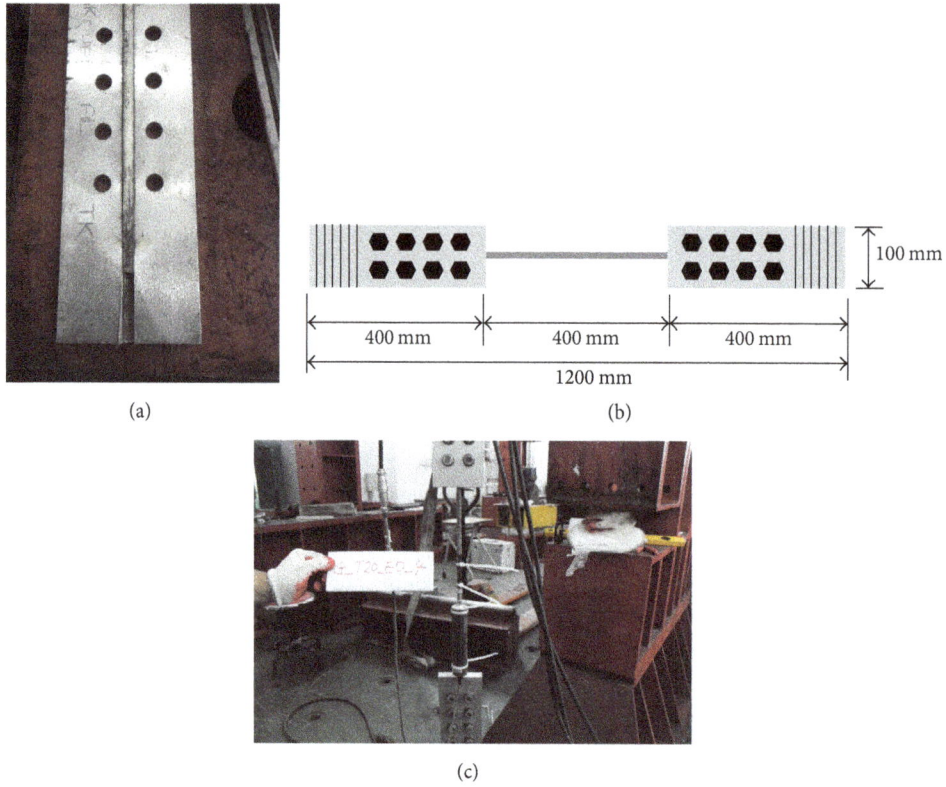

FIGURE 6: Test-setup assembly: (a) gripping fixture, (b) anchoring detail for tensile test, and (c) installation of LVDT.

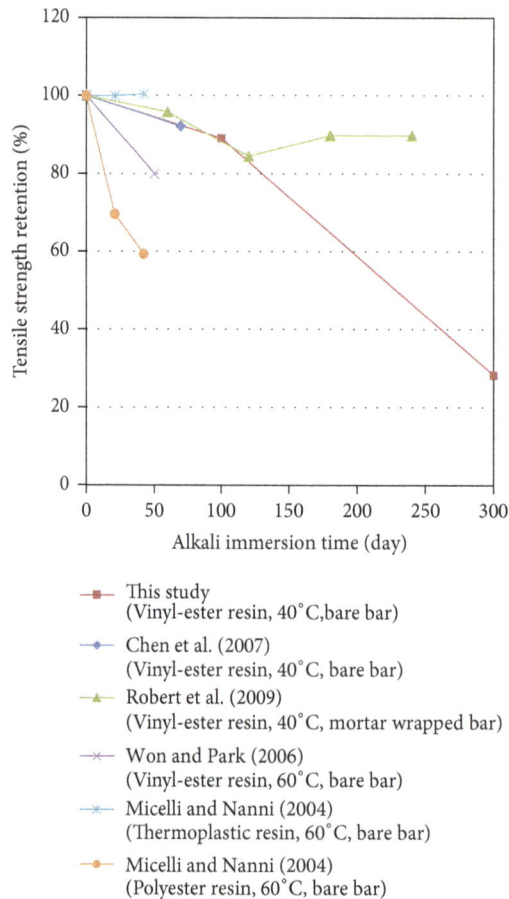

FIGURE 7: Accelerated aging test results of Ref and comparisons with those presented in referred articles.

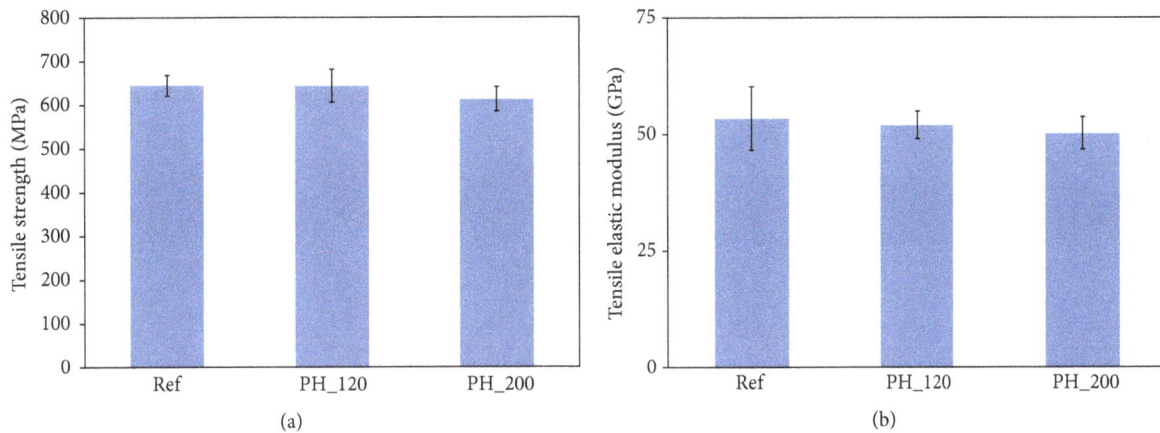

FIGURE 8: Test results for Ref and PH specimens that were not immersed in alkali solution: (a) tensile strength and (b) elastic modulus.

3.2. Alkaline Solution Effects of Preheated GFRP Bar Specimens (PH_120, PH_200). Figure 8 shows the comparison of tensile strength and elastic modulus of the specimens exposed to high temperatures but not exposed to concrete alkalinity (for zero-day case in Table 1). PH_120 and Ref specimens provide almost the same tensile strength and elastic modulus. However, small reduction in the tensile strength is found in PH_200 specimens. The results indicate that the tensile properties that might be degraded upon exposure to high temperatures are recovered fully for PH_120 specimens during cooling down for 3 hours, while recovering about 95% for PH_200 specimens. Such results are similar to those of the postheating tension test performed in the cited previous study [1].

Figure 9 shows the representative tensile stress-strain curves of the specimens. Despite different preheating temperature levels and different immersion days in alkaline solution, the stress-strain relationships are almost linear in all cases. Figures 9(a) and 9(b) show that the stress-strain curves of Ref and PH_120 specimens obtained at zero days and 100 alkali immersion days are very similar. However, as shown in Figure 9(c), the slope of the tensile stress-strain curve of the PH_200 specimens for 100 immersion days shows slightly lower relationship than that of the zero immersion day (preheated only). For longer immersion periods like 300 days and 660 days, the tensile stress-strain curves are much decreased, compared to those of the zero and 100 alkali immersion days. Also, they show almost identical slope. From the test results, it can be recognized that the elastic modulus (slope of the stress-strain curves) of the GFRP bars is closely related to alkali immersion days, not much related to the preheating levels. Regarding the tensile strength shown in Figure 10, the tensile strength decreases for all testing cases as the immersion time to the alkaline solution increases. Also, there is a big drop in the tensile strength until 300 alkali immersion days. However, after 300 alkali immersion days the tensile strength is slightly decreased. More specifically, the specimen Ref and PH_120 exhibit almost the same tensile strength for all immersion days considered here, while PH_200 specimens provide much less tensile strength than those of Ref and PH_120 specimens until 100 alkali immersion

days. After 300 alkali immersion days, however, their tensile strength is almost the same as the other cases.

In order to highlight the relative effects of preexposure to high temperature, the tensile strength and elastic modulus of PH_120 and PH_200 specimens are normalized with the corresponding properties of the Ref specimens and the results are plotted in Figures 11(a) and 11(b). As shown in Figure 11(a), the tensile strength of the preheated specimens is about 90% of Ref specimen tensile strength for 300 alkali immersion days. However, for the specimens tested after 660 immersion days, the tensile strength of the preheated specimens is slightly higher than that of the Ref specimens. Such results indicate that the tensile strength is much affected by the preheating until 300 alkali immersion days, but for longer immersion days the alkaline exposure provides more effect on the tensile strength than the preheating. Figure 11(b) shows that the preheated specimens provide slightly lower elastic modulus than Ref specimens, presenting only 8% maximum difference.

3.3. Failure of Test Specimens. Failure of the specimens without immersing into the alkaline solution is shown in Figure 12. The specimens exhibit a typical tensile failure or rupture induced by tensile forces. When the specimens are ruptured, the delamination of the glass fiber strands is observed in almost the same locations where the heating is applied. The failure of the specimens initiates on the resin matrix at the preheating location and followed through the fiber splitting on the outer layer of the specimens. The fibers are completely ruptured at the preheating location at the end of the testing. Figure 13 shows the tensile failure of the specimens immersed in the alkaline solution for 300 days. The specimens are ruptured at their center near to the preheating location. The delamination or separation of the glass fiber strands is developed with a lesser extent (i.e., smaller area), compared to the rupture length of the specimens preheated only. The lesser extent of separation is developed possibly due to the uneven distribution of the tensile force applied to the specimens. This uneven distribution of the tensile force might be caused by the damage of the resin matrix resulting from the long immersion days in alkaline solution.

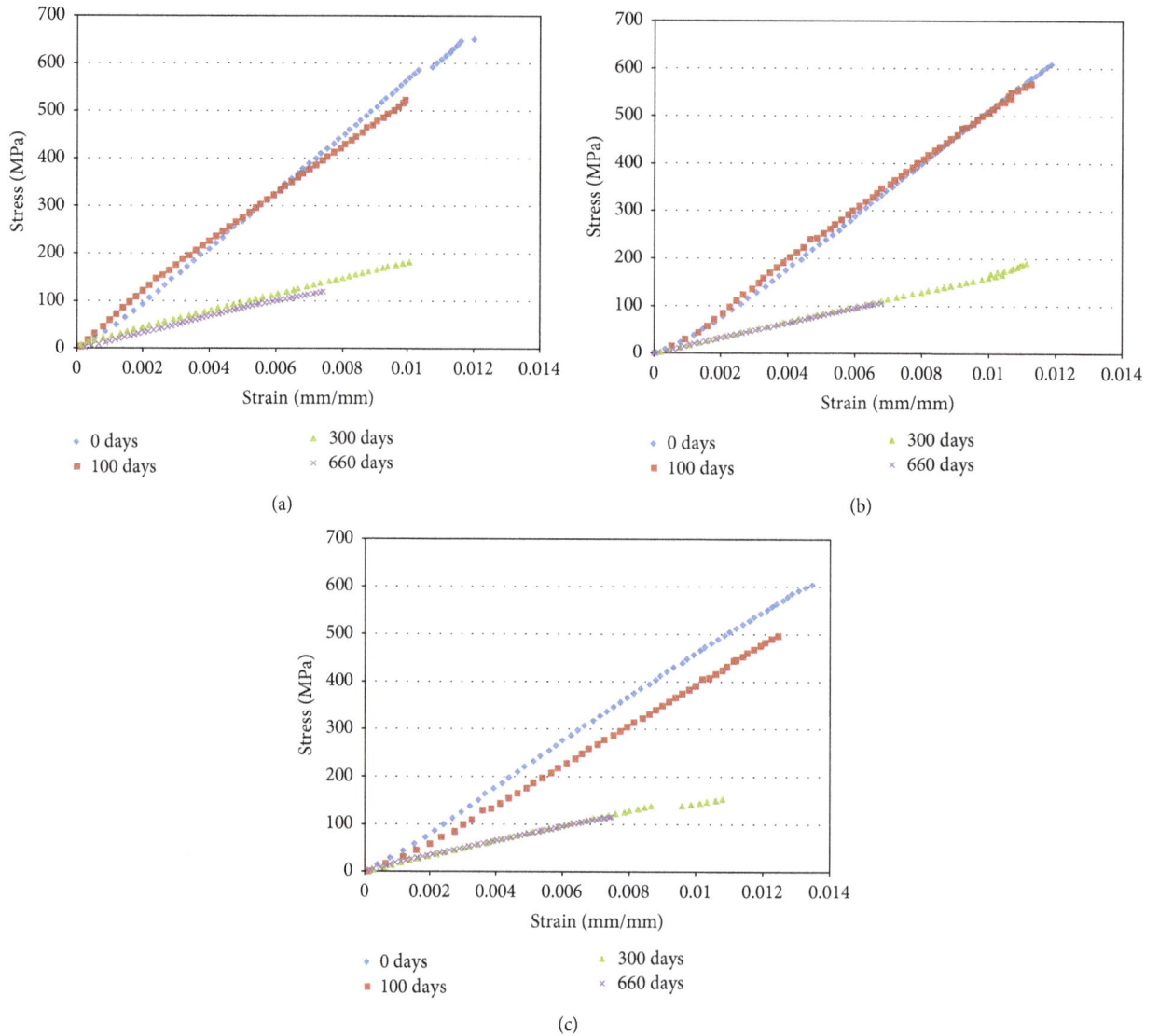

FIGURE 9: Results of tensile stress-strain curve from experimental test: (a) Ref specimen, (b) PH_120 specimen, and (c) PH_200 specimen.

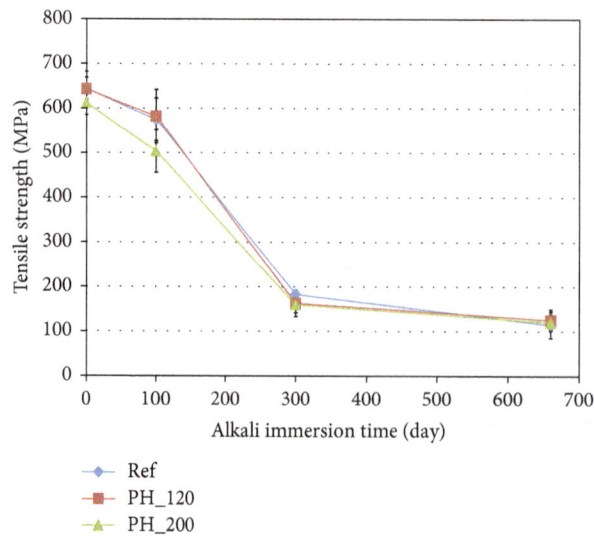

FIGURE 10: Tensile strength reduction according to alkali immersion time.

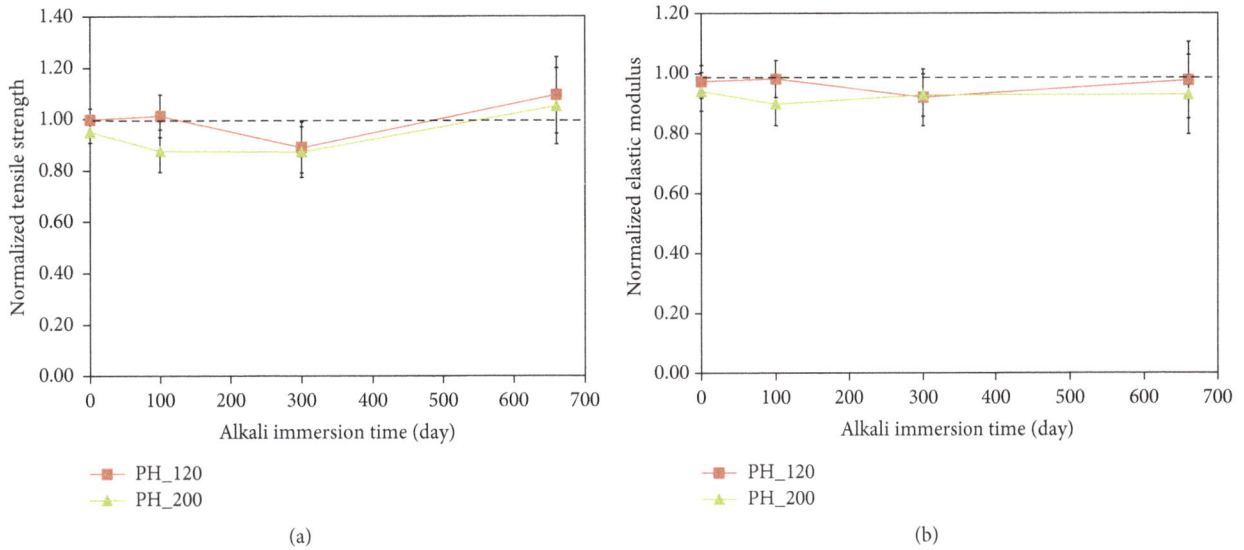

(a)

(b)

FIGURE 11: Comparison of tensile properties: (a) tensile strength and (b) elastic modulus (normalized with Ref specimens).

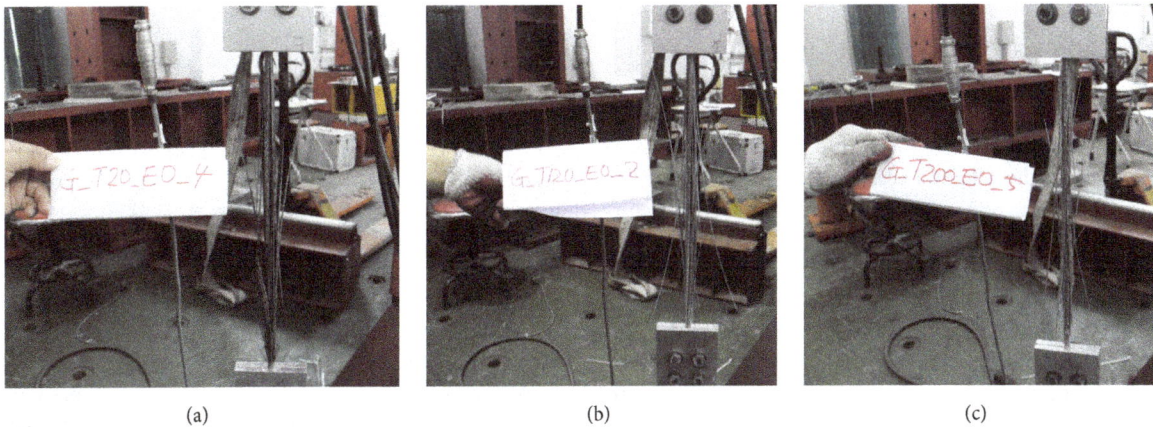

(a)   (b)   (c)

FIGURE 12: Failure of testing specimens without alkaline solution immersion: (a) exposed to room temperature, (b) exposed to 120°C, and (c) exposed to 200°C.

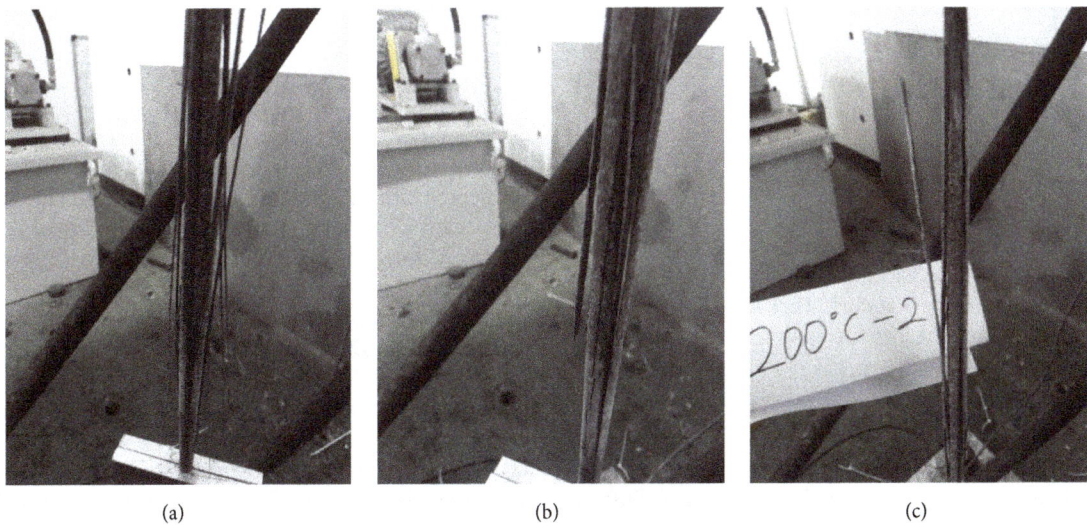

(a)   (b)   (c)

FIGURE 13: Failure of specimens immersed into alkaline solution for 300 days: (a) exposed to room temperature, (b) exposed to 120°C, and (c) exposed to 200°C.

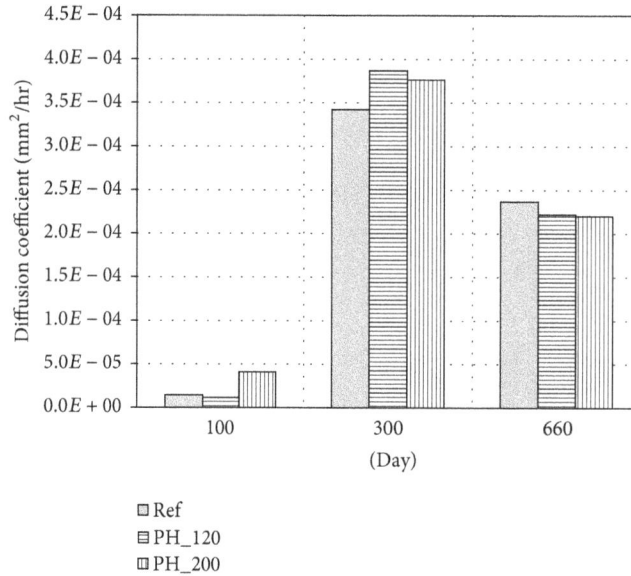

FIGURE 14: Alkaline solution diffusion coefficient results.

4. Diffusion Coefficient of Alkaline Solution

In severe alkaline concrete environment, the mechanical performance degradation of the GFRP reinforcing bars is closely related to the alkaline solution diffusion coefficient [21–26]. Katsuki and Uomoto [21] suggest the relationship between the tensile strength variation and the alkaline diffusion coefficient of FRP reinforcing bars. The relationship is shown in (3) and (4) which are based on Fick's first law. Putting the tensile strength retention obtained from the test into the left side of (3), the parameter x is determined. Herein, the parameter R_0 is the radius of FRP rebar. The alkaline diffusion coefficient D (mm²/hr) is determined by substituting the result of x into (4).

$$\frac{\sigma_t}{\sigma_0} = \left(\frac{1-x}{R_0}\right)^2 \tag{3}$$

$$D = \frac{x^2}{2Ct}, \tag{4}$$

where σ_0 and σ_t indicate tensile strength (MPa) before immersing and at certain immersion time, respectively. The parameter x is the depth of penetration from the surface (mm); the parameter C is a relative term describing the alkaline concentration [(mol/liter)/(mol/liter)]. The parameter t is the alkali immersion time (hr) in 40°C with one mol/liter aqueous NaOH. From the analysis results shown in Figure 14, the diffusion coefficients are 1.1×10^{-5} mm²/hr~4.1×10^{-5} mm²/hr for 100 immersion days, 3.4×10^{-4} mm²/hr~ 3.9×10^{-4} mm²/hr for 300 immersion days, and 2.2×10^{-4} mm²/hr~2.3×10^{-4} mm²/hr for 660 immersion days. In the previous researches, the alkaline solution diffusion coefficients were 2.8×10^{-4} mm²/hr, 3.2×10^{-3} mm²/hr, and 3.3×10^{-3} mm²/hr for 100, 300, and 660 immersion days, respectively, using the same diameter of the GFRP bar

specimens [21–23]. The diffusion coefficient of 100 alkali immersion days is significantly less than the results of previous researches. However, the diffusion coefficients for 300 and 660 immersion days are similar to those of the FRP rebars presented by Katsuki and Uomoto [21], but those are much less than the results of Sen et al. [22] and Trejo et al. [23]. For the 100 immersion days of the testing specimens, the diffusion coefficient of PH_200 specimen is 2.8 to 3.0 times greater than that of other specimens, which indicates the effect of preheating. The diffusion coefficients of 300 and 660 immersion days, however, are not apparently different from each other. Trejo et al. [23] report that more consistent diffusion coefficient can be evaluated from a long term test in which the rebar should be immersed into alkali solution for at least 120 days. This report supports that the diffusion coefficients obtained from the 300 and 660 alkali immersion days are more reliable than the result of the 100 immersion days.

5. Optical Microscopic Analysis

The microscopy images for Ref specimen which are not immersed into alkaline solution are shown in Figure 15. Many voids with a diameter smaller than 10 μm are observed. The voids are formed by air bubbles entrapped between fibers during the pultrusion. However, cracks are not found in the fibers and resin matrix. Figure 16 shows the microscopy images of PH_120 and PH_200 specimens which are not immersed into alkaline solution. Pores with diameters larger than 40 μm are visible between the fibers and in resin-rich areas; also the fibers are cracked as shown in Figure 16. However, the cracks in the resin matrix are found in PH_200 specimen, but not in PH_120 specimen. Figure 17 shows the microscopy images of PH_120 and PH_200 specimens immersed in the alkaline solution for 30 days. For PH_200 specimen, more pores and cracks in the resin and fibers and

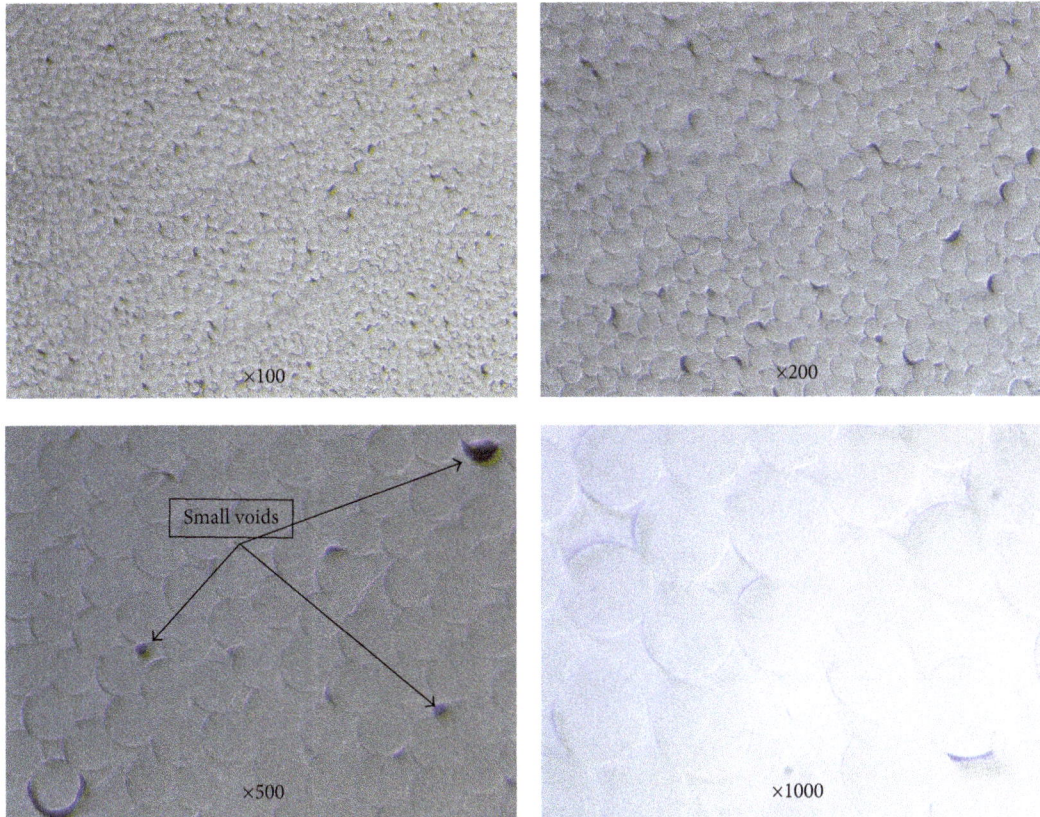

FIGURE 15: Optical microscopic images of Ref specimen.

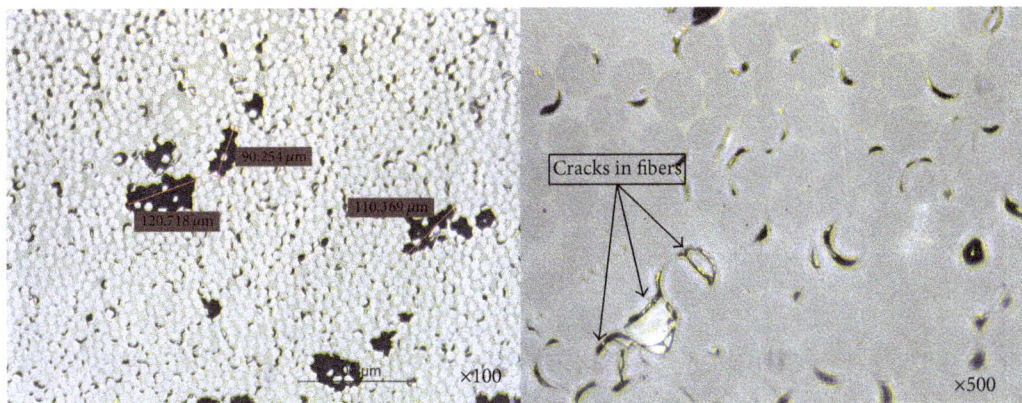

(a)

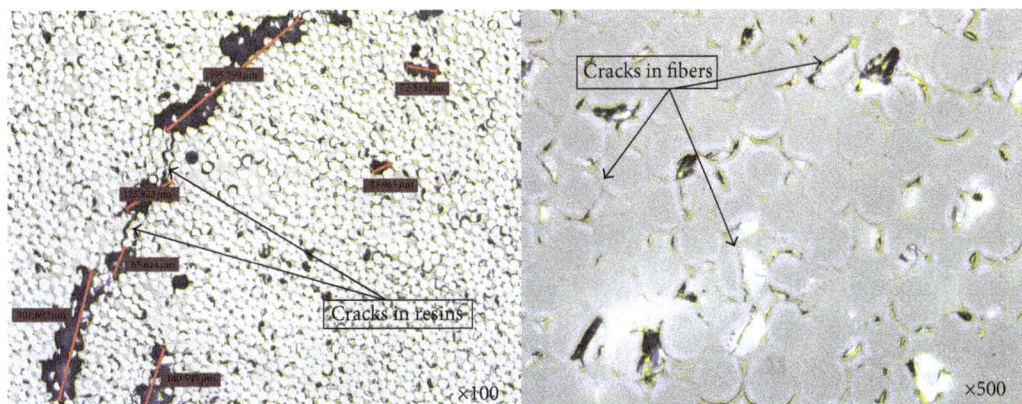

(b)

FIGURE 16: Optical microscopic images before immersing into alkaline solution: (a) PH_120 and (b) PH_200 specimens.

(a)

(b)

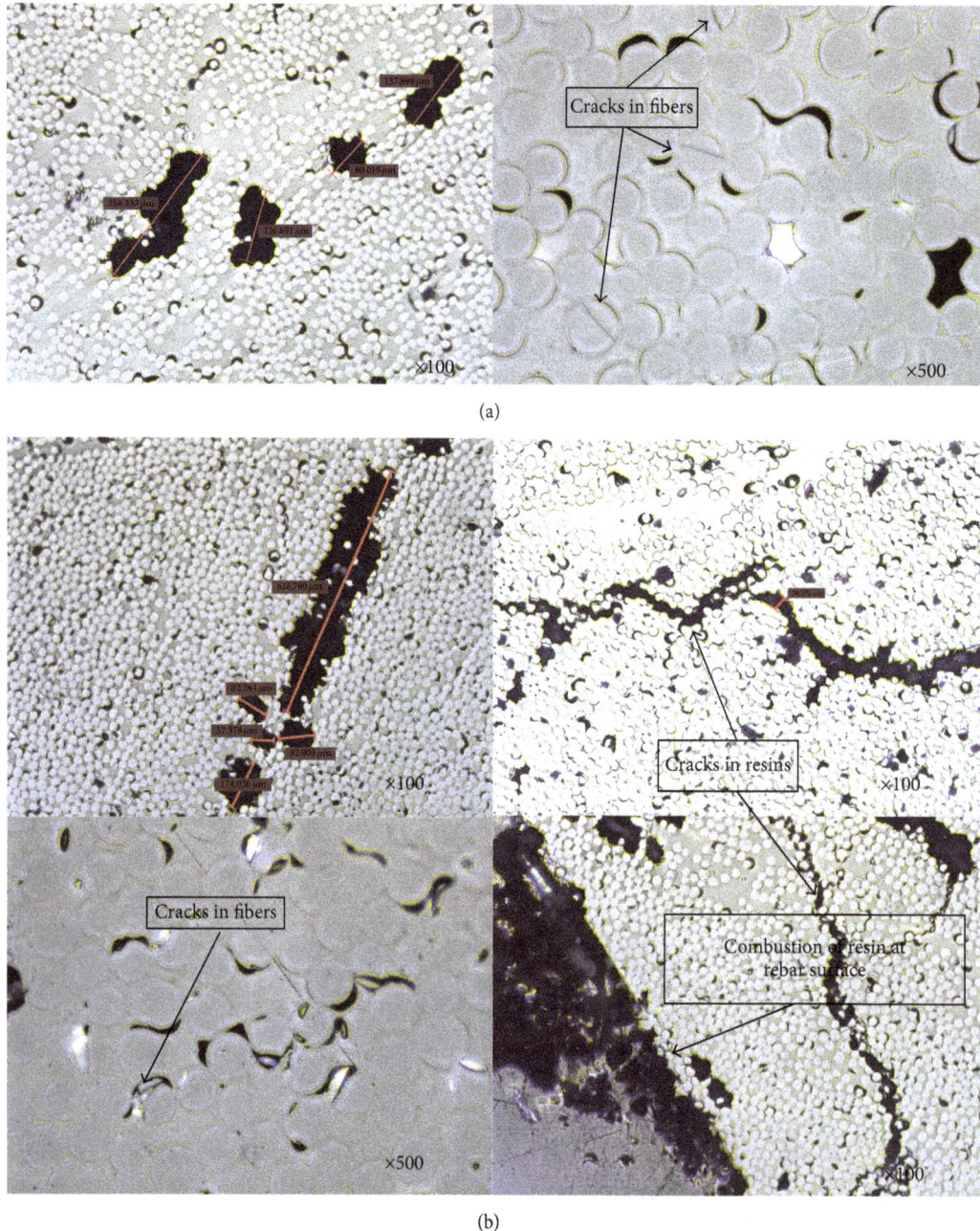

FIGURE 17: Optical microscopic images after immersing into alkaline solution for 30 days: (a) PH_120 and (b) PH_200 specimens.

even the combustion on the rebar surface are found. However, there is no resin crack developed on PH_120 specimen. Also, the pore sizes of PH_120 specimen are not changed compared to the case without immersing into alkaline solution.

6. Remarks and Conclusions

The purpose of this study is to identify the variation in the tensile properties of the structural GFRP reinforcing bars exposed to high temperature. The tensile tests are performed on the 1.2 m long specimens on the UTM and the total number of the specimens tested is 105. The GFRP bar specimens are preexposed to high temperature (120°C and

200°C) and cooled down to room temperature level. After the process, the specimens are immersed in 1 Mole NaOH alkaline solution to represent the concrete environment for long periods of time (100 days, 300 days, 660 days). From the test results, the preheated specimens provide slightly lower elastic modulus than those of unpreheated specimens and the maximum difference is only 8%. The tensile strength also decreases for all testing cases as the increase of the alkaline immersion time. There is a big drop in the tensile strength until 300 alkali immersion days. The tensile strength of the preheated specimens is about 90% of the tensile strength of the unpreheated specimen. After 300 alkali immersion days, however, the tensile strengths are almost identical to each other. The overall results indicate that the tensile strength and elastic modulus of the GFRP bars are closely related to alkali immersion days, not much related to the preheating levels. It leads to the conclusion that the limited damage caused to the GFRP reinforcing bars by the high temperature applied in this study is not expected to induce a significant deterioration in the tensile performance of the GFRP reinforcing bars as that damage is caused by the infiltration of alkaline components. Therefore, the damage caused by such temperature condition may not necessarily be considered for the strength evaluation of FRP rebars. Incidentally, based on the conclusions herein, it can be recommended that the environmental reduction factor specified in ACI 440.1R-06 [27], with design FRP material properties without considering the exposed temperature conditions, is still available for the strength evaluation of the GFRP rebars when they remained in concrete alkalinity environment after fire accident. One note is that if the sustained tension load is considered during preheating process to represent more real situation, the test results obtained in this study will be varied since the fact that the sustained load increases the micro cracks or voids in the specimens during their exposure to high temperature may happen. The sustained tensile loading effects need to be investigated in the future study.

Conflicts of Interest

The authors declare that there are no conflicts of interest regarding the publication of this paper.

Acknowledgments

This research was supported by a grant from the Information & Communication Technology (ICT) Program (code 15SCIP-B066018-03) and Technology Advancement Research Program funded by the Ministry of Land, Infrastructure and Transport of the Korean government (Grant 17CTAP-C117247-02).

References

[1] S. Kumahara, Y. Masuda, H. Tanano, and A. Shimizu, "Tensile strength of continuous fiber bar under high temperature," *ACI Special Publication*, vol. 138, pp. 731–742, 1993.

[2] A. Abbasi and P. J. Hogg, "Temperature and environmental effects on glass fibre rebar: modulus, strength and interfacial bond strength with concrete," *Composites Part B: Engineering*, vol. 36, no. 5, pp. 394–404, 2005.

[3] L. A. Bisby, M. F. Green, and V. K. R. Kodur, "Response to fire of concrete structures that incorporate FRP," *Progress in Structural Engineering and Materials*, vol. 7, no. 3, pp. 136–149, 2005.

[4] Y. C. Wang and V. Kodur, "Variation of strength and stiffness of fibre reinforced polymer reinforcing bars with temperature," *Cement and Concrete Composites*, vol. 27, no. 9-10, pp. 864–874, 2005.

[5] E. J. Bosze, A. Alawar, O. Bertschger, Y.-I. Tsai, and S. R. Nutt, "High-temperature strength and storage modulus in unidirectional hybrid composites," *Composites Science and Technology*, vol. 66, no. 13, pp. 1963–1969, 2006.

[6] M. Robert and B. Benmokrane, "Behavior of GFRP reinforcing bars subjected to extreme temperatures," *Journal of Composites for Construction*, vol. 14, no. 4, pp. 353–360, 2010.

[7] J. R. Correia, M. M. Gomes, J. M. Pires, and F. A. Branco, "Mechanical behaviour of pultruded glass fibre reinforced polymer composites at elevated temperature: Experiments and model assessment," *Composite Structures*, vol. 98, pp. 303–313, 2013.

[8] G. A. Kashwani and A. K. Al-Tamimi, "Evaluation of FRP bars performance under high temperature," *Physics Procedia*, vol. 55, pp. 296–300, 2014.

[9] S. K. Foster and L. A. Bisby, "High temperature residual properties of externally-bonded FRP systems," *ACI Special Publication*, vol. 230, pp. 1235–1252, 2005.

[10] M. Al-Zahrani, S. U. Al-Dulaijan, S. H. Al-Idi, and M. H. Al-Methel, *High temperature effect on tensile strength of GFRP bars and flexural behavior of GFRP reinforced concrete beams*, FRPRCS-8, University of Patras, Patras, Greece, 2007.

[11] Y.-N. Fu, J. Zhao, and Y.-L. Li, "Research on tensile mechanical properties of GFRP rebar after high temperature," *Advanced Materials Research*, vol. 181-182, pp. 349–354, 2011.

[12] S. Alsayed, Y. Al-Salloum, T. Almusallam, S. El-Gamal, and M. Aqel, "Performance of glass fiber reinforced polymer bars under elevated temperatures," *Composites Part B: Engineering*, vol. 43, no. 5, pp. 2265–2271, 2012.

[13] F. Micelli and A. Nanni, "Durability of FRP rods for concrete structures," *Construction and Building Materials*, vol. 18, no. 7, pp. 491–503, 2004.

[14] J.-P. Won and C.-G. Park, "Effect of environmental exposure on the mechanical and bonding properties of hybrid FRP reinforcing bars for concrete structures," *Journal of Composite Materials*, vol. 40, no. 12, pp. 1063–1076, 2006.

[15] Y. Chen, J. F. Davalos, I. Ray, and H. Y. Kim, "Accelerated aging tests for evaluations of durability performance of FRP reinforcing bars for concrete structures," *Composite Structures*, vol. 78, no. 1, pp. 101–111, 2007.

[16] M. Robert, P. Cousin, and B. Benmokrane, "Durability of gfrp reinforcing bars embedded in moist concrete," *Journal of Composites for Construction*, vol. 13, no. 2, pp. 66–73, 2009.

[17] A. Katz, N. Berman, and L. C. Bank, "Effect of high temperature on bond strength of FRP rebars," *Journal of Composites for Construction*, vol. 3, no. 2, pp. 73–81, 1999.

[18] S. U. Al-Dulaijan, M. M. Al-Zahrani, A. Nanni, and T. E. Boothby, "Effect of environmental pre-conditioning on bond of FRP reinforcement to concrete," *Journal of Reinforced Plastics and Composites*, vol. 20, no. 10, pp. 881–900, 2001.

[19] ASTM D 3916-02, *Test Method for Tensile Properties of Pultruded Glass Fiber-Reinforced Plastic Rod*, vol. 08.02, West Conshohocken, PA, USA, 2002.

[20] ACI 440.3R-04, "Guide test methods for fiber-reinforced polymers (FRPs) for reinforcing or strengthening concrete structures," in *Proceedings of the ACI Committee 440*, American Concrete Institute, Farmington Hills, Mich, USA, 2004.

[21] F. Katsuki and T. Uomoto, "Prediction of deterioration of FRP rods due to alkali attack," in *Proceedings of the Second International RILEM Symposium (FRPRCS-2*, pp. 82–89, Ghent, Belgium, 1995.

[22] R. Sen, G. Mullins, and T. Salem, "Durability of E-glass/vinylester reinforcement in alkaline solution," *ACI Structural Journal*, vol. 99, no. 3, pp. 369–375, 2002.

[23] D. Trejo, F. Aguiniga, R. Yuan, R. W. James, and P. B. Keating, "Characterization of design parameters for fiber reinforced polymer composite reinforced concrete systems. Research Report 9-1520-3," Tech. Rep., Texas A&M Univ., College Station, Tex, USA, 2005.

[24] Y. Chen, J. F. Davalos, and I. Ray, "Durability prediction for GFRP reinforcing bars using short-term data of accelerated aging tests," *Journal of Composites for Construction*, vol. 10, no. 4, pp. 279–286, 2006.

[25] P. Gardoni, D. Trejo, and Y. H. Kim, "Time-variant strength capacity model for GFRP bars embedded in concrete," *Journal of Engineering Mechanics*, vol. 139, no. 10, pp. 1435–1445, 2013.

[26] M. Robert and B. Benmokrane, "Combined effects of saline solution and moist concrete on long-term durability of GFRP reinforcing bars," *Construction and Building Materials*, vol. 38, no. 1, pp. 274–284, 2013.

[27] ACI 440.1R-06, "Guide for the Design and Construction of Structural Concrete Reinforced with FRP Bars," in *Proceedings of the ACI Committee 440*, American Concrete Institute, Farmington Hills, Mich, USA, 2006.

16

Experimental Study of the Basic Mechanical Properties of Directionally Distributed Steel Fibre-Reinforced Concrete

Fang-Yuan Li[ID], Cheng-Yuan Cao, Yun-Xuan Cui, and Pei-Feng Wu

Department of Bridge Engineering, College of Civil Engineering, Tongji University, Shanghai 200092, China

Correspondence should be addressed to Fang-Yuan Li; fyli@tongji.edu.cn

Academic Editor: Enzo Martinelli

Directionally distributed steel fibre-reinforced concrete (SFRC) cannot be widely applied due to the limitations of current construction technology, which hinders research on its mechanical properties. With the development of new construction technologies, such as self-compacting concrete or 3D printing, directionally distributed SFRC has found new developmental opportunities. This study tested, compared, and analysed the basic mechanical properties of ordinary concrete, randomly distributed SFRC, and directionally distributed SFRC. The differences between the damage patterns parallel and perpendicular to the direction of the steel fibres were evaluated in directionally distributed SFRC. When the fibre volume fraction is high and the compression is applied perpendicular to the fibre direction, as the loading increases, the transverse deformation of the specimen is constrained by the fibres. When the compression is applied parallel to the fibre direction, the fibre cannot effectively constrain the transverse deformation of the specimens. When the volume fraction of directionally distributed steel fibres was 1.6%, the elastic modulus of the directionally distributed steel fibres was 39% higher than that of ordinary concrete. Comparison of the experimental values of the elastic modulus with those estimated by existing calculation methods revealed that a modification of the current calculation theories may be required to calculate the changes in the elastic modulus of directionally distributed SFRC with a high volume fraction of steel fibres.

1. Introduction

Steel fibre-reinforced concrete (SFRC) was first introduced in the early 20th century. From 1907 to 1908, Некрасоъ, an expert from the former Soviet Union, incorporated metal fibres into concrete, thus heralding the development of SFRC [1]. The earliest study on SFRC was published in 1910; in the United States, Porter envisaged the incorporation of short steel fibres into the concrete matrix to reinforce the concrete materials. In 1911, Graham proposed the incorporation of steel fibres into ordinary reinforced concrete to improve its strength and volume stability.

In the 1940s, scholars and engineers in the United States, Britain, France, Germany, and other countries applied for and were awarded a series of patents on methodologies of improving the performance of concrete by incorporating steel fibres, enhancing the manufacturing process of steel fibres and improving the shape of steel fibres to enhance the bonding strength with the concrete matrix. In 1963, a series of papers on strength-enhancing mechanism of steel fibres were published by Romualdi and Batson [2, 3]. Since then, research and applications of SFRC have developed rapidly. Scholars and engineers in Europe and the United States applied for and were awarded numerous patents on SFRC. Scholars also published many research papers that better clarified the fibres' reinforcement mechanism in concrete. After the 1970s, the United States developed a melt pumping technology to manufacture low-cost steel fibres, which made practical applications of SFRC feasible. Over the next 20 years, the application of steel fibres has drawn widespread attention.

In China, research on and application of SFRC began in the 1970s, and this work was led by the Commission for Science, Technology and Industry for National Defense and the China Building Materials Academy. Dalian University of Technology of China first derived a formula to calculate the tensile strength of randomly distributed SFRC based on fracture mechanics theory. The results were consistent with those given by the theory of composite materials. Zhao's team analysed the reinforcement mechanism and failure pattern of SFRC and systematically conducted theoretical studies of SFRC [4]. Scholars from the former Harbin Institute of Architectural Engineering conducted experimental tests on the basic mechanical properties of SFRC [4]. At the same time, the Committee on Fibre-Reinforced Concrete was set up in the China Civil Engineering Society. The committee compiled and published the "Design and construction procedures of steel fibre-reinforced concrete structure" (CECS 38:93) and the "Test method of steel fibre-reinforced concrete" (CECS 13:89). These works laid the foundation for the application and development of SFRC in China. Since 2000, the committee has organized the revision of design and construction procedures and compiled the "Technical specifications for fibre-reinforced concrete structure" (CECS 38:2004) for the Association of Chinese Construction Standardization.

All theoretical assumptions show that the fibres' distribution in different directions in concrete greatly affects its reinforcement efficiency in the concrete matrix [5, 6]. In the late 1970s, scholars from Italy, Sweden, and other countries announced that by utilizing magnetic devices, the steel fibres' orientation and concentration could be satisfactorily controlled during the vibration processes. These scholars also reported that the bending and shear strengths of directionally distributed SFRC were twice that of randomly distributed SFRC with the same fibre volume fraction. If the fibres were concentrated in a certain position, the enhancement effect could be further improved [6]. However, the original reference could not be found. In addition, the improvements of the compressive and tensile properties were not mentioned.

Before the 21st century, scholars outside China studied several properties of directionally distributed fibre-reinforced concrete. However, due to the difficulty of directionally orienting steel fibres, only a few verification tests were conducted to analyse the advantages of directionally distributed fibre concrete compared with ordinary concrete and randomly distributed SFRC. Few in-depth and comprehensive studies were conducted [7–9], and few studies were published outside China on this topic in the 21st century.

The directions of sustained loading applied to most structural components are usually fixed and thus do not change. For example, beam-style components and the bottoms of beams are subjected to sustained tensile stresses for long periods of time. Therefore, under loading in the same direction in randomly distributed SFRC, only the steel fibres aligned with or are at small angles to the loading direction reinforce the components and resist crack propagation. The other embedded steel fibres either do not contribute or do not fully contribute to the reinforcement of the concrete components. This results in wasted steel fibres [10–14]. Placing directionally oriented steel fibres in the predicted direction of tensile stress to control their angular distribution in the concrete would achieve a "precise design" and "flexible design" for the materials. In this way, the reinforcement of the components' strength and toughness by

the fibres can be fully utilized, while the fibre fraction can be reduced. This procedure would decrease the construction cost while simultaneously increasing the steel fibres' utilization efficiency to 100%. If this goal is achieved, the application of SFRC will have significant advantages in practical engineering. Related studies have also shown that if the fibres' reinforcement effect in unidirectional fibre-reinforced concrete is 1, the reinforcement efficiency of the bidirectional fibre- (such as steel wire mesh-) reinforced concrete is only approximately 0.4 to 0.5. The reinforcement effect in three-directional randomly distributed concrete is even lower [6, 12].

In this paper, ordinary concrete with the same mixtures was used as a reference, and a comparative test of ordinary randomly distributed SFRC and directionally distributed SFRC was conducted. These tests yielded quantitative property information and laid a foundation for future studies of directionally distributed SFRC. The results of this paper are expected to provide technical support for developing seamless widening and splicing technology for bridges, installing expansion joints in bridges, reducing beam end expansion joints and mitigating cracking of the bridge structure. When randomly oriented steel fibres are included in a bridge's non-structural components, where cracks are not a concern, and directionally oriented steel fibres are included in the bridge's structural components, where cracks tend to occur, the issue of cracking that has long-puzzled engineers and operators will be mitigated, and the bridge's durability will be greatly improved. The results will also provide support for the application of self-compacting or 3D concrete printing technology based on directionally distributed SFRC.

2. Preparation of Directionally Distributed SFRC

Currently, most studies on directionally distributed SFRC focus on directionally orienting steel fibres utilizing a magnetic field. The testing requires particularly complex and expensive magnetic field generating equipment that is difficult to operate. In this paper, directionally distributed SFRC cubic specimens were cast by manually placing the steel fibres layer by layer. During the casting process, the uniformity and directional accuracy were carefully monitored at all times, and the deviations between the test pieces were strictly controlled. The specimen casting operation was relatively simple and easy to control.

2.1. Test Materials. For the tests, a concrete mixture was designed based on GB/T 50080-2016 "Standard for Test Methods of Performance of Ordinary Concrete Mixtures" [15] and JGJ 55-2011 "Specification for mix proportion design of ordinary concrete" [16]. The appropriate materials were selected, and a concrete mixture ratio that met the testing requirements was employed.

(1) *Cement*: ordinary Portland cement of Hailuo brand P.O42.5 (low alkali).

(2) *Fine aggregate*: ordinary river sand. It was zone III fine sand with a fineness modulus of 1.73.

TABLE 1: Mixing ratio of benchmark concrete.

Cement	Water	Sand	Coarse aggregates	Water reducer SH-II	C (kg/m^3)	Slump (mm)
1.0	0.40	0.92	2.04	0.0027	550	48

(3) *Coarse aggregate*: ordinary gravel with sizes of 5–20 mm and a continuous grain size distribution. The grades met the requirements.

(4) *Additive*: SH-II-type naphthalene high-efficiency water reducer made by Shanghai Yaoqian Architectural Painting Co., Ltd.

(5) *Steel fibre*: milling-type fibres made by Shanghai Harex Steel Fibre Technology Co., Ltd. with the following dimensions: a length of 32.0 ± 2.0 mm, a width of 2.6 ± 1.2 mm, a thickness of 0.4 ± 0.05 mm, and a length-diameter ratio of 35–45. These fibres had a tensile strength of ≥ 700 MPa.

2.2. Mixing Ratio of Benchmark Concrete. Mixture ratios that met the specified requirements (good viscosity, cohesion, and water retention) were obtained after trial-and-error experimentation. The specific mixture ratios were as follows:

(1) *Water-cement ratio*: the maximum content of steel fibres in this test was 144 kg/m^3. As the compressive strength of concrete was not obviously improved by the incorporation of the steel fibres, the water-cement ratio of the benchmark concrete was determined according to the method in [16].

The test was conducted using grade C50 concrete. The water-cement ratio was determined to be 0.40.

(2) *Water usage*: to accommodate the directional placement of the fibres, the concrete should have a large slump to ensure good viscosity and to reduce the vibration casting time. Additionally, the fine aggregates used in the experiment were fine sand, so the water usage per cubic metre of the specimen was increased by 5–10 kg. Therefore, 220 kg/m^3 of water was used in these specimens.

(3) *Sand ratio*: in this study, the sand ratio of the benchmark concrete specimens was initially determined to be 31%, as per the method in [16].

(4) *The amount of coarse and fine aggregates*: a mass calculating method was used to determine the appropriate amount of coarse and fine aggregates, which was 2500 kg/m^3.

(5) *Water reducer*: trial-and-error casting was conducted using the initially determined mixture ratios as discussed above. Using the workability of the concrete as the main index, the amount of the water reducer to be added was determined to be 1.5 kg/m^3.

The mixture ratios can be also seen in Table 1.

The strength measured after 28 days, that is, the strength of the finally determined mixture, was 49.2 MPa, which met the expected requirements (Table 2). The increase in the

TABLE 2: Cubic compressive strength with different ages of the benchmark ordinary concrete.

Age (d)	3	7	28
Average compressive strength (MPa)	23.6	36.4	49.2

FIGURE 1: Directionally distributed steel fibres in a concrete specimen.

strength of the concrete was rapid at first before gradually slowing in the later period, which agrees with the theoretical predictions.

2.3. Specimen of Directionally Distributed SFRC. Based on the benchmark concrete for normal concrete, the directionally distributed SFRC was made by casting layer by layer and adding directionally oriented steel fibres, as seen in Figure 1.

The cement was evenly mixed with sand, gravel, and water reducer in advance. Then, water was added to the mixture three times until the mixture was completely mixed. The mixture was then poured into an oiled mould layer by layer, each with a height of 20 mm, and each 100 mm cubic specimen included five uniform layers. The steel fibres for each specimen were weighed and divided equally into four batches. After pouring each layer of concrete, one batch of steel fibres was placed in a particular orientation. The even distribution of the steel fibres across the cross section was confirmed. When adding the steel fibres, the fibre spacing should not be too small with respect to the particle size of the coarse aggregate.

3. Experimental Study of the Mechanical Properties of Directionally Distributed SFRC Specimens

Directionally distributed SFRC is a unique composite composed of concrete and unidirectionally embedded steel

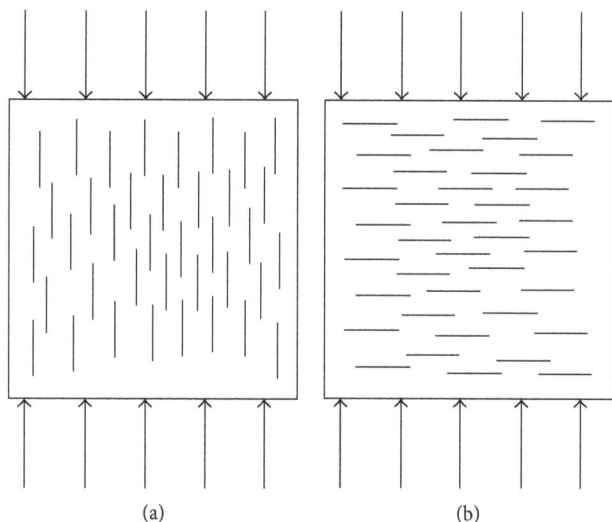

FIGURE 2: Diagrams of loading directions in the split tensile strength test. Loading applied (a) parallel and (b) perpendicular to the fibre direction.

FIGURE 3: Failure pattern of a cubic ordinary concrete specimen under compression.

fibres. Its stress bearing capacity and failure patterns are significantly different from those of ordinary concrete and randomly distributed SFRC [17]. Because of the directional distribution of steel fibres in the concrete, its stress characteristics are anisotropic and more complex than those in ordinary concrete or randomly distributed SFRC, and its test methods and future design and construction methods are unique. Therefore, it is necessary to conduct a comprehensive comparative test of their stress characteristics.

In practical engineering, a concrete structure is rarely subjected to uniaxial loading. However, because the mechanical properties under multiaxial loading are complex, it is difficult to establish a strength theory that can perfectly explain the failure pattern of concrete under different stress conditions. Therefore, most experimental studies are based on the uniaxial strength, which is used to evaluate the basic mechanical properties of concrete structures [18]. In this study, based on relevant research results [19–21] and by using the testing method for ordinary concrete and randomly distributed SFRC, the basic mechanical properties of directionally distributed SFRC were quantitatively studied.

Several types of SFRC specimens were assessed in this study. The concrete's strength grade was C50, and the only variable was the volume fraction of steel ingot-milled fibres. A total of 72 cubic specimens with the dimensions of 100 mm × 100 mm × 100 mm and seven steel fibre volume fractions were made. The compressive strengths of these 72 specimens were tested. The reinforcement effects of different volume fractions of randomly and directionally distributed steel fibres on ordinary concrete were compared and analysed. The elastic modulus was measured through static compression tests using 30 prism specimens with the dimensions of 100 mm × 100 mm × 300 mm. The influence of different volume fractions of randomly distributed and directionally distributed steel fibres on the elastic modulus of ordinary concrete was compared and analysed. The comprehensive analysis of the basic mechanical properties of steel ingot-milled fibre-reinforced

concrete conducted in this paper provides a theoretical basis and guidance for further study of SFRC and its applications in practical engineering.

The mechanical tests in this paper were carried out in accordance with the provisions in two Chinese national standards: "Standard for test method of mechanical properties on ordinary concrete" GB/T 50081-2002 and "Steel fibre-reinforced concrete" JG/T 472-2015 [15, 22].

3.1. Compressive Strength Test. The compressive strength of ordinary SFRC is not significantly affected by the steel fibre properties and volume fraction. Compared with ordinary concrete, the compressive strength of cubic speciments of SFRC is far less important than its tensile strength, but to coordinate the design of SFRC structures with other design specifications, China's "Specification for design and construction of SFRC structures" [22] stipulates that the strength grade of SFRC should be determined based on the standard cubic compressive strength by following the pertinent specifications for the design of concrete structures. Therefore, the compressive strength of cubic specimens of SFRC is still an important index for evaluating the durability and performance of the concrete matrix.

In this study, compressive tests were conducted with cubic specimens of ordinary concrete, ordinary randomly distributed SFRC, and directionally distributed SFRC with seven different fibre volume fractions. The compressive strength tests of the directionally distributed SFRC were conducted both parallel and perpendicular to the fibre direction (Figures 2(a) and 2(b), resp.). When the compressive test was conducted perpendicular to the fibre direction, the load was applied perpendicular to the casting surface.

3.1.1. Failure Pattern

(1) During the test, upon failure of the ordinary concrete specimen (Figure 3), the concrete specimen's side surface perpendicular to the loading surface broke and collapsed to the ground. A "bang" noise was heard when it failed. The concrete's load-bearing capacity rapidly dropped to zero. Cracks propagated through

FIGURE 4: Failure patterns of cubic specimens of randomly distributed SFRC under compression. (a) 0.4%; (b) 0.8%; (c) 1.5%.

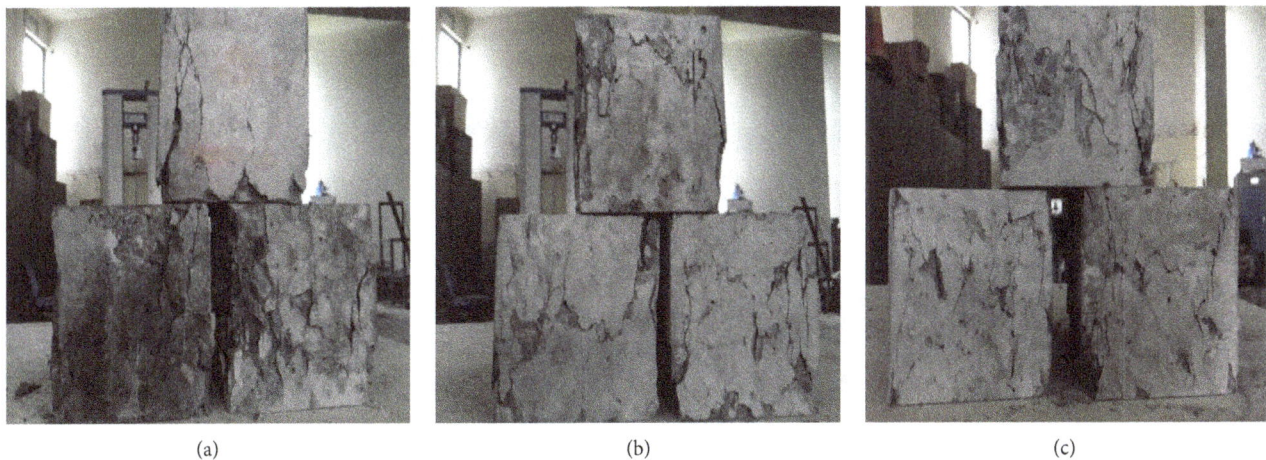

FIGURE 5: Failure patterns of cubic specimens of directionally distributed SFRC under (parallel) compressive loading. (a) 0.4%; (b) 0.8%; (c) 1.5%.

the specimens along several paths, and the specimens failed suddenly.

(2) In the ordinary randomly distributed SFRC specimen (Figure 4), small cracks first appeared on the surface. As the load increased, cracks propagated along many paths but were not connected. The concrete surface was damaged, but the specimen's integrity was maintained. The specimens cracked but did not fail. As the load continued to increase until the specimen failed, the specimen showed obvious signs of failure.

When the fibre volume fraction was low (0.4%), a small amount of concrete peeled from the specimen's surface. The damage pattern was similar to that observed in the ordinary concrete specimens but was not identical. At higher fibre volume fractions (0.8% and 1.5%), the specimens maintained good integrity under compression; only a few discontinuous cracks formed on the specimens' surfaces.

(3) In the failure tests on directionally distributed SFRC specimens under compressive loading, the load can be applied either parallel or perpendicular to the fibre direction.

When the compressive load was applied parallel to the fibre direction (Figures 5 and 6), in specimens with a low fibre volume fraction (0.4%), the failure pattern was similar to that of ordinary concrete, and the concrete in the specimen collapsed at a large area of the surfaces. Cracks propagated from the top to the bottom of the specimen. As the fibre volume fraction increased (to 0.8%), the failure pattern changed. The surface integrity increased, and no through-going cracks formed. As the fibre volume fraction continued to increase (1.5%–1.8%), small-scale areas on the specimens' surfaces started to collapse, and large cracks formed, which compromised the overall integrity.

When the compressive load was applied perpendicular to the fibre direction (Figure 7), in the specimen with a low fibre volume fraction (0.4%), a small amount of concrete collapsed from the specimen's surface, and many non-through-going cracks formed. As the fibre volume fraction increased (to 0.8% and 1.5%), the specimen failure pattern continued to change until no concrete collapsed from the surface, and only nonthrough-going cracks formed. Crack propagation was hindered, and the specimens maintained good integrity.

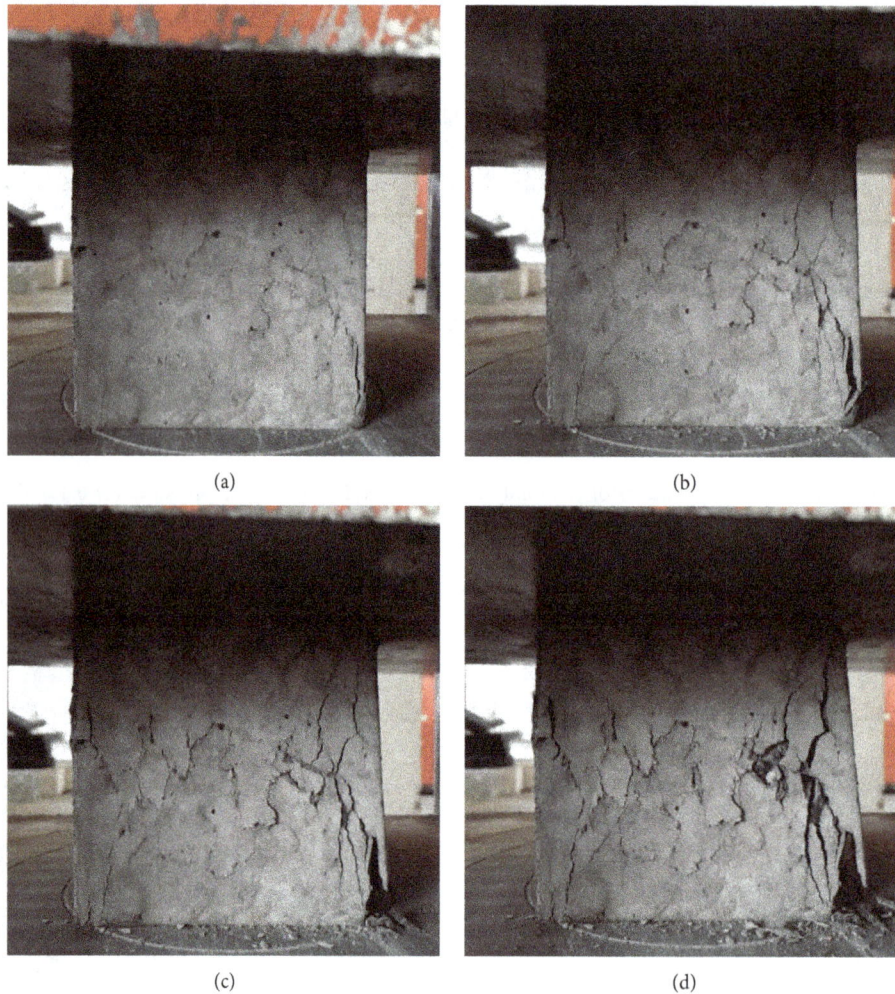

FIGURE 6: Failure process of a cubic specimen of directionally distributed SFRC under (parallel) compressive loading.

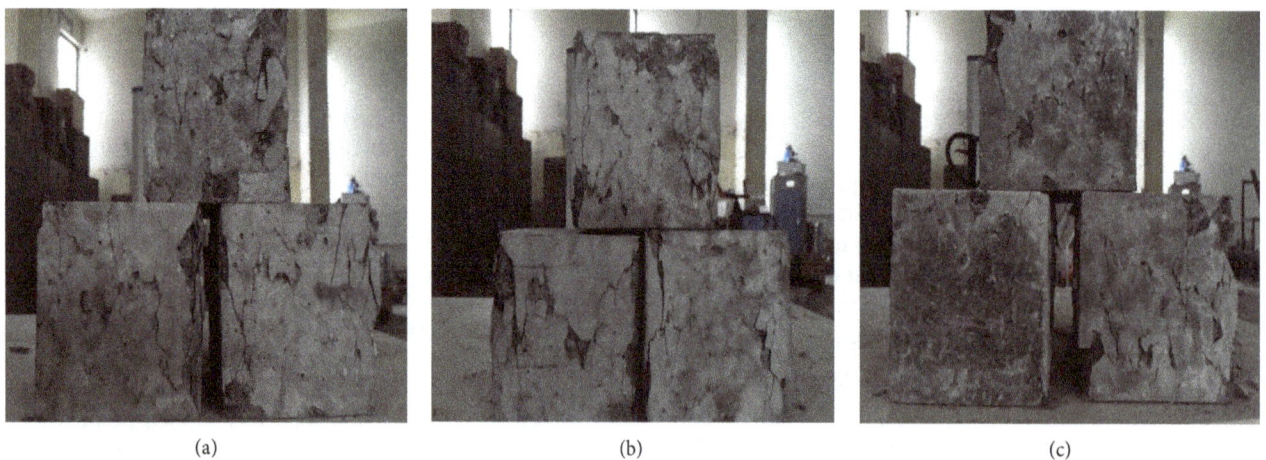

FIGURE 7: Failure pattern of a cubic specimen of directionally distributed SFRC under (perpendicular) compressive loading. (a) 0.4%; (b) 0.8%; (c) 1.5%.

3.1.2. Test Results. The compressive test results are shown in Table 3 and Figure 8, in which benchmark concrete represents the ordinary concrete specimen, "Random steel fibre-reinforced concrete" represents the ordinary randomly distributed SFRC specimen, "Directional SFRC" represents the directionally distributed SFRC specimen, and V_f is the volume fraction of steel fibres in the concrete specimen. When $V_f = 1.0\%$, the converted

TABLE 3: Compressive strength of cubic specimens.

Specimen number	V_f (%)	Compressive strength of cubic specimens (MPa)	Comparison with benchmark concrete
Benchmark concrete	0.0	49.2	1.00
Random SFRC	0.4	48.7	0.99
	0.6	52.2	1.06
	0.8	51.7	1.05
	1.0	52.9	1.08
	1.2	54.8	1.12
	1.5	56.8	1.16
	1.8	61.0	1.24
Directional SFRC (parallel)	0.4	57.3	1.17
	0.6	58.9	1.20
	0.8	55.5	1.13
	1.0	54.1	1.10
	1.2	52.9	1.08
	1.5	54.9	1.12
	1.8	54.3	1.11
Directional SFRC (perpendicular)	0.4	59.5	1.21
	0.6	59.7	1.22
	0.8	55.8	1.14
	1.0	54.7	1.11
	1.2	57.0	1.16
	1.5	60.7	1.24
	1.8	56.7	1.15

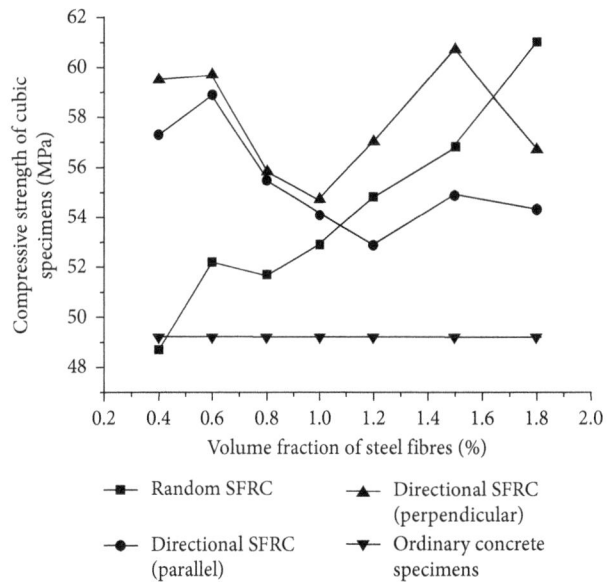

FIGURE 8: Compressive strengths of cubic specimens with different fibre contents.

weight content of the steel fibres embedded in the SFRC is $80\,kg/m^3$.

Table 3 and Figure 8 show the following:

(1) When the fibre volume fraction is below 1.8%, the compressive strength of the ordinary randomly distributed SFRC specimens increases with increasing fibre volume fraction. The maximum strength occurs when the volume fraction is 1.8% and is 24% higher than that of ordinary concrete.

(2) In the directionally distributed SFRC specimens, the compressive strength is higher even when the fibre volume fraction is low (0.6%) (compared with the compressive strength of ordinary concrete, the compressive strengths in the two loading directions are 20% higher (parallel) and 22% higher (perpendicular)). As the fibre volume fraction increases, the compressive strength begins to decrease. When compression is applied parallel to the fibre direction, the compressive strength is lowest when the fibre volume fraction is 1.2% and is lower than that of the ordinary randomly distributed SFRC. The strength increases slightly as the fibre volume fraction increases, but it is always lower than that of the ordinary randomly distributed SFRC. When the compression is applied perpendicular to the fibre direction, the compressive strength is the lowest when the fibre volume fraction is 1.0%, and it increases gradually as the fibre ratio increases. When the fibre volume fraction is 1.5%, the compressive strength of the SFRC reaches its maximum (24% higher than that of ordinary concrete). However,

further increases in the steel fibre ratio result in a sharp decrease in strength.

(3) In the directionally distributed SFRC specimens, the compressive strengths of specimens under compression in the two directions (parallel and perpendicular to the fibre direction) are not significantly different when the fibre volume fraction is low (less than 1%). As the fibre volume fraction increases (1.2%–1.8%), the anisotropy of the specimens increases gradually.

3.1.3. Effect of the Fibre Volume Fraction and Direction on the Compressive Strength of Specimens

(1) Effect of the Fibre Volume Fraction on the Compressive Strength. A comparison of the test results discussed above shows that as the fibre volume fraction increases, the compressive strengths of cubic specimens of both ordinary randomly distributed SFRC and directionally distributed SFRC increase. However, a higher fibre volume fraction does not necessarily result in a greater increase in the compressive strength; the compressive strengths of some specimens with high fibre volume fractions are significantly lower. A possible reason is that as more fibres are incorporated into the concrete, the fibres' specific surface area increases; thus, some fibres are not adequately coated by the concrete slurry, and the spaces between the fibres are not completely filled by the concrete slurry. Therefore, the compactness of the fibre-reinforced concrete decreases. In addition, the increased usage of fibres hinders the casting of the specimens. The specimens are not uniformly vibration solidified, so the fibres are unevenly distributed in the specimens [23–26].

The tests in this study clearly show that, in both ordinary randomly distributed SFRC and directionally distributed SFRC specimens, when the fibre volume fraction is approximately 0.8–1.0%, the compressive strength of the

TABLE 4: Elastic modulus of concrete.

Specimen number	V_f (%)	Elastic modulus (MPa)	Compared with benchmark concrete
Benchmark concrete	0.0	41,400	1.00
Random SFRC	0.8	43,200	1.04
	1.6	43,500	1.05
Directional SFRC	0.8	45,700	1.10
	1.6	57,600	1.39

specimens is significantly lower. This phenomenon is most likely caused by the characteristics of the milled steel fibres used in this experiment. The same result was found in other studies [27, 28]. These findings demonstrate that for this kind of milled steel fibre, which has a special shape, the fibre volume faction features an optimum range. Within this range, the steel fibres have the best reinforcement effect on the concrete's compressive strength. For both ordinary randomly distributed SFRC and directionally distributed SFRC, fibre volume fractions of 0.8% to 1.0% should be avoided.

(2) Effect of the Fibre Direction on the Compressive Strength. A comparison of the compressive strengths of ordinary randomly distributed SFRC with those of the directionally distributed SFRC specimens with same fibre volume fraction under compression in two directions shows that when the fibre volume fraction is low, the strength of the directional SFRC in both loading directions is much higher than that of ordinary random SFRC. As the fibre volume fraction increases, the compressive strength of directional SFRC shows clear anisotropy in the two loading directions. When the compression is applied parallel to the fibre direction, the compressive strength is lower than that of ordinary random SFRC; in contrast, when compression is applied perpendicular to the fibre direction, the compressive strength is higher than that of ordinary random SFRC.

Further analysis shows that when the fibre volume fraction is high and the compression is applied perpendicular to the fibre direction, as the loading increases, the transverse deformation of the specimen is constrained by the fibres. The stress state in the specimen is similar to multiaxial compression. Therefore, its compressive strength is much higher than that of the ordinary concrete and random SFRC specimens. When the compression is applied parallel to the fibre direction, the fibres cannot effectively constrain the transverse deformation of the specimens. When the fibre volume fraction is high, the compactness of the concrete is reduced; thus, the compressive strength decreases as the fibre volume fraction increases.

3.2. Static Compression Test of the Elastic Modulus. The elastic modulus is an important mechanical property of concrete that reflects the relationship between the stress and the strain in the concrete. This parameter is necessary for calculating the deformation of the concrete structure, crack development, and thermal stress in concrete. There are three ways to calculate the elastic modulus: (1) the tangent modulus, (2) the secant modulus, and (3) the chord modulus. The static compressive elastic modulus studied in this experiment is actually the concrete's chord modulus.

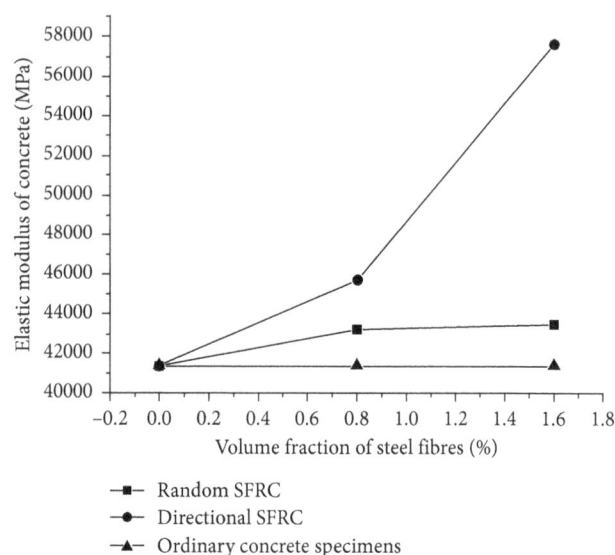

FIGURE 9: Elastic modulus of concrete with different fibre volume fractions.

In accordance with the Chinese "Standard test method for mechanical properties of ordinary concrete" (GB/T 50081-2002), a prism with the dimensions of 150 mm × 150 mm × 150 mm was used as the standard specimen. Six specimens were made for each test. Of the six specimens, three were used to measure the axial compressive strength, and the other three were used to measure the elastic modulus of the concrete.

Static compressive elastic modulus tests were conducted with ordinary concrete specimens, randomly distributed SFRC specimens with two fibre volume fractions (V_f values of 0.8% and 1.6%), and directionally distributed SFRC specimens with two fibre volume fractions (V_f values of 0.8% and 1.6%). In the directional SFRC, the fibres were oriented in the longitudinal direction of the specimen.

3.2.1. Test Results. Table 4 and Figure 9 show the results for the static compressive elastic modulus, in which benchmark concrete represents the ordinary concrete specimen, "Random SFRC" represents the ordinary randomly distributed SFRC specimen, "Directional SFRC" represents the directionally distributed SFRC specimen, and V_f is the volume fraction of steel fibres in the concrete specimens. When $V_f = 1.0\%$, the converted weight content of the steel fibres embedded in the SFRC is 80 kg/m³.

Table 4 and Figure 9 show the following:

(1) The elastic modulus of the ordinary randomly distributed SFRC is higher than that of the ordinary

TABLE 5: Comparison of the elastic moduli of ordinary randomly distributed SFRC (unit: MPa).

Steel fibre volume fraction (%) used in different theories and formulas	Results given by composite materials theory	Results given by the empirical formula proposed by Yuan et al.	Results given by the empirical formula proposed by Gao et al.	Results given by the empirical formula proposed by Wang et al.	Experimental values
0.8	41,400	37,200	44,100	40,800	43,200
1.6	41,500		44,400	45,600	43,500

TABLE 6: Comparison of the elastic moduli of directionally distributed SFRC (unit: MPa).

Steel fibre volume fraction (%) used in different theories and formulas	Results given by composite materials theory	Results given by semiempirical formulas proposed by Balpin-Tsal	Experimental values
0.8	41,800	42,000	45,700
1.6	42,200	42,600	57,600

concrete, but it is only slightly higher. At the maximum fibre volume fraction of 1.6%, the elastic modulus is only 5% higher.

(2) Compared with the ordinary concrete, the elastic modulus of the directionally distributed SFRC increases significantly as the fibre volume fraction increases. When the fibre volume fraction is 0.8%, the elastic modulus of the directionally distributed SFRC is 10% higher than that of the ordinary concrete. As the fibre volume fraction increases to 1.6%, the elastic modulus increases by 39% compared with that of the ordinary concrete. The elastic modulus of directional SFRC clearly increases significantly as the fibre volume fraction increases.

3.2.2. Comparison with the Results from Other Studies.
Based on experiments, many scholars have established empirical or semiempirical formulas to evaluate the elastic modulus of ordinary SFRC. However, because of differences in the concrete mixture, the material properties of the concrete (e.g., the aggregates' elastic modulus, composition, and variety), the amount used, the test methods, and the results of these formulas vary widely. In this study, several representative formulas [29–32] were selected, and the test results were compared with the values produced by these formulas.

(1) Elastic Modulus of Directionally Distributed SFRC. Balpin-Tsal proposed the following semi-empirical formulas [32]:

$$E = \frac{1 + 2\gamma V_f}{1 - \gamma V_f}E_c,$$
$$\gamma = \frac{(E_f/E_c) - 1}{(E_f/E_c) + 2},$$ (1)

where E is the elastic modulus of the SFRC (MPa), V_f is the volume fraction of steel fibres (%), E_c is the elastic modulus of ordinary concrete (MPa), and E_f is the elastic modulus of steel fibres (MPa).

(2) The Elastic Modulus of Ordinary Three-Dimensional Randomly Distributed SFRC. Based on experiments, Yuan et al. proposed the following empirical formula [29]:

$$Y = 35.814 + 0.0306X_1 + 0.0034X_2,$$ (2)

where Y is the elastic modulus of SFRC (GPa), X_1 is the length to diameter ratio of the steel fibres, and X_2 is the tensile strength of the steel fibres (GPa).

Based on experiments, Gao et al. proposed the following empirical formula [31]:

$$E = 34800 f_{cu}^{0.06},$$ (3)

where E is the elastic modulus of the SFRC (MPa) and f_{cu} is the compressive strength of the SFRC (MPa).

Based on experiments, Wang et al. proposed the following empirical formula [30]:

$$y = 26600x^2 - 37.4x + 39.372,$$ (4)

where x is the steel fibre volume fraction (%) and y is the elastic modulus of the SFRC (GPa).

The SFRC elastic modulus values produced by the formula based on the theory of composite materials [27, 28, 33–35] and by the semiempirical and empirical formulas in the studies discussed above were compared with the test results in this study. The results are listed in Tables 5 and 6.

A comparison of the results in Tables 5 and 6 shows the following:

(1) There are differences among the elastic moduli of randomly distributed SFRC given by the theory of composite materials, the empirical formulas [36, 37], and the experimental values. The results given by the empirical formula proposed by Yuan et al. differ the most from the theoretical values (up to 14.5%). The differences among the results given by the other formulas and theoretical values and the experimental values are all approximately 5%. Therefore, the results given by these formulas and the theory of composite materials can be used as a general reference.

(2) The elastic moduli of the directionally distributed SFRC given by the theory of composite materials and the semiempirical formulas all differ from the experimental values. Therefore, the determination of the elastic modulus of directionally distributed SFRC is more complicated than that of ordinary randomly distributed SFRC. The existing theoretical and

semiempirical formulas do not completely satisfy the engineering demands. Consequently, additional studies should be conducted in this area.

4. Conclusions

(1) The properties of directionally distributed SFRC specimens under compression show no obvious anisotropy when the fibre volume fraction is low. When compression is applied perpendicular to the fibre direction, the damage on the specimen's surface when the specimen fails is limited; only small cracks form on the specimen's surface, and the integrity of the specimen is maintained.

(2) When compression is applied parallel to the fibre direction and the fibre volume fraction is high, the specimen's failure pattern resembles that of ordinary concrete. The compressive strength of ordinary randomly distributed SFRC increases steadily as the fibre volume fraction increases. Its strength is between the compression of parallel and perpendicular fibre directions.

(3) The elastic modulus of ordinary randomly distributed SFRC is not affected significantly by changes in the steel fibre volume fraction. However, the elastic modulus of the directionally distributed SFRC specimen is greatly affected by changes in the steel fibre volume fraction. When the fibre volume fraction was 1.6%, the elastic modulus was 39% higher than that of ordinary concrete. It must be quantified as an engineering parameter through experimental studies.

(4) The specimen's elastic modulus of directionally distributed SFRC increased significantly. The main reason is that although the fibre volume fraction is low, as in ordinary randomly distributed SFRC, the fibres' directional distribution may change the elastic deformation properties of the concrete matrix. This change increases the vertical stiffness of the concrete prism in the test, which increases the elastic modulus of the specimen.

(5) The results for directionally distributed SFRC given by existing semiempirical formulas and theoretical calculations are different from the experimental values. The determination of the elastic modulus of directionally distributed SFRC is more complex than that of ordinary randomly distributed SFRC. Therefore, further research in this area is needed for the formulas.

Conflicts of Interest

The authors declare that there are no conflicts of interest regarding the publication of this paper.

Acknowledgments

This research was supported by the National Basic Research Program of China (973 Program) under Grant no. 2013CB036303 and the Zhejiang Provincial Public Interest Technology Research Industrial Project under Grant no. 2016C31099.

References

[1] B. Gangil, A. Patnaik, A. Kumar, and M. Kumar, "Investigations on mechanical and sliding wear behaviour of short fibre-reinforced vinylester-based homogenous and their functionally graded composites," *Proceedings of the Institution of Mechanical Engineers Part L-Journal of Materials-Design and Applications*, vol. 226, no. 4, pp. 300–315, 2012.

[2] J. P. Romualdi and G. B. Batson, "Mechanics of crack arrest in concrete," *Journal of Engineering Mechanics Division*, vol. 89, no. 3, pp. 147–168, 1963.

[3] A. E. Naaman, F. Moavenzadeh, and F. J. Mcgarry, "Probabilistic analysis of fibre reinforced concrete," *Journal of the Engineering Mechanics Division*, vol. 100, no. 2, pp. 397–413, 1974.

[4] G. Zhao and C. Huang, *Summary of Research on Strengthening Mechanism of Steel Fiber Reinforced High-Strength Concrete and Design Method*, Structure Laboratory of Department of Civil Engineering, Dalian University of Technology, Dalian, China, 1993.

[5] J. P. Romualdi and J. A. Mandel, "Tensile strength of concrete affected by uniformly distributed closely spaced short lengths of wire reinforcements," *Journal of the American Concrete Institute*, vol. 61, pp. 657–671, 1964.

[6] J. Zhao and L. Cheng, *Design and Construction of Steel Fiber Reinforced Concrete*, Heilongjiang Science and Technology Press, Harbin, China, 1988.

[7] S. M. Abtahi, M. Sheikhzadeh, and S. M. Hejazi, "Fiber-reinforced asphalt-concrete-a review," *Construction and Building Materials*, vol. 24, no. 6, pp. 871–877, 2010.

[8] J. Schnell, K. Schladitz, and F. Schuler, "Direction analysis of fibres in concrete on basis of computed tomography," *Beton-und Stahlbetonbau*, vol. 105, no. 2, pp. 72–77, 2010.

[9] A. Patnaik, C. MacDonald, M. MacDonald, and V. Ramakrishnan, "Review of ASTM C1399 test for the determination of average residual strength of fiber reinforced concrete," in *Proceedings of the First International Conference on Recent Advances in Concrete Technology*, pp. 687–696, Washington, DC, USA, September 2007.

[10] X. B. Lu and C. T. T. Hsu, "Behavior of high strength concrete with and without steel fiber reinforcement in triaxial compression," *Cement and Concrete Research*, vol. 36, no. 9, pp. 1679–1685, 2006.

[11] T. Fukushima, "Resources circulation-oriented ecomaterials design of continuous fiber reinforced concrete (FRPRC)," in *Proceedings of the International Conference on Processing & Manufacturing of Advanced Materials (Thermec'2003) Pts 1–5*, vol. 426–432, pp. 3323–3328, Madrid, Spain, July 2003.

[12] A. Poitou, F. Chinesta, and G. Bernier, "Orienting fibers by extrusion in reinforced reactive powder concrete," *Journal of Engineering Mechanics*, vol. 127, no. 6, pp. 593–598, 2001.

[13] R. Mu, "Analysis of the distribution of steel fiber in aligned steel fiber reinforced concrete using digital X-ray CT scanning," *Journal of Chinese Electron Microscopy Society*, no. 6, pp. 487–491, 2015.

[14] T. Ponikiewski and J. Katzer, "X-ray computed tomography of fibre reinforced self-compacting concrete as a tool of assessing

its flexural behaviour," *Materials and Structures*, vol. 49, no. 6, pp. 2131–2140, 2016.

[15] Ministry of Housing and Urban-Rural Development of the People's Republic of China, *Standard for Test Method of Performance on Ordinary Fresh Concrete*, China Architecture and Building Press, Beijing, China, 2016.

[16] H. Li, *Investigation on the Preparation and Properties of the Aligned Steel Fiber Reinforced Concrete*, Hebei University of Technology, Tianjin, China, 2013.

[17] A. Najigivi, A. Nazerigivi, and H. R. Nejati, "Contribution of steel fiber as reinforcement to the properties of cement-based concrete: a review," *Computers and Concrete*, vol. 20, no. 2, pp. 155–164, 2017.

[18] T. Ponikiewski and J. Katzer, "Mechanical properties and fibre density of steel fibre reinforced self-compacting concrete slabs by DIA and XCT approaches," *Journal of Civil Engineering and Management*, vol. 23, no. 5, pp. 604–612, 2017.

[19] A. Jansson, I. Lofgren, K. Lundgren, and K. Gylltoft, "Bond of reinforcement in self-compacting steel-fibre-reinforced concrete," *Magazine of Concrete Research*, vol. 64, no. 7, pp. 617–630, 2012.

[20] L. Soufeiani, S. N. Raman, M. Z. B. Jumaat, U. J. Alengaram, G. Ghadyani, and P. Mendis, "Influences of the volume fraction and shape of steel fibers on fiber-reinforced concrete subjected to dynamic loading–a review," *Engineering Structures*, vol. 124, pp. 405–417, 2016.

[21] B. Nepal, C. S. Chin, and S. Jones, "A review on agricultural fibre reinforced concrete," in *Sustainable Buildings and Structures*, pp. 125–130, Taylor & Francis Group, London, UK, 2016.

[22] Ministry of Housing and Urban-Rural Development of the People's Republic of China, *Steel Fiber Reinforced Concrete*, Standards Press of China, Beijing, China, 2015.

[23] D. Daviau-Desnoyers, J. P. Charron, B. Massicotte, P. Rossi, and J. L. Tailhan, "Influence of reinforcement type on macrocrack propagation under sustained loading in steel fibre-reinforced concrete," *Structural Concrete*, vol. 17, no. 5, pp. 736–746, 2016.

[24] H. Ahmad, M. H. M. Hashim, S. H. Hamzah, and A. Abu Bakar, "Steel fibre reinforced self-compacting concrete (SFRSC) performance in slab application: a review," in *Proceeedings of the International Conference on Advanced Science, Engineering and Technology (ICASET 2015)*, p. 1774, Penang, Malaysia, December 2015.

[25] S. Abdallah, M. Z. Fan, X. M. Zhou, and S. Le Geyt, "Anchorage effects of various steel fibre architectures for concrete reinforcement," *International Journal of Concrete Structures and Materials*, vol. 10, no. 3, pp. 325–335, 2016.

[26] S. Mukhopadhyay and S. Khatana, "A review on the use of fibers in reinforced cementitious concrete," *Journal of Industrial Textiles*, vol. 45, no. 2, pp. 239–264, 2015.

[27] R. Mu, Q. M. Zhao, and W. Q. Tian, "Investigation on the preparation and properties of aligned steel fibre reinforce cement paste," *Journal of Hebei University of Technology*, no. 2, pp. 101–104, 2012.

[28] F. Y. Li, "A concrete modeling structure with directional steel fibres," China Patent 201621193064.9, 2017.

[29] H. Wang, *Steel Fiber Reinforced Concrete*, China Water and Power Press, Beijing, China, 1985.

[30] R. Wang, *Mechanical Experiment Research on Steel Fiber Reinforced Concrete and Numerical Simulation of Failure Process*, Northeastern University, Shenyang, China, 2004.

[31] D. Gao, J. Zhao, and J. Tang, "A experimental study on elastic modulus of fiber reinforced high-strength concrete," *Industrial Construction*, vol. 34, no. 10, pp. 47–49, 2004.

[32] Z. Zhang and F. Zhang, "Theoretical calculation of elastic modulus of steel fiber reinforced concrete," in *Proceedings of the 4th National Academic Conference of Fiber Cement and Fiber Concrete*, Nanjing, China, 1992.

[33] X. S. Lin, *High and Ultra High Strength Concrete with Steel Fibres*, Science Press, Beijing, China, 2002.

[34] S. C. Lee, J. Y. Cho, and F. J. Vecchio, "Tension-stiffening model for steel fiber-reinforced concrete containing conventional reinforcement," *ACI Structural Journal*, vol. 111, no. 3, pp. 717–718, 2014.

[35] Z. F. Jiang, "Steel fibre concrete," *Architecture Technology*, no. 1, pp. 56–59, 1986.

[36] M. Gencoglu, T. Uygunoglu, F. Demir, and K. Guler, "Prediction of elastic modulus of steel-fiber reinforced concrete (SFRC) using fuzzy logic," *Computers and Concrete*, vol. 9, no. 5, pp. 389–402, 2012.

[37] E. Shadafza and R. S. Jalali, "The elastic modulus of steel fiber reinforced concrete (SFRC) with random distribution of aggregate and fiber," *Civil Engineering Infrastructures Journal*, vol. 49, no. 1, pp. 21–32, 2016.

17

Influences of Ultrafine Slag Slurry Prepared by Wet Ball Milling on the Properties of Concrete

Yubo Li [ID],[1] **Shaobin Dai,**[1] **Xingyang He** [ID],[2] **and Ying Su**[2]

[1]*Wuhan University of Technology, Wuhan 430070, China*
[2]*Hubei University of Technology, Wuhan 430070, China*

Correspondence should be addressed to Xingyang He; lunwenhe2017@sina.com

Academic Editor: Sverak Tomas

The application of ultrafine ground-granulated blast-furnace slag (GGBFS) in concrete becomes widely used for high performance and environmental sustainability. The form of ultrafine slag (UFS) used in concrete is powder for convenience of transport and store. Drying-grinding-drying processes are needed before the application for wet emission. This paper aims at exploring the performances of concrete blended with GGBFS in form of slurry. The ultrafine slag slurry (UFSS) was obtained by the process of grinding the original slag in a wet ball mill, which was mixed in concrete directly. The durations of grinding were 20 min, 40 min, and 60 min which were used to replace Portland cement with different percentages, namely, 20, 35, and 50, and were designed to compare cement with original slag concrete. The workability was investigated in terms of fluidity. Microstructure and pore structure were evaluated by X-ray diffraction (XRD), scanning electron microscopy (SEM), and mercury intrusion porosimetry (MIP). The fluidity of concrete mixed with UFSS is deteriorated slightly. The microstructure and early strength were obviously improved with the grind duration extended.

1. Introduction

Cement, the main cementitious material used in the concrete, is the base material of construction. Cement manufacture is a highly energy-intensive process. The total energy consumption of the global cement industry is estimated at 2% of global primary energy use, which accounts for approximately 7% of all global carbon dioxide emissions [1, 2]. The development of ecofriendly concrete in the construction industry is gaining deep concern and intensive research worldwide.

Ground-granulated blast-furnace slag (GGBFS) is a by-product of steel making [3]. GGBFS has been found to exhibit excellent cementitious properties when it was finely powdered [4]. The use of GGBFS as supplementary cementitious material not only reduces the usage of cement [5] but also improves the porosity performance of OPC concrete [6]. In addition, GGBFS is one of the major precursor materials used in the production of alkali-activated materials (AAMs) [7]. As a kind of alternative cementitious material, GGBFS shows comparable mechanical performance to blended slag cement and similar or even lower global warming potential compared with the best available concrete technology [8]. The reactivity of GGBFS is considered an important parameter to assess the effectiveness of GGBFS in concrete composites, which varies greatly with the source of slag, types of raw material used, methods of cooling, and the duration of milling [9]. Pal et al. [10] investigated the relationship between the hydraulic index (HI) of slag and the influencing factors of slag, including glass content, fineness, and chemical composition. Zhao et al. [11] studied the particle characteristics and hydration activity of GGBFS containing industrial crude glycerol-based milling aids. The result indicated that means and milling aids strongly affect the activity of GGBFS.

Wet milling is a method of producing powder slurry by milling materials with mediums together, with advantages of uniformity, high milling efficiency, and small noise contrast with dry milling. It is widely used in cement and ceramics industries [12] and rarely in the field of slag powder. Nowadays, the main milling methods of slag are dry milling processes and need to go through drying-milling-drying

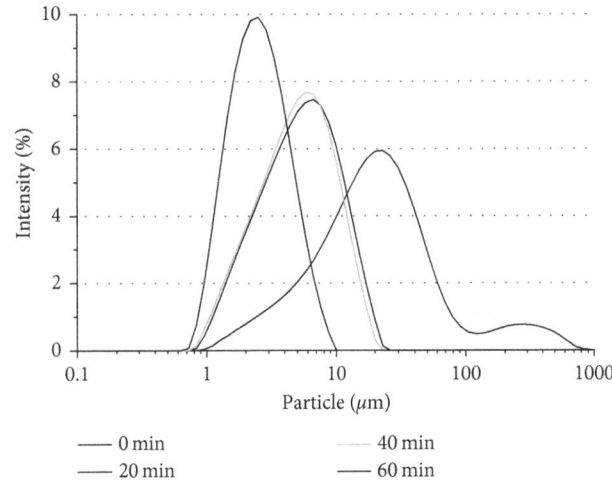

FIGURE 1: Different particle size distributions of slag particles at different wet milling durations.

processes before application for wet emissions. If the dry milling process is replaced by the wet milling process, the drying process can be omitted and be applied to concrete directly.

In this paper, the feasibility of UFSS prepared by the wet ball mill process applied in concrete was investigated. The properties of concrete were tested by kinds of test methods and analysis methods.

2. Materials and Experimental Details

2.1. Materials

2.1.1. Ground-Granulated Blast-Furnace Slag (GGBFS). The GGBFS was from the Capital Iron and Steel Company in the state of Beijing (China). Specific gravity of GGBFS was 2.92. The particle size distribution and microstructure of slag slurry/powder for different wet milling durations and the original slag (0 min) are shown in Figure 1, Table 1, and Figure 2. The chemical composition and physical properties of slag are shown in Table 2. Compared with the original slag, the size range of UFSS was reduced and the morphology was smooth.

2.1.2. Cement. Cement was from Jidong Cement Co., Ltd, the state of Hebei province (China), Portland cement 42.5, and the specific surface area is $316 \, m^2/kg$. The physical and mechanical properties of cements are shown in Table 2.

2.1.3. Other Materials. The fine aggregates and coarse aggregates used were natural river sand and broken stones. Maximum particle size of fine aggregates and coarse aggregates was 2.36 mm and 20 mm, respectively. Fineness modulus of fine aggregates was 2.7. Table 3 shows the properties of coarse and fine aggregates of concrete. The indicator test methods were carried out according to the Chinese national standards GB14684-2011 and GB14685-2011.

TABLE 1: The main integral distribution of slag powder at different wet grinding times.

Sample	$d \, (0.1)$ (μm)	$d \, (0.5)$ (μm)	$d \, (0.9)$ (μm)
0 min	4.32	18.18	70.47
20 min	1.82	5.01	11.41
40 min	1.734	4.73	10.59
60 min	1.21	2.32	4.62

2.2. Experimental Details

2.2.1. Experimental Design. The durations of grinding were 20 min (UFSS20), 40 min (UFSS40), and 60 min (UFSS60) which were used to replace Portland cement with different percentages, namely, 20, 35, and 50, and were designed to compare portland cement with original slag concrete (from B1 to E3, Table 4). Consulting the literature [13–16], 10–60% of slag replace cement in concrete displays good performance, and the fineness of slag can also make a difference. In this study, the specimens of compressive strength and workability were concrete, and that of microstructure tests were corresponding paste. The designations of mixtures are shown in Table 4; the concrete mix proportion is cement : water : sand : stone = 1 : 0.5 : 1.57 : 2.36. The mixture codes in Table 4 are the percentage figures, which indicated the weight percentage between different raw materials.

2.2.2. Wet Ball Milling Details. The ball crusher is YXQM-2L, the grinding speed is 400 r/min, the ratio of water to solid is 0.5, the ratio of ball to materials is 4, and the milling media are Φ8 mm agate ball and Φ3 mm zirconia ball.

2.2.3. Specimen Preparation. GGBFS and water were mixed and ground. Firstly, different UFSS were obtained through controlled grinding duration based on the milling curve. Then, the UFSS was mixed with the remaining water during the preparation of the specimen, and the other conventional steps were carried out according to the Chinese national standard GB/T 50081-2002.

FIGURE 2: SEM images of slag powder at different wet grinding times: (a) raw slag and wet grinding for (b) 20 min, (c) 40 min, and (d) 60 min.

TABLE 2: Chemical composition of GGBFS and cement.

	Oxide composition of GGBFS (wt.%)	Oxide composition of cement (wt.%)
SiO_2	33.50	23.03
Al_2O_3	12.52	5.11
Fe_2O_3	1.10	3.34
CaO	37.90	63.33
MgO	9.29	2.06
SO_3	2.51	2.33

TABLE 3: The properties index of coarse and fine aggregates of concrete.

	Grading	Mud content (%)	Apparent density (kg/m³)	Bulk density (kg/m³)
Coarse	Continuous	0.4	2773	1458
Sand	II level	1.6	2659	1463

2.3. Testing Methods

2.3.1. Flowability. The flowability tests of the fresh mixed concrete were performed in conformity with the Chinese national standard GB/T 2419-2005.

2.3.2. Compressive Strength. The compressive strength tests were performed in conformity with the Chinese national standard GB/T 17671-2005, and the ages were 3 days, 7 days, 28 days, and 90 days.

2.3.3. Scanning Electron Microscopy (SEM). The microstructure of specimens of 28 days was tested by scanning electron microscopy (FEI Quanta 450FEG), with the magnification of ×2000 and ×10000.

2.3.4. X-Ray Diffraction (XRD). The model of laboratory X-ray diffraction used was D/MAX-RB (RIGAKU Corporation, Japan). The test angle range was 5–70°, and the test error was controlled within 0.02° ($\Delta 2\theta \leq \pm 0.02°$). The age of specimens was 28 days.

2.3.5. Mercury Intrusion Porosimetry (MIP). The microporous structure of specimens at 28 days was tested by mercury intrusion porosimetry complying ISO 15901-1:2005. The porosity, median pore diameter of area, and the average pore diameter of hydration production were analyzed using Demo windows 9400 series software.

3. Results and Discussion

3.1. Workability. The workability is the ease of working with a freshly mixed concrete in the stages of handing, placing, compacting, and finishing. Slump is always regarded as an indicator of the workability of concrete [14]. To explore the feasibility of UFSS replace cement in concrete, the slump was tested and is depicted in Figure 3.

The content and grinding duration of UFSS reduced slump. Mixing E3 with 50% UFSS60 showed a slump value of 169 mm as compared to 180 mm and 174 mm showed by mixtures E1 and E2 which had 20% and 35% UFSS60.

TABLE 4: Designations of mixtures in this research.

Number		GGBFS		Cement (wt.%)	Sand (wt.%)	Coarse (wt.%)	Water (wt.%)
		Content (wt.%)	Milling duration (min)				
A	A1	0	—	100	157	236	50
B	B1	20	0	80	157	236	50
	B2	35	0	65	157	236	50
	B3	50	0	50	157	236	50
C	C1	20	20	80	157	236	50
	C2	35	20	65	157	236	50
	C3	50	20	50	157	236	50
D	D1	20	40	80	157	236	50
	D2	35	40	65	157	236	50
	D3	50	40	50	157	236	50
E	E1	20	60	80	157	236	50
	E2	35	60	65	157	236	50
	E3	50	60	50	157	236	50

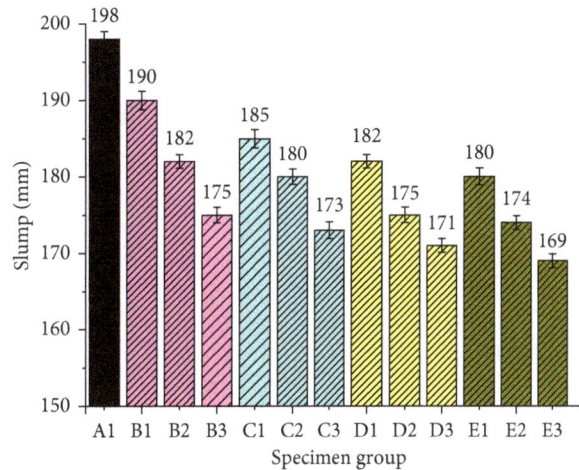

FIGURE 3: Slump of mixture with different contents and milling durations.

Similarly discipline was observed in series A, B, C, and D. The increase of content of slag decreased the fluidity, which confirmed the trend obtained by Deb et al. [14]. Under the condition of same dosage of slag, take 35% as an example, the slump of C2, D2, and E2 were reduced by 5.3%, 7.9%, and 8.4%, respectively, compared to B2. Similarly discipline was observed in 20% and 50% series.

The slump was decreased slightly with the increase of content and milling duration extended for slag. The results of this part were similar to the results of literatures [17, 18]. The reasons for that were that the increases of surface of slag powder led to increased water demand. What is more, the slag was contacted with the aqueous medium directly, and the vitreous network of slag particles is more easily to be dissolved in the process of wet ball milling. The surface of the slag particles becomes rough, leading to poor workability. This may be a further reason for the above results.

3.2. Compressive Strength. The results of compressive strength tests of specimens at different ages are depicted in

Figure 4. UFSS could accelerate the development of compressive strength at the early age of concrete. For E3 with 50% of UFSS60, the 3-day strength reached 60.1% of that of 28 days, while the values of B3, C3 and, D3 were found to be 48.1%, 56%, and 58.7%, respectively. For 50%, the ratios of 7 and 28 days strength of original slag, UFSS20, UFSS40, and UFSS60 were found to be 63.7%, 65.7%, 69.3%, and 74.2%, respectively.

The above results confirmed the trend obtained in dry ground slag by Yan Shi and Arash Aghaeipour [19, 20]. The reasons for the above results may be the disparity of activity and particle size of slag. The average diameter (d (0.5)) of UFSS20, UFSS40, and UFSS60 was 5.01 μm, 4.73 μm, and 2.32 μm, respectively, while that of original slag was 18.18 μm. The compactness of concrete was improved, and the large pores were reduced effectively (specification in Section 3.3 specifically). Sharmila and Dhinakaran [21] studied the compressive strength of commercially available ultrafine slag applied to the concrete. They found that the performance of concrete was improved. The study [22] showed that the activity of slag was lower than that of cement and failed to exhibit considerable reaction in the early age.

(a)

(b)

(c)

(d)

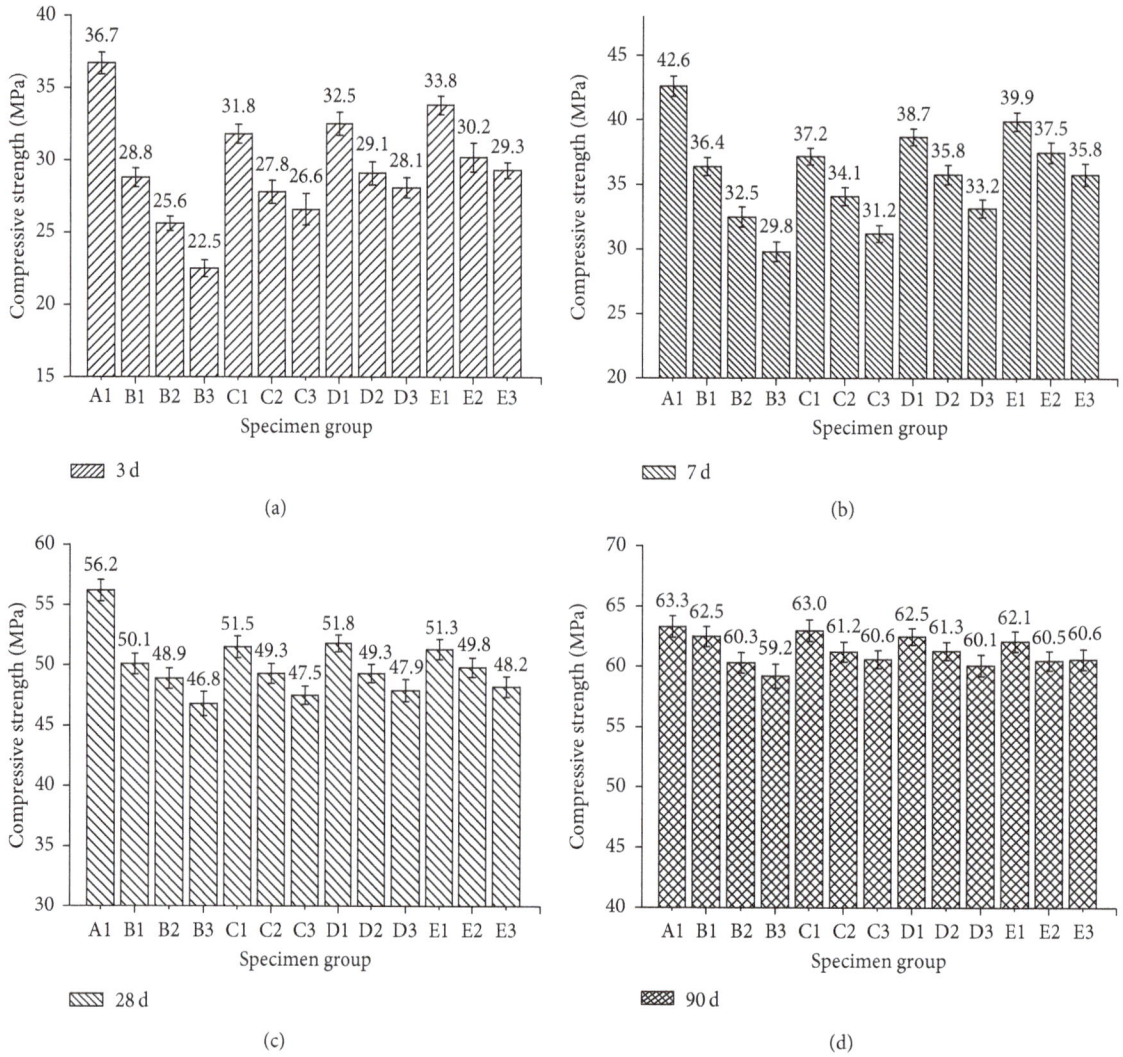

FIGURE 4: Compressive strength of specimens at different ages.

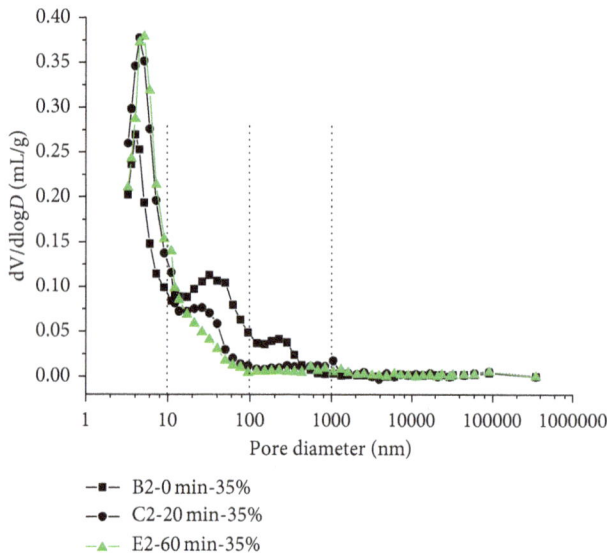

FIGURE 5: Pore size distribution curves of specimens with 35% of UFSS at 28 days.

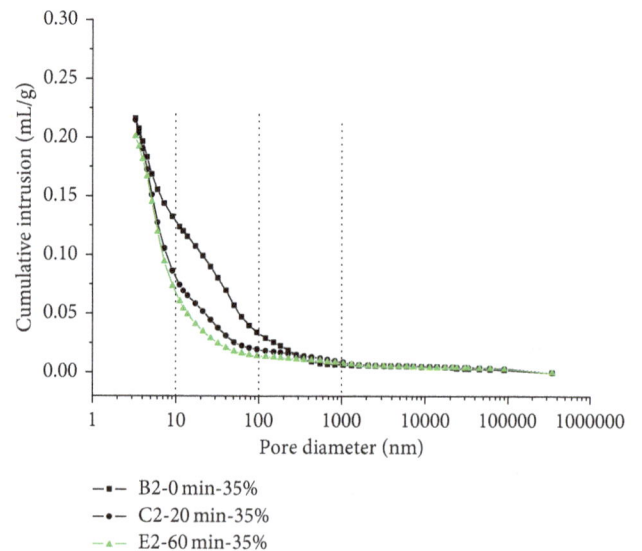

FIGURE 6: Cumulative pore volumes of specimens with 35% of slag cured for 28 days.

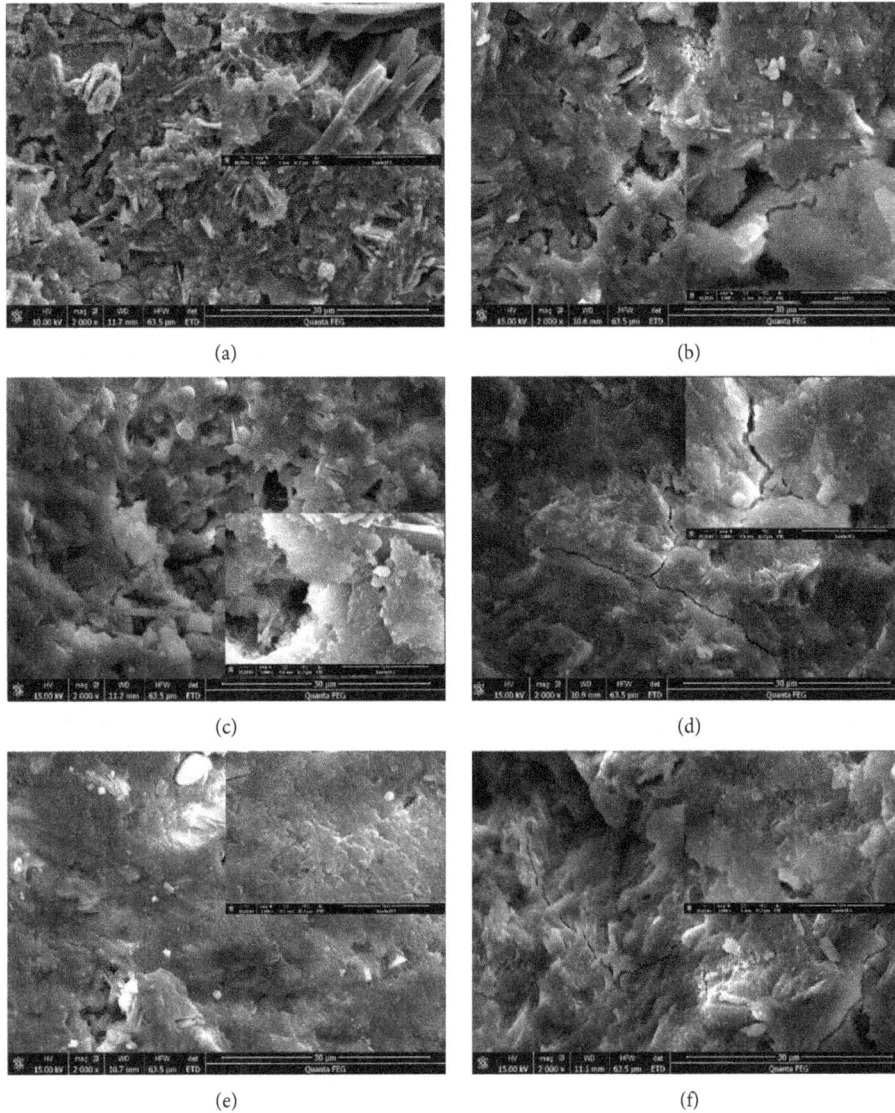

FIGURE 7: SEM images of specimens cured for 28 days: (a) B2-0 min-35%, (b) C2-20 min-35%, (c) D1-40 min-20%, (d) D2-40 min-35%, (e) D3-40 min–50%, and (f) E2-60 min-35%.

The activity of slag disposed by the milling process was improved, which was beneficial for the improvement of compressive strength especially in early ages.

The results of compressive strength at ages 28 days and 90 days showed that the compressive strength of specimens decreased with the increase of slag content and increased with the duration of grinding extend. The gaps of compressive strength of specimens become smaller with the extension of age. Sharmila and Dhinakaran [23] found that the compressive strength of concrete was reduced when the content of ultrafine slag exceeded 15%. The aggregation of fine particles leads to a higher porosity. However, this phenomenon was not observed with slag of 50% in this study. The wet ball mill was carried out in aqueous media, the surface of the particles formed hydrated films under the effect of aqueous media, and the surface energy of particles was relatively lower to the dry ball mill. What is more, the

UFSS was mixed with water first and then mixed together with other raw materials; the particles get a good dispersion in the concrete.

3.3. Pore Structure. The results of pore size distribution (PSD) of B2, C2, and E2 at 28 days are shown in Figure 5. UFSS refined the pores and created a distribution peak for the pore range within 100 nm, and the trend was clearly observed in the specimens with UFSS60. Three characteristic ranges of pore sizes were divided: <10 nm (Part I), 10 nm–100 nm (Part II), and >100 nm (Part III), with various peaks in each range, which represent small pores, middle capillary pores, and larger capillary pores [24].

For specimens incorporating 35% of UFSS, the pattern of PSD curves was significantly different for different grinding durations. B2 samples (original slag, 35%) present triplet

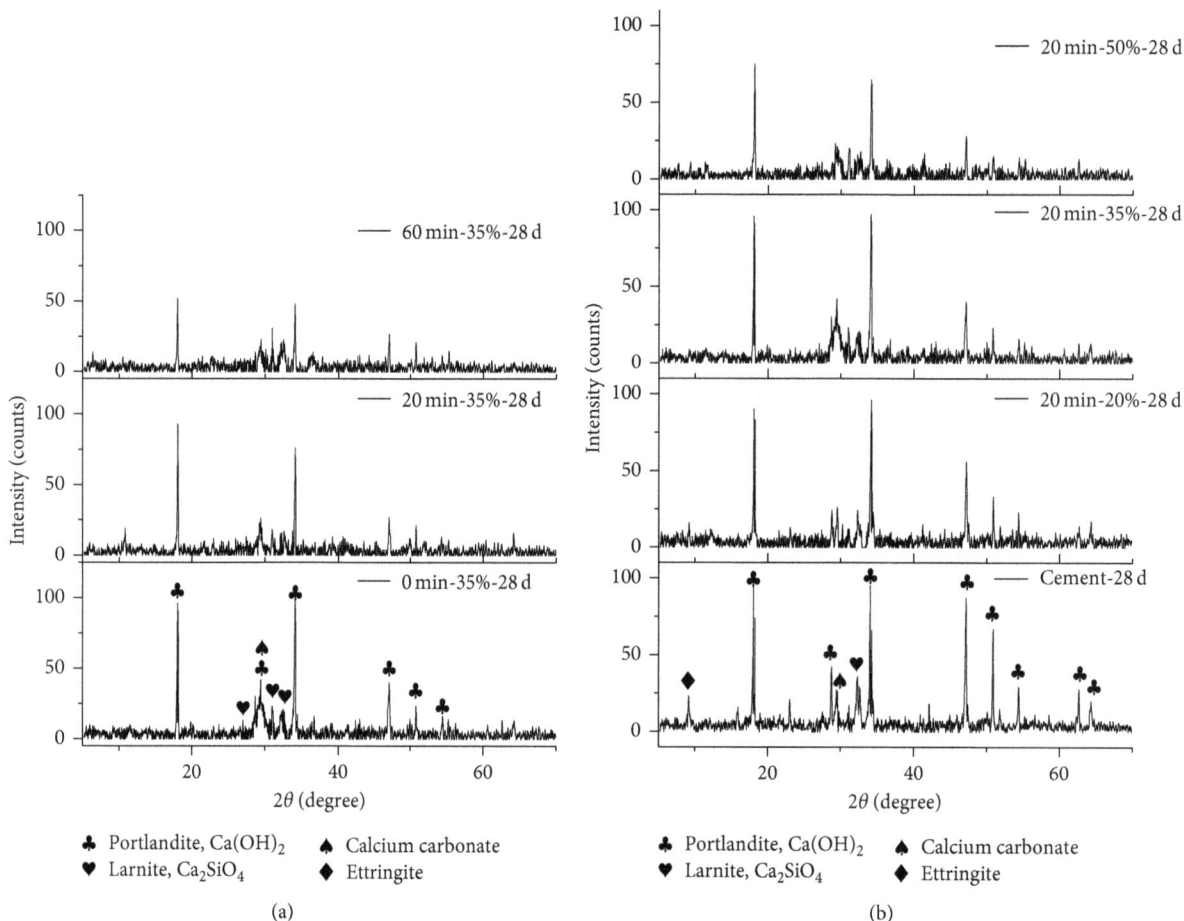

FIGURE 8: XRD of the specimens cured for 28 days.

peak characteristics in Part I, Part II, and III. The pores consisted of small pores, middle capillary pores, and larger capillary pores. C2 samples (UFSS20, 35%) present dual peak characteristics in Part I and Part II. The pores were refined and mainly consisted of small pores and middle capillary pores. E2 samples (UFSS20, 35%) present single peak characteristics in Part I, which mainly consists of small pores.

For the range of Part I, the formation of small pores was mainly the hydration of cementitious materials [24, 25]. The hydration extent of B2 was limited compared to the other two groups. The values of C2 and E2 were close, indicating that the extents of hydration reaction were similar with an age of 28 days. For the range of Part II, the pores >10 nm were formed by the filling effect of particles in concrete [26]. The peak of the PSD curve of B2 was obvious; the compactness of hardened body was poor relatively. The peak of the PSD curve of C2 exits and is weaker than B2, indicating that UFSS20 could improve the compactness of concrete compared to original slag. E2 showed a smooth curve in this part, indicating that the compactness of concrete could be improved obviously. For the range of Part III, only the curves of B2 exit a peak. UFSS obtained from wet milling could improve the compactness of concrete effectively.

Figure 6 shows the cumulative pore volumes of the specimens corresponding to Figure 5. As the particle size of slag

reduces, the pore volume decreased. The porosities of B2, C2, and E2 were 24.7986%, 24.0260%, and 23.7765%, respectively. The porosities decreased with the reduction of sizes of particles. The slag slurry reduced the distribution of larger pores and optimized the porosity distribution. The compactness and compressive strength of concrete were optimized.

3.4. Microstructure Tests. Figure 7 shows the SEM images of specimens incorporating UFSS at 28 days. The larger capillary pores were easily observed in B2. From the images of D1, D2, and D3, the compactness of specimens was improved with the increasing of slag content. Ultrafine slag particles played an important role in the compactness performance of harden body. The compactness of pates was improved with the grind duration extending, which was consistent with the results of MIP tests. This further explained the increase of compressive strength of concrete with ultrafine slag.

The cement particles were wrapped by slag particles, causing delay of the reaction process for cement particles, and the hydrated products of cement particles were not easily observed. However, the activity of slag particles was lower than that of cement particles even through the milling process. This was why the compressive strength decreased with the increased content of slag, even if the porosity was improved.

3.5. X-Ray Diffraction Analysis. XRD analysis was carried out to examine the influences of the content and fineness of slag on hydration and phase. XRD images are shown in Figure 8. From XRD analysis, mineral phases of portlandite, calcium carbonate, larnite, and ettringite were found.

In the cement paste blended with slag, three structure reactions were involved: cement hydration, the pozzolanic reaction of slag, and hydration reaction of slag [27, 28]. However, slag exhibits a pozzolanic reaction in the presence of calcium hydroxide ($Ca(OH)_2$) formed upon cement hydration [20]. The extent of slag involved in the reaction could be reflected by the amount of calcium hydroxide under same content conditions. It could be seen from the XRD images (Figure 8(a)) that the content of calcium hydroxide decreased with the decrease of particle size. The results illustrated that the activity slag treated through wet milling was increased, which accounted for the increase of compressive strength. With different slag dosages, the test results of various contents of slag (Figure 8(b)) complied with the above regularity: the amount of calcium hydroxide reduced with the increase in content.

4. Conclusion

This study investigated the influences of ultrafine slag slurry (UFSS) prepared by the wet ball mill on the properties of cement and concrete. The results obtained were summarized as follows:

(i) Wet ball mill could improve the size distribution of slag particles effectively. The slag decreased the slump of concrete slightly.

(ii) Use of UFSS as a substitute to cement improved the compressive strength of concrete especially at early ages.

(iii) UFSS optimized the pore size distribution of the plaster. The amount of large pores (10–100 nm) were decreased notably and created distribution in the range of small pores (<10 nm).

Conflicts of Interest

The authors declare that there are no conflicts of interest regarding the publication of this article.

Acknowledgments

The authors gratefully acknowledge the support of the National Key Research and Development Plan (no. 2017YFB0310003) and the Science and Technology Support Program of Hubei Province (no. 2017ACA178) for funding this project.

References

[1] G. Tesema and E. Worrell, "Energy efficiency improvement potentials for the cement industry in Ethiopia," *Energy*, vol. 93, no. 2, pp. 2042–2052, 2015.

[2] J. Li, P. Tharakan, D. Macdonald, and X. Liang, "Technological, economic and financial prospects of carbon dioxide capture in the cement industry," *Energy Policy*, vol. 61, no. 10, pp. 1377–1387, 2013.

[3] C. L. H Wang and C. Y. Lin, "Strength development of blended blast furnace slag cement mortars," *Journal of the Chinese Institute of Engineers*, vol. 9, no. 3, pp. 233–239, 1986.

[4] M. C. G. Juenger and R. Siddique, "Recent advances in understanding the role of supplementary cementitious materials in concrete," *Cement and Concrete Research*, vol. 78, pp. 71–80, 2015.

[5] A. A. Ramezanianpour, "Effect of curing on the compressive strength, resistance to chloride-ion penetration and porosity of concretes incorporating slag, fly ash or silica fume," *Cement and Concrete Composites*, vol. 17, no. 2, pp. 125–133, 1995.

[6] M. Mahoutian, Y. Shao, A. Mucci, and B. Fournier, "Carbonation and hydration behavior of EAF and BOF steel slag binders," *Materials and Structures*, vol. 48, no. 9, pp. 3075–3085, 2015.

[7] M. Iqbal Khan, G. Fares, and S. Mourad, "Optimized fresh and hardened properties of strain hardening cementitious composites: effect of mineral admixtures, cementitious composition, size, and type of aggregates," *Journal of Materials in Civil Engineering*, vol. 29, no. 10, p. 04017178, 2017.

[8] F. Puertas and A. Fernández-Jiménez, "Mineralogical and microstructural characterization of alkali-activated fly ash/slag pastes," *Cement and Concrete Composites*, vol. 25, no. 3, pp. 287–292, 2003.

[9] S. Kumar, R. Kumar, A. Bandopadhyay et al., "Mechanical activation of granulated blast furnace slag and its effect on the properties and structure of portland slag cement," *Cement and Concrete Composites*, vol. 30, no. 8, pp. 679–685, 2008.

[10] S. C. Pal, A. Mukherjee, and S. R. Pathak, "Investigation of hydraulic activity of ground granulated blast furnace slag in concrete," *Cement and Concrete Research*, vol. 33, no. 9, pp. 1481–1486, 2003.

[11] J. Zhao, D. Wangb, and P. Yana, "Particle characteristics and hydration activity of ground granulated blast furnace slag powder containing industrial crude glycerol-based grinding aids," *Construction and Building Materials*, vol. 104, pp. 134–141, 2016.

[12] H. Goudarzi and S. Baghshahi, "PZT ceramics prepared through a combined method of B-site precursor and wet mechanically activated calcinate in a planetary ball mill," *Ceramics International*, vol. 43, no. 4, pp. 3873–3878, 2017.

[13] A. Nazari, M. H. Rafieipour, and S. Riahi, "The effects of CuO nanoparticles on properties of self compacting concrete with GGBFS as binder," *Materials Research*, vol. 14, no. 3, pp. 307–316, 2011.

[14] P. S. Deb, P. Nath, and P. K. Sarker, "The effects of ground granulated blast-furnace slag blending with fly ash and activator content on the workability and strength properties of geopolymer concrete cured at ambient temperature," *Materials and Design*, vol. 62, pp. 32–39, 2014.

[15] J. Qiu, H. S. Tan, and E.-H. Yang, "Coupled effects of crack width, slag content, and conditioning alkalinity on autogenous healing of engineered cementitious composites," *Cement and Concrete Composites*, vol. 73, pp. 203–212, 2016.

[16] A. Allahverdi and M. Mahinroosta, "Mechanical activation of chemically activated high phosphorous slag content cement," *Powder Technology*, vol. 245, pp. 182–188, 2013.

[17] Y. Tang, X. Zuo, S. He, O. Ayinde, and G. Yin, "Influence of slag content and water-binder ratio on leaching behavior of cement pastes," *Construction and Building Materials*, vol. 129, pp. 61–69, 2016.

[18] P. Nath and P. K. Sarke, "Effect of GGBFS on setting, workability and early strength properties of fly ash geopolymer concrete cured in ambient condition," *Construction and Building Material*, vol. 66, pp. 163–171, 2016.

[19] A. Aghaeipour and M. Madhkhan, "Effect of ground granulated blast furnace slag (GGBFS) on RCCP durability," *Construction and Building Materials*, vol. 141, pp. 533–541, 2017.

[20] Y. Shi, H. Chen, J. Wang, and Q. Feng, "Preliminary investigation on the pozzolanic activity of superfine steel slag," *Construction and Building Materials*, vol. 82, pp. 227–234, 2015.

[21] P. Sharmila and G. Dhinakaran, "Compressive strength, porosity and sorptivity of ultra fine slag based high strength concrete," *Construction and Building Materials*, vol. 120, pp. 48–53, 2016.

[22] A. Karimpour, "Effect of time span between mixing and compacting on roller compacted concrete (RCC) containing ground granulated blast furnace slag (GGBFS)," *Construction and Building Materials*, vol. 24, no. 11, pp. 2079–2083, 2010.

[23] P. Sharmila and G. Dhinakaran, "Strength and durability of ultra fine slag based high strength concrete," *Structural Engineering and Mechanics*, vol. 55, no. 3, pp. 675–686, 2015.

[24] K. Li, Q. Zeng, M. Luo, and X. Pang, "Effect of self-desiccation on the pore structure of paste and mortar incorporating 70% GGBS," *Construction and Building Materials*, vol. 51, pp. 329–337, 2014.

[25] Q. Zeng, K. Li, T. Fen-Chong, and P. Dangla, "Pore structure characterization of cement pastes blended with high-volume fly ash," *Cement and Concrete Research*, vol. 42, no. 1, pp. 194–204, 2012.

[26] Y. C. Choi, J. Kim, and S. Choi, "Mercury intrusion porosimetry characterization of micropore structures of high-strength cement pastes incorporating high volume ground granulated blast-furnace slag," *Construction and Building Materials*, vol. 137, pp. 96–103, 2017.

[27] X. Feng, E. J. Garboczi, D. P. Bentz, P. E. Stutzman, and T. O. Mason, "Estimation of the degree of hydration of blended cement pastes by a scanning electron microscope point-counting procedure," *Cement and Concrete Research*, vol. 34, no. 10, pp. 1787–1793, 2004.

[28] Y. C. Ding, T. W. Cheng, P. C. Liu, and W. H. Lee, "Study on the treatment of BOF slag to replace fine aggregate in concrete," *Construction and Building Materials*, vol. 146, pp. 644–651, 2017.

18

Optical-Fiber-Based Smart Concrete Thermal Integrity Profiling: An Example of Concrete Shaft

Ruoyu Zhong, Ruichang Guo, and Wen Deng ⓘ

Missouri University of Science and Technology, Rolla, MO, USA

Correspondence should be addressed to Wen Deng; wendeng@mst.edu

Academic Editor: Andrey E. Miroshnichenko

Concrete is currently the most widely used construction material in the world. The integrity of concrete during the pouring process could greatly affect its engineering performance. Taking advantage of heat production during the concrete curing process, we propose an optical-fiber-based thermal integrity profiling (TIP) method which can provide a comprehensive and accurate evaluation of the integrity of concrete immediately after its pouring. In this paper, we use concrete shaft as an example to conduct TIP by using the optical fiber as a temperature sensor which can obtain high spatial resolution temperature data. Our method is compared with current thermal infrared probe or embedded thermal sensor-based TIP for the concrete shaft. This innovation makes it possible to detect defects inside of the concrete shaft with thorough details, including size and location. First, we establish a 3D shaft model to simulate temperature distribution of concrete shaft. Then, we extract temperature distribution data at the location where the optical fiber would be installed. Based on the temperature distribution data, we reconstruct a 3D model of the concrete shaft. Evaluation of the concrete integrity and the existence of the potential defect are shown in the paper. Overall, the optical-fiber-based TIP method shows a better determination of defect location and size.

1. Introduction

Concrete is currently the most widely used construction material all over the world. Concrete consists of both fine and coarse aggregates that are bonded by cement paste. Hydration reaction occurs when cement is blended with water. This hydration is an exothermal reaction, which means it generate heat and results in temperature rise during concrete curing. At the beginning, tricalcium aluminate (C_3A) reacts with water and generates a large amount of heat, but the reaction would not last long. It is followed with a short period that release less heat called the dormant phase. After a short period of dormancy, the alite and belite start to react and continuously generate heat. The maximum heat generation would last 10 to 20 hours after pouring. Since cementitious material can generate a large amount of heat, defects of concrete would lead to temperature divergence in the concrete structure. By taking advantage of such a feature of concrete, we propose an innovative method of optical fiber-based thermal integrity profiling (TIP) to inspect the concrete integrity and use the concrete shaft as an example to demonstrate the method.

Since the concrete shaft serves as the deep foundation, the quality of the concrete shaft is critical for the safety of superstructures. Defects within the concrete shaft would degrade the shaft performance (Figure 1). The existence of defects within the concrete shaft is mainly due to some problems of construction and design deficiencies [1]. Among 5,000 to 10,000 shafts tested, 15% of shafts showed the deviation from ideal signal and 5% of tested shafts showed indisputable defect indication [2, 3]. Since both excavation and concreting are blind processes when building drilled shafts, it is impossible to completely prevent defects from happening during construction. Determination of whether defects exist in the concrete shaft and how severe the defects are, are crucial to evaluate whether the concrete shaft would satisfy its design purpose. Among the existing methods, nondestructive testing is a widely accepted method for shaft integrity testing.

Currently, major nondestructive testing methods include low strain integrity test and cross-hole sonic logging (CSL). The low strain integrity test, also known as sonic pulse echo method, uses light hammer impacts and evaluates the

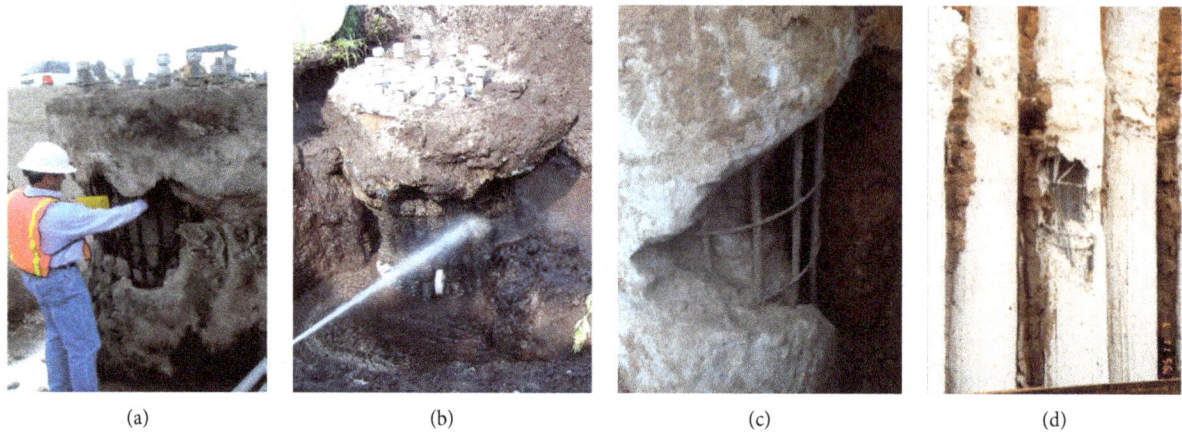

(a)   (b)   (c)   (d)

FIGURE 1: Different kinds of defects (images are from an online source).

collected force and velocity recordings to evaluate shaft integrity [4–6]. The low strain integrity test is cost-efficient and effective. However, this method has limitations including operator's familiarity and experience, and length/width ratio of the concrete shaft. CSL is a widely used nondestructive integrity test method. For CSL, 3–8 access tubes must be installed within a shaft cross section [7]. Then, a signal generator coupled with receiver is lowered, maintaining a consistent elevation, to test the integrity of concrete shaft. This method has higher accuracy, but is limited within the reinforcement cage. The CSL method only tests the integrity of concrete shaft between the access tubes, while outside of that zone is left untested. However, the bending capacity of the concrete shaft is mainly dependent on the outer part. The core of the concrete shaft has little contribution to bending capacity [8, 9]. The integrity of the outer part of concrete shaft should be evaluated as well.

TIP, which is a new nondestructive testing method, makes use of the hydration heat generated during concrete curing to determine whether defects exist and estimates their size and location according to the temperature distribution along the concrete shaft [10–12]. Temperature distribution is measured by lowering a thermal probe with infrared thermocouples into access tubes or by an embedded thermal sensor during the curing process. Inverse modeling of temperature distribution would provide information on whether the reinforcement cage has been misplaced or improper formation has happened. In addition, the location and type of defect would be indicated from the data. A relatively cool region indicates a shortage of concrete at that particular location, whereas a relatively warm region indicates extra concrete. Compared with previous methods, TIP covers a larger area and provides a more comprehensive result. However, due to the limited amount of access tubes, temperature data for inverse modeling could be insufficient to accurately predict temperature distribution of the concrete shaft which could limit further development of this method [11, 13].

Referring to advancements in optical fiber studies, Rayleigh scattering caused by local refractive index fluctuations along the glass fiber can be used to measure strains and temperature. Every point on the optical fiber can send a different Rayleigh scattering signal when subjected to temperature change, and therefore, every point along the fiber acts as a temperature sensor [14–18]. This feature of optical fibers makes it an ideal temperature sensor to measure high spatial resolution temperature distribution. Currently, this technology has been applied to measuring and recording temperature data, for example, in car engines, microwave ovens, or large furnaces for the steel industry. Advances in the research on Rayleigh scattering-based optical fiber make its application on TIP possible. By applying this optical fiber to TIP as a temperature sensor, more comprehensive and consistent temperature data can be provided. The conventional method sets an access tube every 300 mm diameter, and a measurement point within access tubes with vertical intervals less than 500 mm. In this optical-fiber based method, the optical fiber would be wrapped around the reinforcement cage spirally and densely with negligible cost of the fiber itself. Even if the vertical interval is the same as the conventional method when wrapping the optical fiber, the horizontal interval would still be significantly smaller. Temperature data measured by optical-fiber-based TIP would have high spatial resolution. The measured temperature data interval can be as small as 4 mm. Thus, the inverse modeling of temperature distribution can produce a more reliable integrity report [19–22].

The objective of this paper is to address the advantages of our proposed optical-fiber-based TIP method regarding its inverse modeling of temperature distribution of defected concrete shaft by having high resolution spatial temperature data. We used finite element method (FEM) to simulate temperature distribution of a defective concrete shaft. Temperature data were extracted in two different ways based on the concepts of our new method and conventional infrared thermal probe method. Based on the temperature distribution data, we reconstructed the 3D geometry of concrete shaft based on two methods. The impact of the size and location of the defect on temperature distribution is discussed in this paper.

2. Methodology

2.1. Governing Equation.
The principle of TIP is to take advantage of the correlation between the shape of the concrete shaft and temperature distribution. The temperature distribution is simulated using FEM. The governing equation of temperature (T) distribution in concrete shaft is as follows:

$$\rho C_p \frac{\partial T}{\partial t} = \left[\frac{\partial}{\partial x}\left(k\frac{\partial T}{\partial x} \right) + \frac{\partial}{\partial y}\left(k\frac{\partial T}{\partial y} \right) + \frac{\partial}{\partial z}\left(k\frac{\partial T}{\partial z} \right) \right] + Q, \tag{1}$$

where C_p represents the heat capacity of the material, k is the thermal conductivity of the material, and Q is the heat source of the material.

2.2. Heat Generation.
The total amount of heat production and the rate of heat production are two important factors of temperature distribution. These two factors determine the temperature of the concrete shaft and the timing for TIP to be performed. The amount of heat and heat production rate are related to the ingredients of the concrete. Concrete with different proportions would generate different amounts of heat. The total heat production can be determined by using the following equations [23]:

$$\begin{aligned} Q_0 &= Q_{cem}p_{cem} + 461p_{slag} + Q_{FA}p_{FA}, \\ Q_{cem} &= 500p_{C_3S} + 260p_{C_2S} + 866P_{C_3A} + 420p_{c_4AF} \\ &\quad + 624p_{SO_3} + 1186p_{FreeCaO} + 850p_{MgO}, \\ Q_{FA} &= 1800p_{FACaO}. \end{aligned} \tag{2}$$

The degree of hydration can be determined by the following equations [10, 23]:

$$\begin{aligned} \alpha(t) &= \alpha_u \exp\left(-\left[\frac{\tau}{t_e} \right]^\beta \right), \\ \alpha_u &= \frac{(1.031w/cm)}{(0.194 + w/cm)} + 0.5p_{FA} + 0.3p_{SLAG} < 1, \\ \beta &= p_{C_3S}^{0.227} \cdot 181.4 \cdot p_{C_3A}^{0.146} \cdot Blaine^{-0.535} \cdot p_{SO_3}^{0.558} \\ &\quad \cdot \exp\left(-0.647p_{SLAG} \right), \\ \tau &= p_{C_3S}^{-0.401} \cdot 66.78 \cdot p_{C_3A}^{-0.154} \cdot Blaine^{-0.804} \cdot p_{SO_3}^{-0.758} \\ &\quad \cdot \exp\left(2.187 \cdot p_{SLAG} + 9.5 \cdot p_{FA} \cdot p_{FACaO} \right), \end{aligned} \tag{3}$$

where $\alpha(t)$ represents the degree of hydration of cement at time t, w/cm is the water-cement ratio, and β and τ are determined by the cementitious constituent fractions. According to ASTM D7949-14, the recommended timing to perform TIP would be 12 hours after concrete placement until the number of days is equivalent to the foundation diameter in meters divided by 0.3 m.

2.3. Heat Transport.
Heat transport is another important factor for temperature evolution within the concrete shaft. Heat is dissipated into the surrounding soil simultaneously after heat is generated due to hydration. Heat transport includes three mechanisms: conduction, convection, and radiation. In this situation, heat conduction is the predominant mechanism in heat transport. Heat conduction in the material is represented by thermal conductivity k.

Soil consists of solids, air, and water. The specific value of thermal conductivity of soil is determined by the constitution of soil and the thermal conductivity of each phase. The thermal conductivity can be determined by using the following equation [24–26]:

$$k_1 = k_s - n[k_s - S_w k_w - (1 - S_w) k_a], \tag{4}$$

where n denotes porosity and S_w represents the degree of saturation.

However, this model does not consider the effect caused by the shape of the void inside of soil. Thus, they introduce a shape factor $\chi = \sqrt{S_w}$ into the equation to represent the effect caused by the shape of the void. Then, the equation becomes

$$k = \sqrt{S_w}\{k_s - n[k_s - S_w k_w - (1 - S_w) k_a]\} + (1 - \sqrt{S_w}) k_a. \tag{5}$$

2.4. Heat Capacity.
We assume that the temperature of the soil is the same among the three phases and that the heat capacity of the soil is also related to the three phases. The heat required to raise the temperature of soil one degree can be calculated by $C_s m_s + C_w m_w + C_g m_g$. The total weight of the soil is $m_s + m_w + m_g$. Therefore, the value of soil heat capacity can be determined as follows:

$$C_p = \frac{C_s m_s + C_w m_w + C_g m_g}{m_s + m_w + m_g}. \tag{6}$$

Considering that the mass of air is negligible, the equation can be simplified as

$$C_p = \frac{C_s + C_w w}{1 + w}, \tag{7}$$

where w is water content.

2.5. Simulation Parameters.
A common concrete shaft consists of two parts: concrete and reinforcement cage. To get the data of temperature distribution, the sensor must be deployed inside of concrete shaft. As mentioned above, the optical fiber would be wrapped around the reinforcement cage spirally so that temperature along the fiber would be obtained. When it comes to conventional TIP, temperature can only be measure through access tubes or at the points where embedded sensors are set.

In order to simulate the temperature evolution and distribution within the shaft, a 3D model is established as shown in Figure 2. The model consists of four parts: concrete inside reinforcement cage, concrete outside reinforcement cage, soil surrounding shaft, and soil below shaft. In this

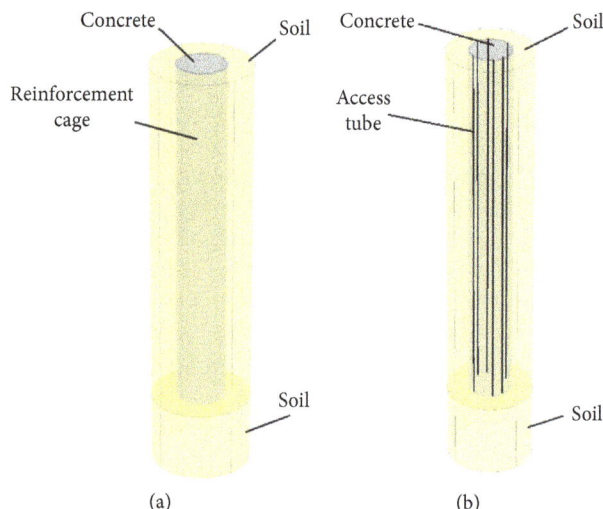

FIGURE 2: Shaft-soil model.

TABLE 1: Soil properties.

Properties	Unit	Value
Density	kg/m^3	1800
Soil solid thermal conductivity	W/m·K	5
Water thermal conductivity	W/m·K	0.5
Air thermal conductivity	W/m·K	0.05
Soil solid heat capacity	J/(kg·K)	850
Water heat capacity	J/(kg·K)	4190
Porosity	%	51.1
Water content	%	39.8
Saturation	%	97

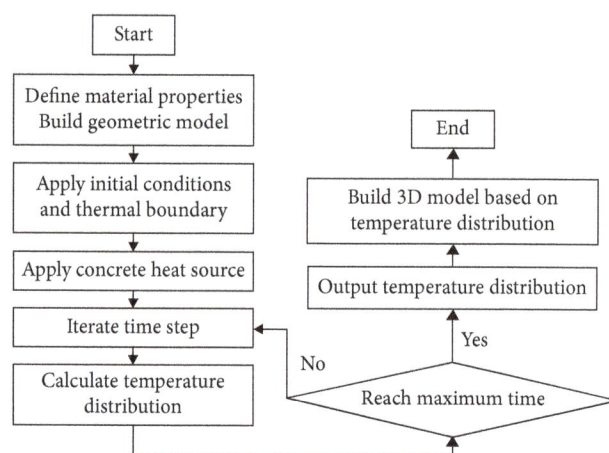

FIGURE 3: Flow chart.

case, heat transfer into the reinforcement cage has been neglected since the reinforcement cage has a low heat capacity, high thermal conductivity, and relatively small volume. However, simulation of the reinforcement cage surface is still necessary because the reinforcement cage is where the optical sensor and access tubes are deployed. The location of the reinforcement cage surface would be the interface between the core concrete cylinder and the concrete cover. To get as much data as possible for high spatial resolution temperature distribution data, optical fiber is chosen to be deployed spirally. The pitch of optical fiber is 300 mm. To simulate access tubes in the conventional TIP method, we use vertical lines to represent the access tubes. According to ASTM D7949-14, one access duct should be placed every 300 mm in diameter. Therefore, there would be 6 vertical lines on the reinforcement cage in our setting.

The thickness of soil outside the concrete shaft is chosen based on the distance between two concrete shafts. In this case, the thickness is equal to the diameter of the concrete shaft. The properties of soil are listed in Table 1.

This simulation was conducted using FEM. The mesh type is free tetrahedral, with the minimum element size of 0.21 m. Several defects would be set on the shaft. Size and location are important factors we will inspect when evaluating the quality of the concrete shaft. The flow chart of the simulation can be found in Figure 3. In the simulation, the ability of the optical-fiber-based TIP and the conventional TIP to detect defect size and location will be compared and discussed.

3. Result and Discussion

In this section, we discuss the result of simulations using two different ways to extract data based on the concept of different TIP methods. Location and size of the defects are considered. First, we compare the results from two methods regarding how defect location will affect the result. Then, we investigate how defect size will affect the result.

3.1. Location Prediction. When performing TIP, we consider the location of the peak value as the location of the defect. To compare the accuracy of both methods, numerical simulation of a 6-foot diameter concrete shaft with a 12-inch sized cubic defect at selected locations is conducted. The location and size of defects are listed in Table 2. The location selected would be: defect is exactly at the measurement point of access tube method, defect is shifted from measurement point of access tube method, and defect is between measurement points of access tubes, separately (Figure 4). The result is shown in Figure 5; the concave region indicates the region that has negative value of temperature divergence. The region with dark blue color is the determination of defect by each method. The area of dark blue region indicates the size of the defect, whereas the location of that region indicates the location of the defect.

When the defect is located at the position where the infrared thermal probe measures temperature, both methods provide accurate determination of location. Furthermore, the temperature distribution can roughly indicate the shape of the defect. When the defect is shifted from the position where the infrared thermal probe measures temperature, although both methods can detect the defect, the result from conventional TIP method deviates from the actual location of the defect. At the same time, the optical-fiber-based TIP method can still have an accurate prediction of defect location. When the defect happened to be located between two measurement points of the conventional TIP method, although both methods can detect the defect, the result from

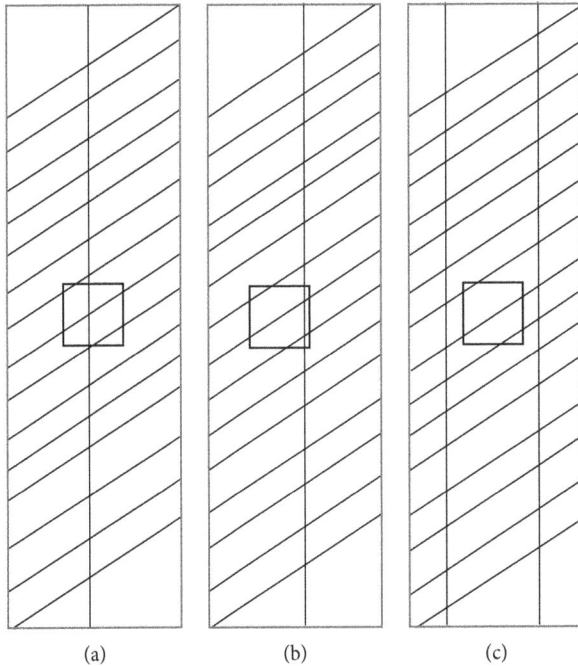

FIGURE 4: Schematic of defect location. (a) Defect located on access tube, (b) defect shifted from access tube, and (c) defect located between access tubes.

the conventional TIP method can hardly predict the location of the defect. The temperature distribution between access tubes does not show a significant peak value. The location of the defect can be anywhere within the low temperature region. Optical-fiber-sensor-based TIP, on the other hand, can still have an accurate determination of the defect.

According to the result presented above, in all three situations, the optical-fiber-based TIP method has great outcome despite the location of the defect. The even distribution of measurement points and relatively small interval not only increase the possibility of the defect being located at the measurement point, but also diminish the effect when the peak value is not located at the measurement point, which contributes to more accurate reconstructed temperature distribution.

Considering that in most situations, the defect is not located exactly at the measurement point, we could draw a conclusion that the optical-fiber-based TIP method would always have the same or better determination of defect location.

3.2. Size Sensitivity. The size of the defect is also a significant factor that needs to be considered for shaft integrity test. The size of defect is related to the magnitude of temperature divergence. The peak value of temperature distribution is crucial to determine the size of the defect. To compare the accuracy of both methods, numerical simulation of a 6-foot concrete shaft is conducted with different defects of different sizes located between access tubes (Table 3). The size of the defects is 18 inches, 15 inches, and 10 inches (Figure 6). An anomaly that has 12% of area reduction is an anomaly

needed for further evaluation. Both methods should have the ability to detect defects at this size.

The result is shown in Figure 7; the concave region indicates that the region that has a negative value of temperature divergence. The region with dark blue color is the determined defect by each method. The area of dark blue region indicates the size of the defect, whereas the location of that region indicates the location of the defect.

When the defect is a 15-inch cube at the lateral surface of the concrete shaft, both methods can detect the existence of the defect. However, the optical fiber method has a larger temperature divergence, closer to the actual temperature distribution in that region. When the defect is a 10-inch cube at the lateral surface of concrete shaft, the infrared thermocouple probe or embedded sensor-based TIP cannot detect the defect between access tubes. The temperature divergence caused by the defect would only be maintained within a certain zone. Once the effect zone is located totally between access tubes, conventional TIP may miss the existing defect which may have a negative effect on the performance of the concrete shaft. The optical-fiber-based TIP method can still detect defects. The area of defect is smaller compared with the result shown in Figure 7, indicating that the size of the defect is smaller than the result shown in Figure 7. This demonstrates that the TIP method has the ability to measure size based on the temperature distribution result.

Since the size of defect is related to the magnitude of temperature divergence, accurate temperature distribution is crucial to defect size evaluation. The value of peak temperature divergence decreases from the center of defect outward. As distance from center of defect increase, the temperature distribution would be closer to the intact part. The closer the measurement point is to the center of the defect, the higher the accuracy of temperature distribution measurement would be. The conventional TIP has no measurement point between access tubes, which limits the minimum size of defect that can be detected. Optical-fiber-based TIP, on the other hand, has the ability to detect smaller defects due to high spatial resolution. However, if the size of the defect is too small, even optical-fiber-based TIP will not be able to detect it.

4. Conclusion

In this paper we proposed an optical-fiber-based TIP method. This method can be an improvement of infrared thermocouple probes and embedded sensors which are applied to conventional TIP by having high spatial resolution temperature data. The method also changes the way to deploy sensors from vertically deployed to spirally deployed around the reinforcement cage. These two changes enable TIP to measure a high resolution and consistent temperature distribution within the concrete shaft, leading to more accurate determination of integrity of the concrete shaft.

To verify the advantages of optical-fiber-based TIP, we investigate two factors of defects: location and size. In the location section, we set three situations: defect on access tube, shift from access tube, and between access tubes. When

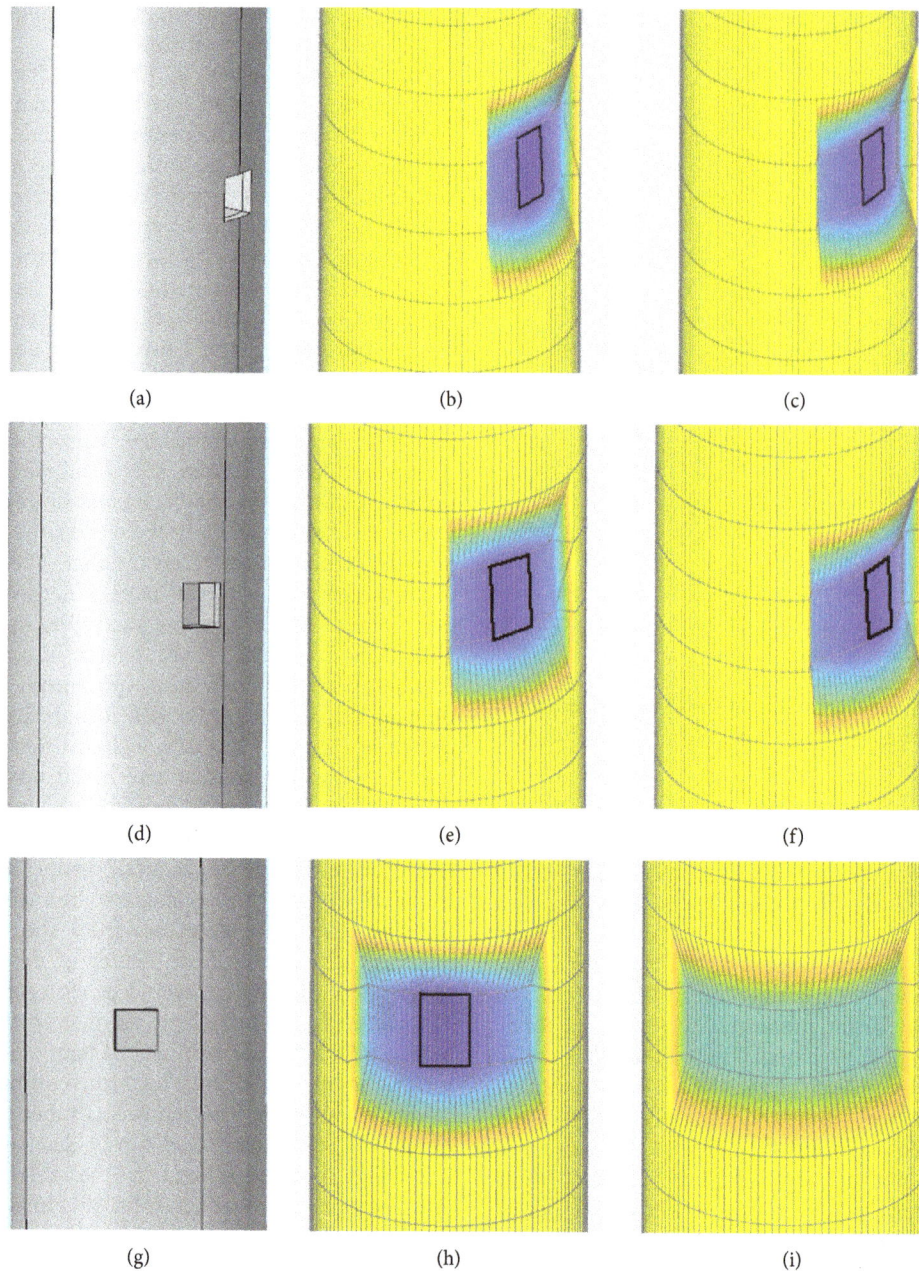

FIGURE 5: Result of simulation with different locations. (a, d, g) Actual location of defect, (b, e, h) location determination by the optical fiber method, and (c, f, i) location determination by the conventional method.

TABLE 2: Location determination by different methods.

Situation	Actual location			Optical fiber method			Conventional method		
	Depth	θ	r	Depth	θ	r	Depth	θ	r
On tube	7.5 m	180	0.6096	7.58 m	179.7	0.6148	7.58 m	181	0.6148
Around tube	7.5 m	171	0.6096	7.575 m	169	0.6175	7.575 m	179.7	0.619
Between tubes	7.5 m	150	0.6096	7.576 m	149.1	0.6096	N/A	N/A	0.774

the defect is located exactly on the access tube, both methods have an accurate determination of the location. However, when the defect is located between access tubes where conventional TIP does not have a measurement point,

optical-fiber-based TIP shows higher accuracy on location determination. We also simulate three situations with different size defects located between access tubes. Since optical-fiber-based TIP has measurement points evenly

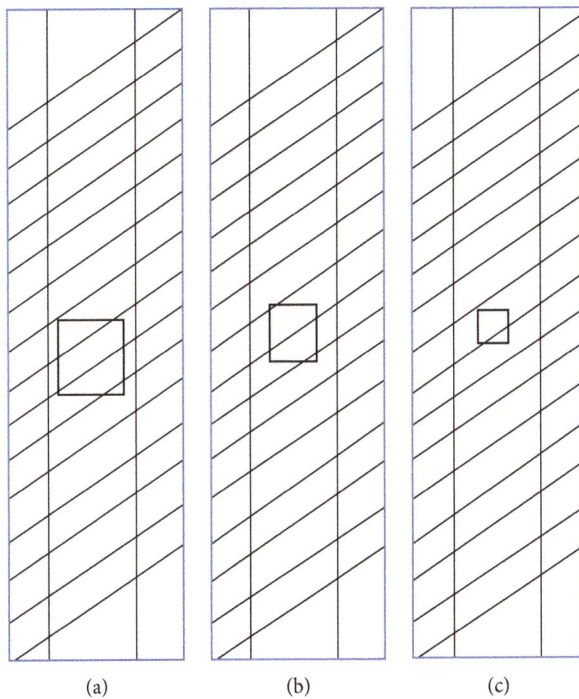

FIGURE 6: Schematic of defect location. (a) 18-inch cube, (b) 15-inch cube, and (c) 10-inch cube.

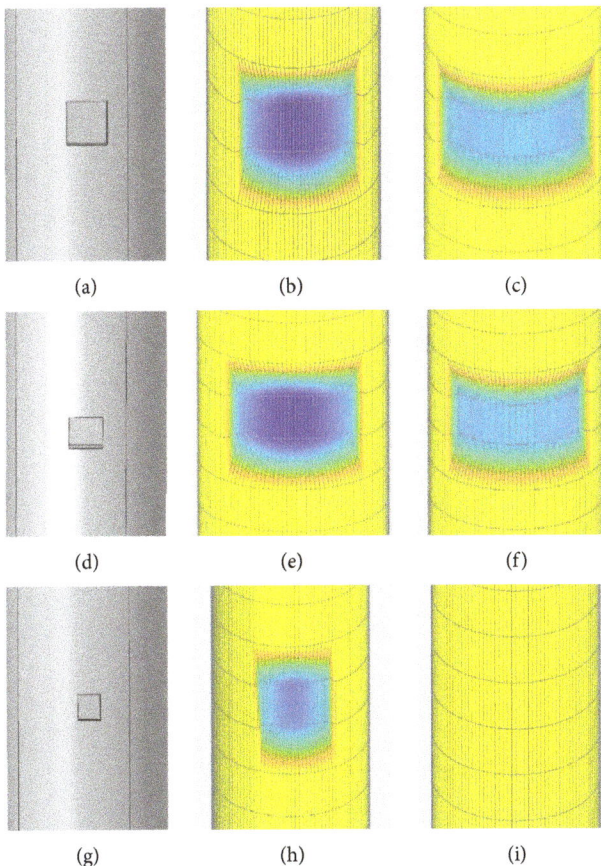

FIGURE 7: Result of simulation with different sizes. (a, d, g) Actual size of defect, (b, e, h) size determination by the optical fiber method, and (c, f, i) size determination by the conventional method.

TABLE 3: Size determination.

Actual size	Optical fiber method	Conventional method
18-inch cube	17.60 inches	13.34 inches
15-inch cube	14.68 inches	10.90 inches
10-inch cube	9.84 inches	5.45 inches

distributed at the surface, the sensitivity of optical-fiber-based TIP to the size of the defect is significantly higher. In the simulation regarding the shape of the defect, since measurement points of optical-fiber-based TIP distribute evenly within the defect, a more precise outline of the defect is depicted by optical-fiber-based TIP. Overall, optical-fiber-based TIP shows higher accuracy in prediction of shaft defects.

Conflicts of Interest

The authors declare that they have no conflicts of interest.

Acknowledgments

This material is based on the work supported as part of the University of Missouri Research Board at the University of Missouri system. Additional support was provided by the Geotechnical Engineering program and Center for Advancing Faculty Excellence of Missouri University of Science and Technology. The authors would like to thank Dr. Jie Huang's valuable discussion on this problem.

References

[1] M. O'Neill, *Integrity Testing of Foundations*, Transportation Research Board, Washington, DC, USA, 1991.

[2] O. Klingmuller and F. Kirsch, *Current Practices and Future Trends in Deep Foundations*, American Society of Civil Engineers, Reston, VA, USA, 2004.

[3] D. Brown and A. Schindler, "High performance concrete and drilled shaft construction," in *Proceedings of GSP 158 Contemporary Issues in Deep Foundations*, Denver, CO, USA, February 2007.

[4] N. Massoudi and W. Teffera, "Non-destructive testing of piles using the low strain integrity method," in *Proceedings of 5th International Conference on Case Histories in Geotechnical Engineering*, pp. 13–17, Missouri University of Science and Technology, New York, NY, USA, April 2004.

[5] J. C. Ashlock and M. K. Fotouhi, "Thermal integrity profiling and crosshole sonic logging of drilled shafts with artificial defects," in *Proceedings of Geo-Congress 2014 Technical Papers, GSP 234 © ASCE*, Atlanta, Georgia, February 2014.

[6] S.-J. Hsieh, R. Crane, and S. Sathish, "Understanding and predicting electronic vibration stress using ultrasound excitation, thermal profiling, and neural network modeling," *Nondestructive Testing and Evaluation*, vol. 20, no. 2, pp. 89–102, 2005.

[7] D. Li, L. Zhang, and W. Tang, "Closure to "reliability evaluation of cross-hole sonic logging for bored pile integrity" by

D. Q. Li, L. M. Zhang, and W. H. Tang," *Journal of Geotechnical and Geoenvironmental Engineering*, vol. 133, no. 3, pp. 343-344, 2007.

[8] K. Johnson, G. Mullins, and D. Winters, "Concrete temperature control via voiding drilled shafts," in *Proceedings of Contemporary Issues in Deep Foundations, ASCE Geo Institute, GSP 158*, no. 1, pp. 1–121, Denver, CO, USA, October 2007.

[9] K. R. Johnson, "Temperature prediction modeling and thermal integrity profiling of drilled shafts," in *Proceedings of Geo-Congress 2014 Technical Papers, GSP 234 © ASCE*, Atlanta, Georgia, February 2014.

[10] G. Mullins, "Thermal integrity profiling of drilled shafts," *DFI Journal-The Journal of the Deep Foundations Institute*, vol. 4, no. 2, pp. 54–64, 2010.

[11] G. Mullins, "Advancements in drilled shaft construction, design, and quality assurance: the value of research," *International Journal of Pavement Research and Technology*, vol. 6, no. 2, pp. 993–999, 2013.

[12] K. R. Johnson, "Analyzing thermal integrity profiling data for drilled shaft evaluation," *DFI Journal-The Journal of the Deep Foundations Institute*, vol. 10, no. 1, pp. 25–33, 2016.

[13] A. G. Davis, "Assessing reliability of drilled shaft integrity testing," *Transportation Research Record*, vol. 1633, pp. 108–116, 1998.

[14] D. Samiec, *Distributed Fiber Optic Temperature and Strain Measurement with Extremely High Spatial Resolution*, Polytech GmbH, Waldbronn, Germany, 2011.

[15] W. Pei, W. Yua, S. Li, and J. Zhou, "A new method to model the thermal conductivity of soil–rock media in cold regions: an example from permafrost regions tunnel," *Cold Regions Science and Technology*, vol. 95, pp. 11–18, 2013.

[16] T. J. Moore, M. R. Jones, D. R. Tree, and D. D. Allred, "An inexpensive high-temperature optical fiber thermometer," *Journal of Quantitative Spectroscopy and Radiative Transfer*, vol. 187, pp. 358–363, 2017.

[17] H. Su, H. Jiang, and M. Yang, "Dam seepage monitoring based on distributed optical fiber temperature system," *IEEE Sensors Journal*, vol. 15, no. 1, pp. 9–13, 2015.

[18] R. K. Palmer, K. M. McCary, and T. E. Blue, "An analytical model for the time constants of optical fiber temperature sensing," *IEEE Sensors Journal*, vol. 17, no. 17, pp. 5492–5511, 2017.

[19] A. Leal-Junior, A. Frizera-Neto, C. Marques, and M. Pontes, "A polymer optical fiber temperature sensor based on material features," *Sensors*, vol. 18, no. 2, p. 301, 2018.

[20] Y. Rui, C. Kechavarzi, O'. L. Frank, C. Barker, D. Nicholson, and K. Soga, "Integrity testing of pile cover using distributed fibre optic sensing," *Sensors*, vol. 17, no. 12, p. 2949, 2017.

[21] Y. Cardona-Maya and J. F. Botero-Cadavid, "Refractive index desensitized optical fiber temperature sensor," *Revista Facultad de Ingeniería*, no. 85, pp. 86–90, 2017.

[22] J. Huang, X. Lan, M. Luo, and H. Xiao, "Spatially continuous distributed fiber optic sensing using optical carrier based microwave interferometry," *Optics Express*, vol. 22, no. 15, pp. 18757–18769, 2014.

[23] A. Schindler and K. Folliard, "Heat of hydration models for cementitious materials," *ACI Materials Journal*, vol. 102, no. 1, pp. 24–33, 2005.

[24] W. Liu, P. He, and Z. Zhang, "A calculation method of thermal conductivity of soils," *Journal of Glaciology and Geocryology*, vol. 24, no. 6, pp. 770–773, 2002.

[25] C. S. Blázquez, A. F. Martín, I. M. Nieto, and D. Gonzalez-Aguilera, "Measuring of thermal conductivities of soils and rocks to be used in the calculation of a geothermal installation," *Energies*, vol. 10, no. 795, pp. 1–19, 2017.

[26] D. Barry-Macaulay, A. Bouazza, R. M. Singh, B. Wang, and P. G. Ranjith, "Thermal conductivity of soils and rocks from the Melbourne (Australia) region," *Engineering Geology*, vol. 164, pp. 131–138, 2013.

19

The Role of Phosphorus Slag in Steam-Cured Concrete

Jin Liu and Dongmin Wang

School of Chemical and Environmental Engineering, China University of Mining and Technology, Beijing 100083, China

Correspondence should be addressed to Dongmin Wang; wangdongmin-2008@163.com

Academic Editor: Xiao-Jian Gao

Steam curing is an effective method to increase the hydration degree of binder containing phosphorus slag. The role of phosphorus slag in steam-cured concrete was investigated by determining the hydration heat, hydration products, nonevaporable water content, pore structure of paste, and the compressive strength and chloride ion permeability of concrete. The results show that elevated steam curing temperature does not lead to new crystalline hydration products of the composite binder containing phosphorus slag. Elevating steam curing temperature enhances the early hydration heat and nonevaporable water content of the binder containing phosphorus slag more significantly than increasing steam curing time, and it also results in higher late-age hydration degree and finer pore structure. For steam-cured concrete containing phosphorus slag, elevating curing temperature from 60°C to 80°C tends to decrease the late-age strength and increase the chloride permeability. However, at constant curing temperature of 60°C, the steam-cured concrete containing phosphorus slag can achieve satisfied demoulding strength and late-age strength and chloride permeability by extending the steam curing duration.

1. Introduction

With the development of the construction of concrete engineering, an increasing number of precast concrete elements are applied to modern architectures. Compared with pumping concrete, precast concrete has many advantages, such as more efficient manufacture process, more stable quality, shorter building time, lower cost, and safer and cleaner environment of construction [1–4]. Steam curing is the most popular method in the production of precast concrete elements, which makes great contribution to construction industrialization [5, 6]. The pressure of steam curing includes high pressure, normal atmospheric pressure, and no pressure [7]. Due to economic consideration, steam curing with normal atmospheric pressure is widely used. Presetting period, heating up period, constant temperature period, and cooling period constitute a steam curing process [8, 9]. The temperature of constant period is one of the most important parameters of steam curing regimes, which is usually at 40~90°C [10–14]. It is notable that steam-cured concrete normally has a high permeability and a low strength gain rate at late ages due to the nonuniformly distributed hydration

products and loose pore structure of paste caused by high curing temperature [15–18].

Mineral admixtures such as fly ash and ground granulated blast furnace slag (GGBS) are widely used in modern concrete. Application of mineral admixtures to concrete might improve workability of fresh concrete, reduce hydration heat of cement, and enhance late strength and durability of concrete at late ages. The traditional mineral admixtures are becoming increasing scarce, so new kinds of mineral admixtures such as steel slag, limestone powder, and phosphorus slag are gradually used in concrete production [19–21]. Phosphorus slag is a by-product in the production of yellow phosphorus by electric furnace method in industry. It is reported that about 8 to 10 tons of phosphorus slag are produced for 1 ton of yellow phosphorus production [22]. More than 8 million tons of phosphorus slag are generated every year in China and the utilization ratio is very low [23]. The major chemical compositions of phosphorus slag are CaO and SiO_2, which normally accounts for over 80% [24]. Based on different nature of phosphate ores, the minor compositions of phosphorus slag are 2.5~5% Al_2O_3, 1~5% P_2O_5, 0.5~3% MgO, 0.2~2.5% Fe_2O_3, and 0~2.5% F [25].

TABLE 1: The chemical compositions of the raw materials%.

Sample	SiO_2	Al_2O_3	Fe_2O_3	CaO	MgO	SO_3	Na_2O_{eq}	Loss	P_2O_5	F
Cement	22.36	7.73	3.66	57.21	3.10	3.54	0.73	2.31	—	—
Phosphorus slag	38.27	5.33	0.29	43.12	1.69	1.30	1.75	2.16	4.62	2.46

Note. $Na_2O_{eq} = Na_2O + 0.685\,K_2O$.

Researches showed that phosphorus slag could refine the late-age pore structure of hardened paste, reduce early hydration heat of cement, and enhance durability of concrete [26, 27]. It is well accepted that phosphorus slag has a strong retarding effect on the early hydration of cement [28–30]. The concrete incorporating phosphorus slag achieves very low early strength, so the application of phosphorus slag to concrete is restricted to a certain extent.

Due to low reactivity or lack of sufficient alkali activation, mineral admixtures usually exhibit a low reaction degree at early ages [31, 32]. Correspondingly, the concrete containing mineral admixtures usually achieves a lower early strength compared with the plain cement concrete [33, 34]. However, the hydration degree of mineral admixtures increases significantly with the increase of curing temperature and steam curing time at early ages, which contributes to the strength development of concrete [35–38]. Therefore, steam curing is an effective method to improve the strength of concrete containing mineral admixtures at early ages [39–41]. It is a potential method to improve the late-age pore structure of steam-cured concrete by using mineral admixture.

In this paper, phosphorus slag was used in steam-cured concrete. In order to enhance the demoulding strength of steam-cured concrete containing phosphorus slag, two methods were employed: increasing the steam curing temperature and extending the steam curing duration. The effects of these two methods on the properties of steam-cured concrete containing phosphorus slag were compared.

2. Raw Materials and Test Methods

2.1. Raw Materials. The cement used was Ordinary Portland cement with the specific surface area of $350\,m^2kg^{-1}$ and the strength grade of 42.5 complying with the Chinese National Standard GB175-2007. The chemical compositions of the cement and the phosphorus slag are provided in Table 1. Figure 1 shows the scanning electron microscope (SEM) image of phosphorus slag, which indicates that the particles of phosphorus slag have an irregular shaped morphology. Figure 2 shows the XRD patterns of the phosphorus slag, which indicates that most of the mineral phases of the phosphorus slag are amorphous. The particle size distributions of the cement and the phosphorus slag are shown in Figure 3. Coarse and fine aggregates were crushed limestone between 5 mm and 25 mm and natural river sand smaller than 5 mm, respectively.

2.2. Test Methods. Table 2 lists the mix proportions of concrete. Concrete C and concrete CC were plain cement concrete with the water-to-binder ratios (W/B) of 0.4 and 0.32, respectively. Concrete N1 and concrete NN1 were the

FIGURE 1: Scanning electron microscope image of the phosphorus slag.

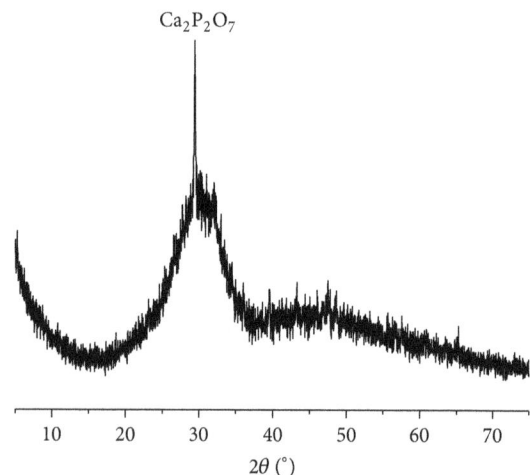

FIGURE 2: XRD pattern of the phosphorus slag.

ones containing 15% phosphorus slag with the W/B of 0.4 and 0.32, respectively. Concrete N2 and concrete NN2 were the ones containing 30% phosphorus slag with the W/B of 0.4 and 0.32, respectively. Table 3 lists the mix proportions of pastes. The composition of binder and W/B of the pastes corresponds to those of the concrete.

Two curing methods for the concrete were set in this study: (1) standard curing: concrete was cured in a room with a temperature of $20 \pm 1°C$ and a relative humidity higher

TABLE 2: Mix proportions of concrete/kg·m^{-3}.

Sample	Cement	Phosphorus slag	Sand	Stone	Water
C	380	0	785	1083	152
N1	323	57	785	1083	152
N2	266	114	785	1083	152
CC	450	0	780	1076	144
NN1	382.5	67.5	780	1076	144
NN2	315	135	780	1076	144

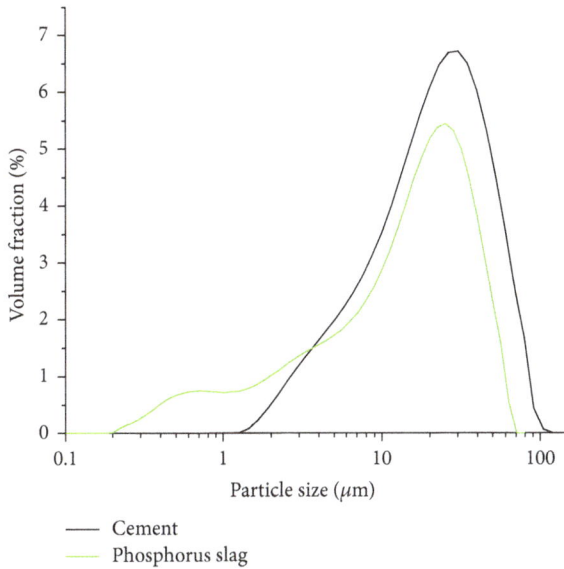

FIGURE 3: Particle size distributions of the cement and the phosphorus slag.

TABLE 3: Mix proportions of pastes/g.

Sample	Cement	Phosphorus slag	Water
C	100	0	40
N1	85	15	40
N2	70	30	40
CC	100	0	32
NN1	85	15	32
NN2	70	30	32

than 95% and (2) steam curing: the concrete was placed in a standard curing room for a precuring period of 3 h after casting and then cured in a curing box. For plain cement concrete, the temperature of the constant period was 60°C and duration of constant period was 8 h. For the concrete containing phosphorus slag, duration of constant period was extended to 11 h when the temperature of the constant period was 60°C, and the temperature of the constant period was increased to 80°C when the duration of constant period was 8 h. Both of the times for heating up period and cooling period were 2 h. The concrete was placed in the standard curing room after steam curing.

Concrete of $100 \times 100 \times 100$ mm was cast. The compressive strength of concrete was tested at the ages of demoulding time, 3 d, 28 d, and 90 d. For the concrete cured under standard condition, the demoulding time was 18 h. The chloride ion permeability of concrete was tested at the ages of 28 d and 90 d. The chloride ion permeability was evaluated by measuring the charge passed of concrete according to ASTM C 1202 "Standard Test Method for Electrical Indication of Concrete's Ability to Resist Chloride Ion Penetration."

The pastes were cast in plastic sealed tubes after preparation and cured under the same conditions with concrete. Hardened pastes were extracted and then immersed into absolute alcohol to prevent further hydration at testing ages. The pore characteristics of pastes were determined by mercury intrusion porosimetry (MIP). X-ray diffraction (XRD) was used to determine the mineral phases of hydration products. The nonevaporable water (w_n) content of paste was calculated as the mass difference between the samples dried at 105°C and heated at 1000°C normalized by the mass after being dried at 105°C and correcting for the loss on ignition of unhydrated samples [42]. The hydration heat evolution of the binder at the W/B of 0.4 and 0.32 within 24 h was tested by using an isothermal calorimeter at constant temperatures of 60°C and 80°C, respectively.

The temperature of constant period is denoted with suffix "-60" or "-80." Additionally, the time of constant period is denoted with suffix "-8" or "-11." For example, "C-60-8" represents the sample C cured at 60°C for 8 h.

3. Results and Discussion

3.1. Hydration Heat. Figure 4(a) shows the exothermic rates during the hydration of samples C, N1, and N2 at 60°C within 24 h. At W/B of 0.4, with the increase of phosphorus slag addition, the dormant period of binder is prolonged. In addition, the exothermic peak of binder is postponed and the exothermic peak value of binder decreases. These results indicate that the addition of phosphorus slag tends to decrease the exothermic rate of binder at early ages. Figure 4(b) shows the exothermic rates during the hydration of samples CC, NN1, and NN2 at 60°C within 24 h. At W/B of 0.32, the influence of phosphorus slag on the exothermic rate of binder is basically the same with that of binder at W/B of 0.4.

Figures 5(a) and 5(b) show the exothermic rates during the hydration of samples C, N1, and N2 as well as samples CC, NN1, and NN2 at 80°C within 24 h, respectively. When the

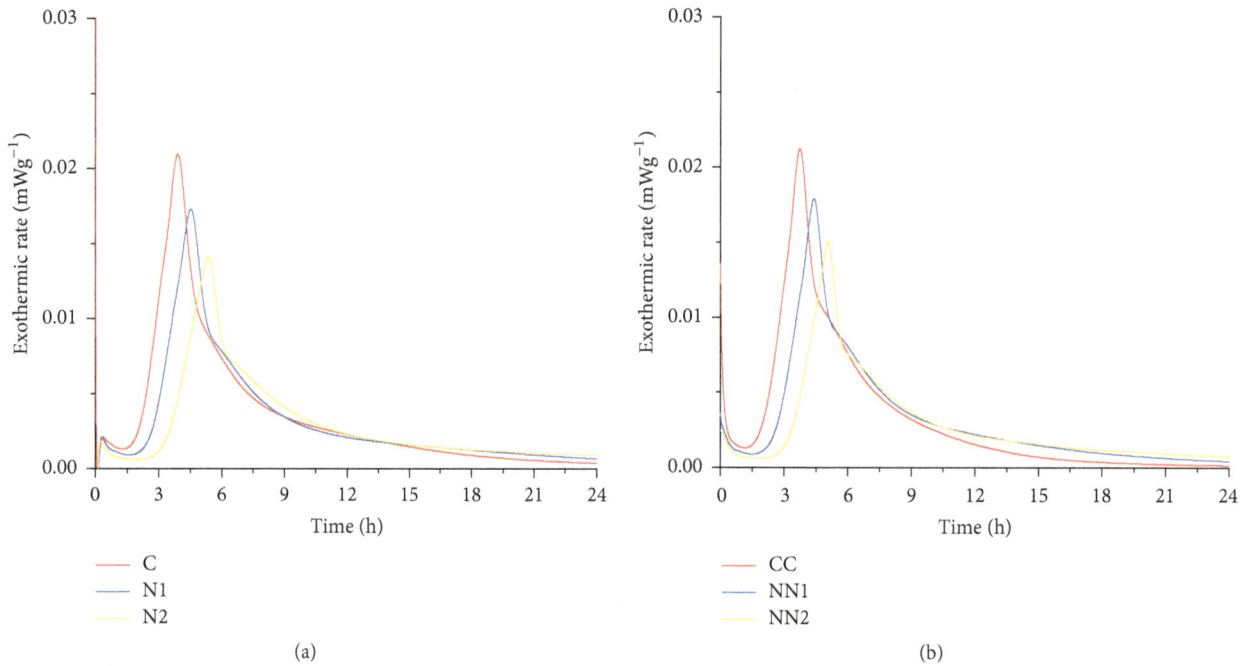

FIGURE 4: (a) Exothermic rates during the hydration of samples C, N1, and N2 at 60°C. (b) Exothermic rates during the hydration of samples CC, NN1, and NN2 at 60°C.

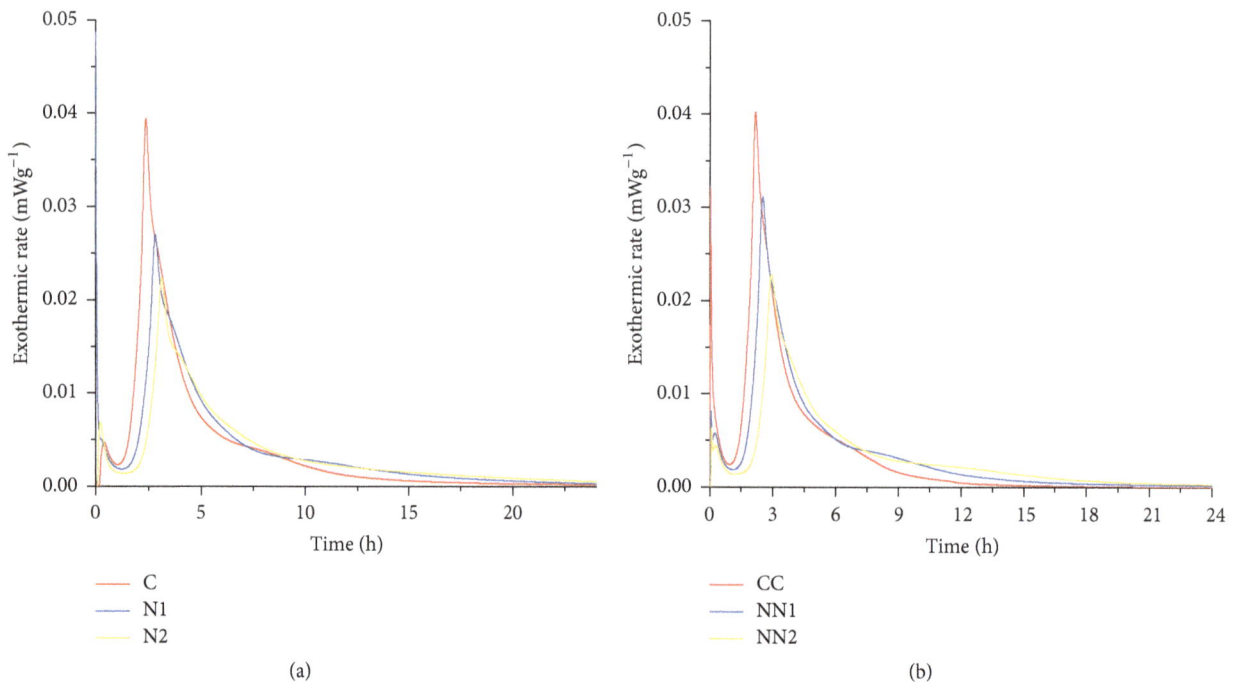

FIGURE 5: (a) Exothermic rates during the hydration of samples C, N1, and N2 at 80°C. (b) Exothermic rates during the hydration of samples CC, NN1, and NN2 at 80°C.

temperature is elevated from 60°C to 80°C, the exothermic peak of binder occurs earlier and the exothermic peak value of binder increases. These results indicate that the exothermic rate of binder increases significantly with an elevated temperature from 60°C to 80°C. What is more, the addition of

phosphorus slag also tends to decrease the exothermic rate of binder at early ages at 80°C.

The accumulative hydration heats of the plain cement and the binders containing phosphorus slag are shown in Table 4. At W/B of 0.4, under constant temperature of

TABLE 4: Accumulative hydration heats of the plain cement and the binders containing phosphorus slag ($J \cdot g^{-1}$).

Sample	60°C		80°C
	8 h	11 h	8 h
C	222.1	257.3	288.5
N1	183.3	217.8	258.2
N2	141	181.9	210.6
CC	230.6	262.2	290.6
NN1	189.2	224.7	248.2
NN2	148.4	185.2	207.8

60°C, the increase ratios of accumulative hydration heats of samples C, N1, and N2 are 15.8%, 18.8%, and 29.0% by increasing curing time from 8 h to 11 h, respectively. At constant curing time of 8 h, the increase ratios of accumulative hydration heats of samples C, N1, and N2 are 29.9%, 40.9%, and 49.4% by increasing curing temperature from 60°C to 80°C, respectively. These results indicate that both extended duration and elevated temperature promote the hydration degree of cement and the binders containing phosphorus slag significantly, but the promoting effect of elevated temperature is greater. What is more, the promoting effect of extended duration and elevated temperature on hydration degree of the binders containing phosphorus slag is greater at higher phosphorus slag replacement. This is because both the hydration of cement and reaction of phosphorus slag are promoted in the composite binder and the reaction degree of phosphorus slag is promoted more significantly than the hydration degree of cement. At W/B of 0.32, under constant temperature of 60°C, the increase ratios of accumulative hydration heats of samples CC, NN1, and NN2 are 13.7%, 18.8%, and 24.8% by increasing curing time from 8 h to 11 h, respectively. At constant curing time of 8 h, the increase ratios of accumulative hydration heats of samples CC, NN1, and NN2 are 26.0%, 31.2%, and 40.0% by increasing curing temperature from 60°C to 80°C, respectively. The influence of extended duration and elevated temperature on hydration degree of plain cement and the binders containing phosphorus slag is basically the same with that of binder at W/B of 0.4.

3.2. XRD Results. Figure 6 shows the XRD results of the plain cement paste and the paste containing 30% phosphorus slag under the standard curing condition at the age of 90 d. The XRD pattern only displays the crystalline phases in the hydration products of the binder. However, gel is amorphous and therefore the characteristic peak of gel cannot be found in the XRD pattern. It can be seen that the angle positions of the characteristic peaks of two kind of pastes are identical at the age of 90 d. This indicates that there is no new crystalline substance in the hydration products of the composite binder containing phosphorus slag, and the reaction products of phosphorus slag are amorphous gel. Figure 7 shows the XRD results of the paste containing 30% phosphorus slag under different curing temperatures at the age of 90 d. It can be seen that the angle positions of the characteristic peaks of the pastes containing 30% phosphorus slag under different

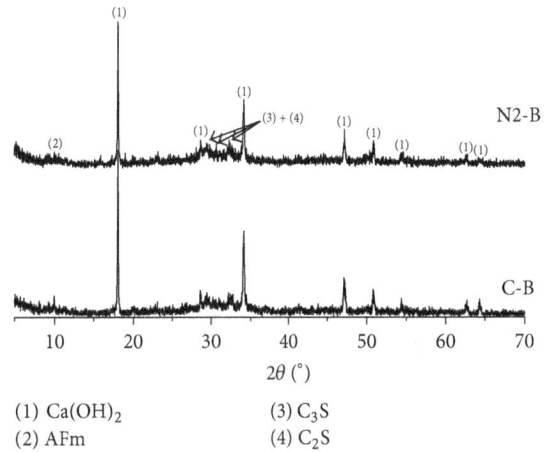

(1) Ca(OH)$_2$
(2) AFm
(3) C$_3$S
(4) C$_2$S

FIGURE 6: XRD patterns of the standard-cured paste at the age of 90 d.

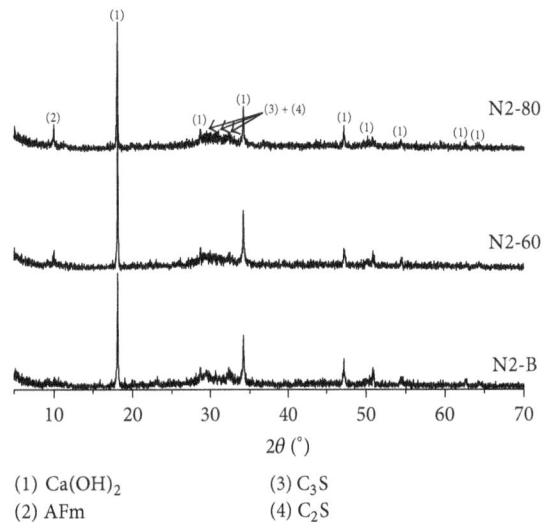

(1) Ca(OH)$_2$
(2) AFm
(3) C$_3$S
(4) C$_2$S

FIGURE 7: XRD patterns of the paste containing phosphorus slag at the age of 90 d.

steam curing temperatures at the age of 90 d are identical. This indicates that elevated steam curing temperature does not lead to new crystalline hydration products of the composite binder containing phosphorus slag.

3.3. Nonevaporable Water Content. The nonevaporable water (w_n) content represents the amount of hydration products. The w_n content can be used to determine the hydration degree of binder. Figures 8 and 9 show the w_n contents of the pastes containing phosphorus slag with W/B of 0.4 and 0.32 at the age of demoulding time, respectively. It is clear that the w_n contents of the pastes containing phosphorus slag under the steam curing of 60°C for 11 h and 80°C for 8 h are significantly higher than that under standard curing at the age of demoulding time. It is believed that elevated temperature accelerates the hydration of cement as well as the reaction of phosphorus slag. Meanwhile, the increased amplitude of the w_n content of each sample under the steam curing of 80°C

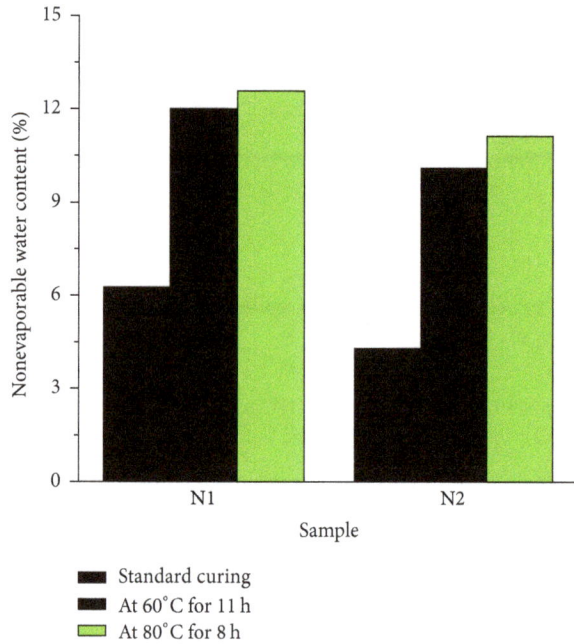

FIGURE 8: Nonevaporable water contents of the pastes containing phosphorus slag with W/B of 0.4 at the age of demoulding time.

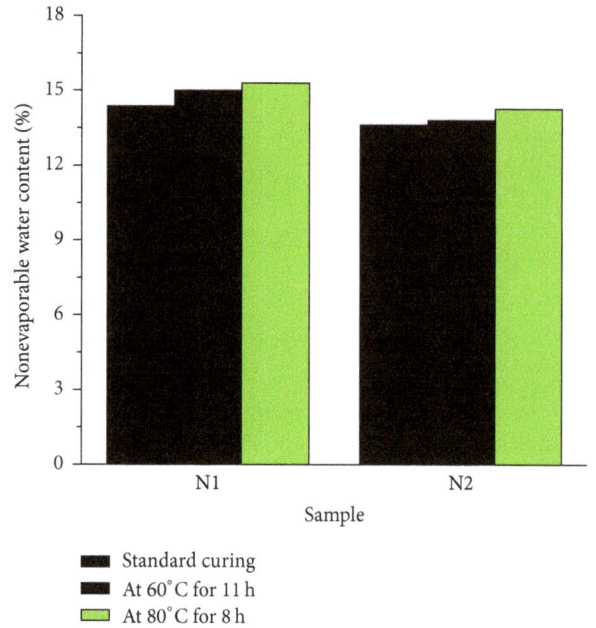

FIGURE 10: Nonevaporable water contents of the pastes containing phosphorus slag with W/B of 0.4 at the age of 90 d.

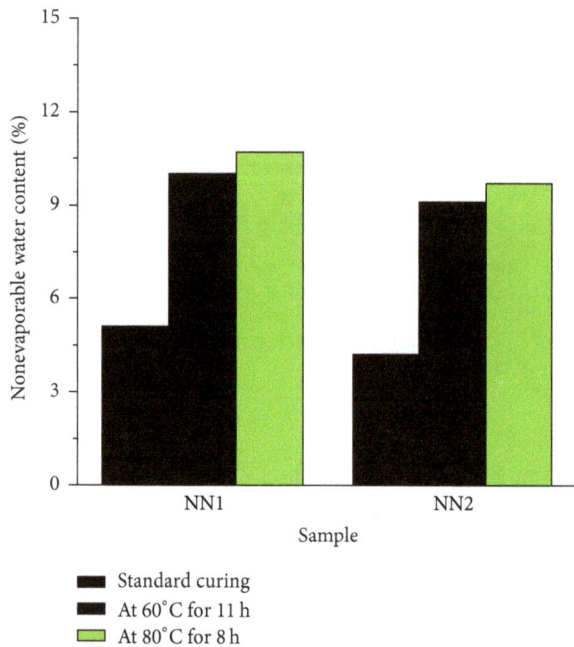

FIGURE 9: Nonevaporable water contents of the pastes containing phosphorus slag with W/B of 0.32 at the age of demoulding time.

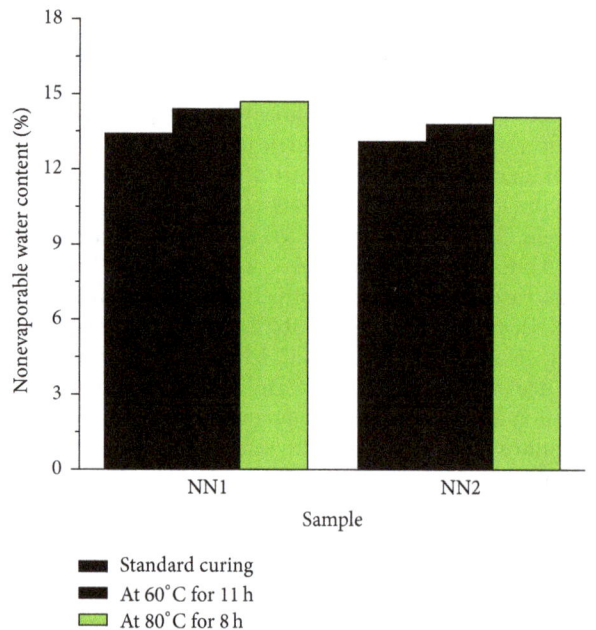

FIGURE 11: Nonevaporable water contents of the pastes containing phosphorus slag with W/B of 0.32 at the age of 90 d.

for 8 h is larger than that under the steam curing of 60°C for 11 h. This indicates that elevated curing temperature tends to accelerate the early hydration rate of the binder containing phosphorus slag more effectively than extended steam curing duration.

Figures 10 and 11 show the w_n contents of the paste containing phosphorus slag with W/B of 0.4 and 0.32 at

the age of 90 d, respectively. In general, dense and thick C-S-H layer grows around cement grains if the binder is cured under high temperature at early ages, which tends to hinder the hydration of cement at late ages [43]. However, it is interesting to find that the w_n contents of the pastes containing phosphorus slag under the curing condition of 60°C for 11 h and 80°C for 8 h are also a little higher than that

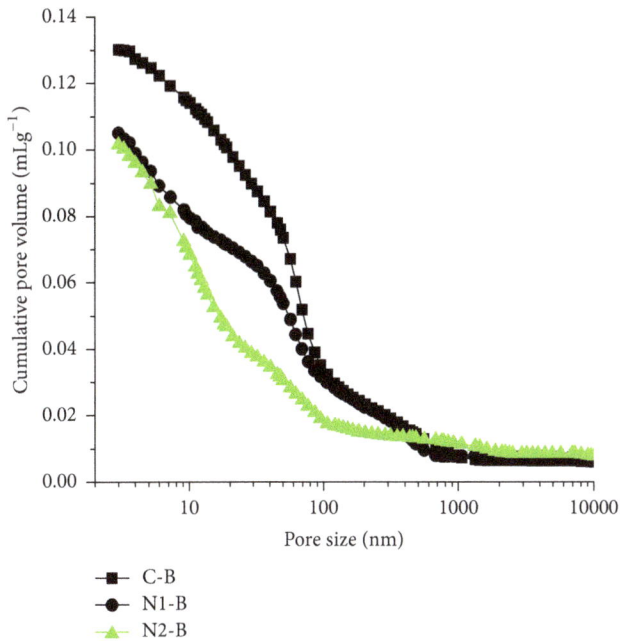

FIGURE 12: Pore structures of the standard-cured hardened paste with W/B of 0.4 at the age of 90 d.

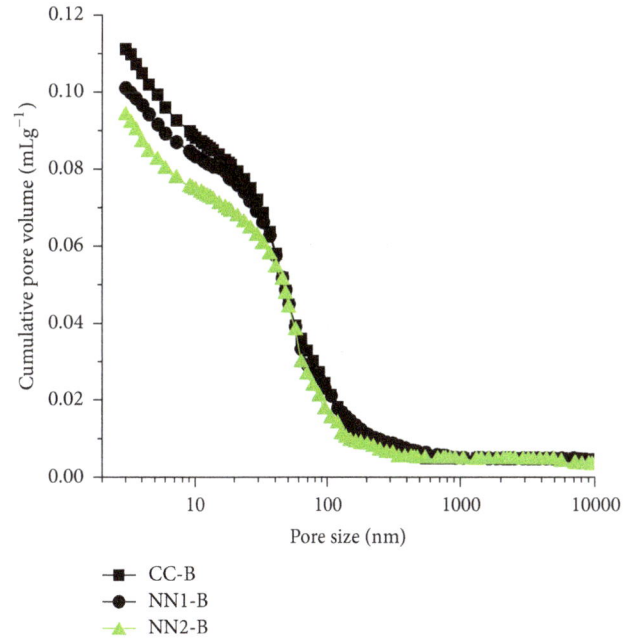

FIGURE 13: Pore structures of the standard-cured hardened paste with W/B of 0.32 at the age of 90 d.

under standard curing at the age of 90 d. The reasons may be as follows: (1) the sample experienced a precuring period of 3 h and the duration of the steam curing is not longer than 11 h, so the adverse effect of elevated temperature on the late hydration is not so obvious; (2) an elevated steam curing temperature promotes the reaction degree of phosphorus slag resulting in more reaction products. In addition, the w_n content of the paste containing phosphorus slag under the steam curing of 80°C for 8 h is higher than that under the steam curing of 60°C for 11 h at the age of 90 d. This is probably due to the fact that an elevated steam curing temperature tends to promote the reaction degree of phosphorus slag more significantly than an extended steam curing duration.

It can be concluded from Figures 8–11 that the w_n content of the paste containing phosphorus slag under the steam curing of 80°C for 8 h is higher than that under the steam curing of 60°C for 11 h at the ages of demoulding time as well as 90 d. The influences of elevating curing temperature and increasing curing time on the early nonevaporable water content of the paste containing phosphorus slag are consistent with those on the early hydration heats of the binder containing phosphorus slag.

3.4. Pore Structure. The pore size distributions of the standard-cured hardened pastes with W/B of 0.4 and 0.32 at the age of 90 d are depicted in Figures 12 and 13, respectively. For the samples with W/B of 0.4, it is evident that the cumulative pore volume of the hardened paste containing 15% phosphorus slag is lower than that of the hardened plain cement paste. Both the cumulative pore volume and the proportion of pores larger than 100 nm of the hardened paste containing 30% phosphorus slag are lower than that of the plain cement paste. For samples with W/B of 0.32,

the cumulative pore volume decreases with the increase of phosphorus slag addition. However, at such a low W/B of 0.32, though the addition of phosphorus slag tends to reduce porosity, its influence on the pore structure is limited to some extent. In conclusion, the addition of phosphorus slag can optimize the pore structure of hardened paste under standard curing condition, and the optimization effect becomes more obvious with the increase of the replacement ratio of phosphorus slag within the replacement limit of 30%.

The pore size distributions of the steam-cured hardened pastes with W/B of 0.4 at the age of 90 d are depicted in Figure 14. The hardened plain cement paste cured at 60°C for 8 h is employed as the control group. It is clear that both the cumulative pore volume and the proportion of pores larger than 100 nm of the hardened pastes containing phosphorus slag under the steam curing of 60°C for 11 h and 80°C for 8 h are lower than those of control group. Meanwhile, both the cumulative pore volume and the proportion of pores larger than 100 nm of the hardened pastes containing phosphorus slag under the steam curing of 80°C for 8 h are the least, which indicates that elevated curing temperature tends to optimize the late-age pore structure of the hardened paste containing phosphorus slag more effectively than extended steam curing duration.

The pore size distributions of the steam-cured hardened pastes with W/B of 0.32 at 90 d are depicted in Figure 15. It is clear that both the cumulative pore volume and the proportion of pores larger than 100 nm of the hardened pastes containing phosphorus slag under the steam curing of 60°C for 11 h are lower than those of control group. The proportion of pores larger than 100 nm of the hardened paste containing phosphorus slag under the steam curing of 80°C for 8 h

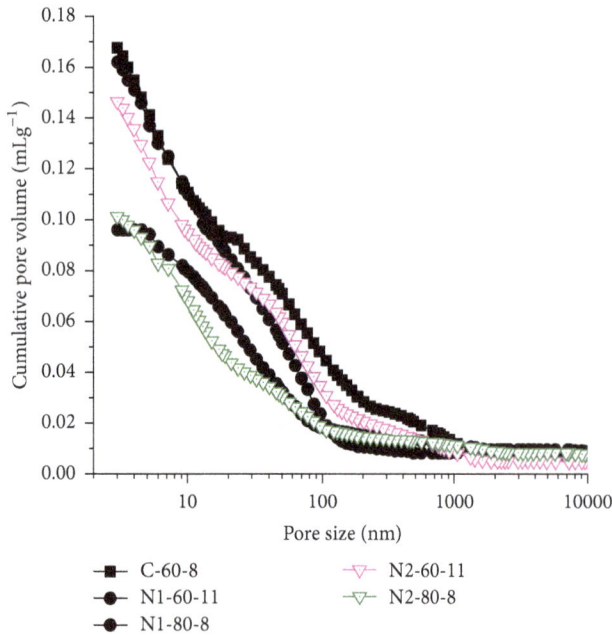

FIGURE 14: Pore structures of the steam-cured hardened paste with W/B of 0.4 at the age of 90 d.

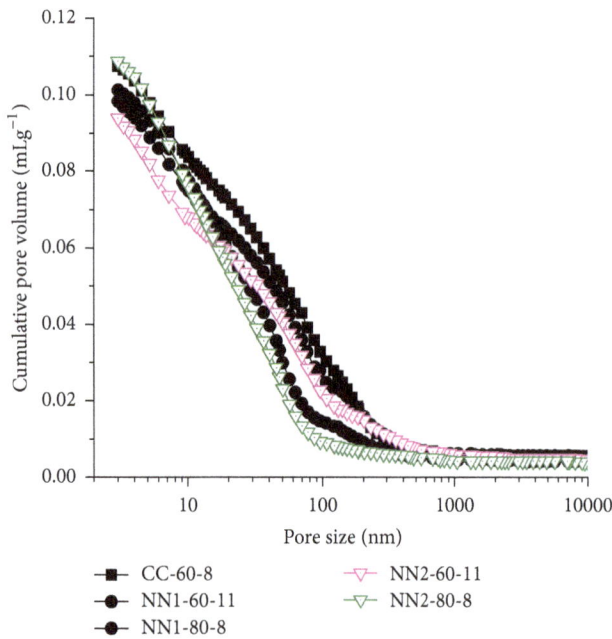

FIGURE 16: Compressive strength of the standard-cured concrete with W/B of 0.4.

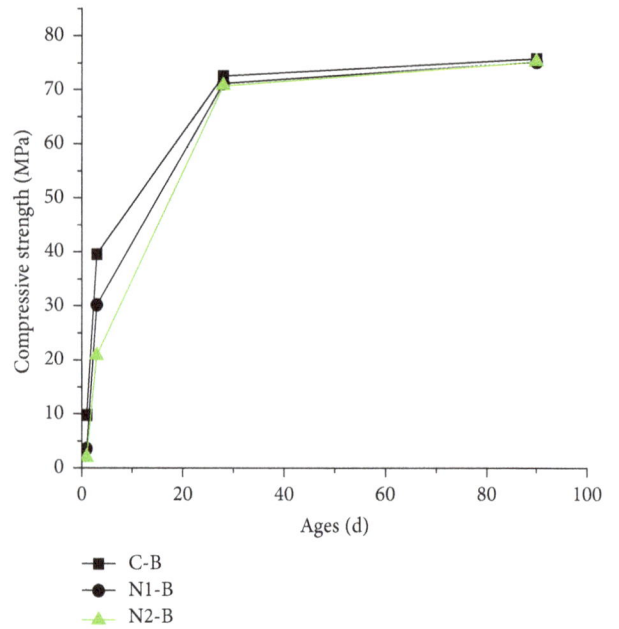

FIGURE 15: Pore structures of the steam-cured hardened paste with W/B of 0.32 at the age of 90 d.

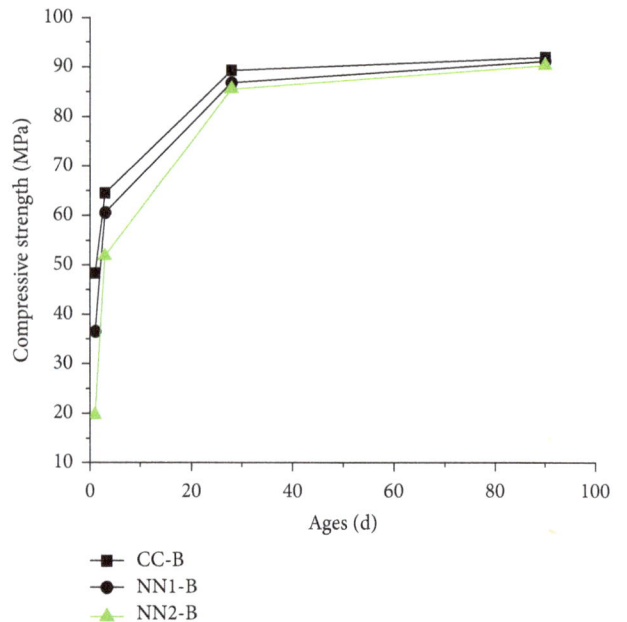

FIGURE 17: Compressive strength of the standard-cured concrete with W/B of 0.32.

is also the least, and the proportion of pores smaller than 20 nm of the hardened paste containing phosphorus slag under the steam curing of 80°C for 8 h is higher than that under the steam curing of 60°C for 11 h. This indicates that elevated curing temperature tends to refine the late-age pore structure of the hardened paste containing phosphorus slag more effectively than extended steam curing duration.

In conclusion, the paste containing phosphorus slag can achieve denser pore structure than the plain cement paste

at the late ages by elevating steam curing temperature or extending steam curing duration. The late-age pore structure of the hardened paste containing phosphorus slag under the condition of elevating curing temperature is finer than that under the condition of extending steam curing duration, which corresponds to the trends of nonevaporable water content results.

3.5. *Compressive Strength.* Figures 16 and 17 show the compressive strength of standard-cured concrete with the W/B

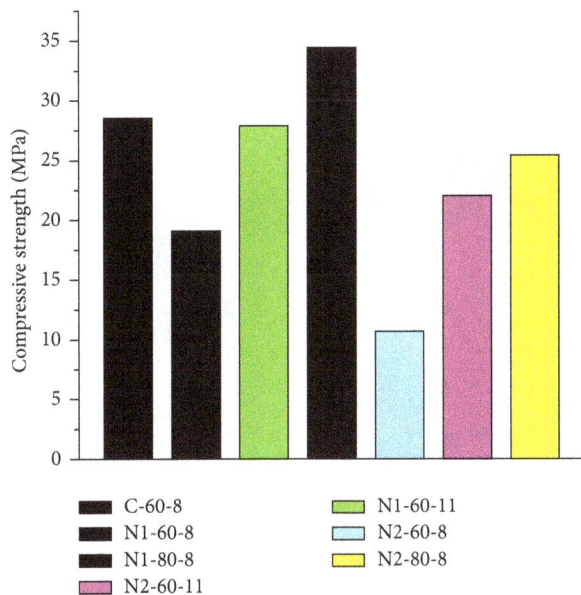

FIGURE 18: Demoulding strength of the steam-cured concrete with W/B of 0.4.

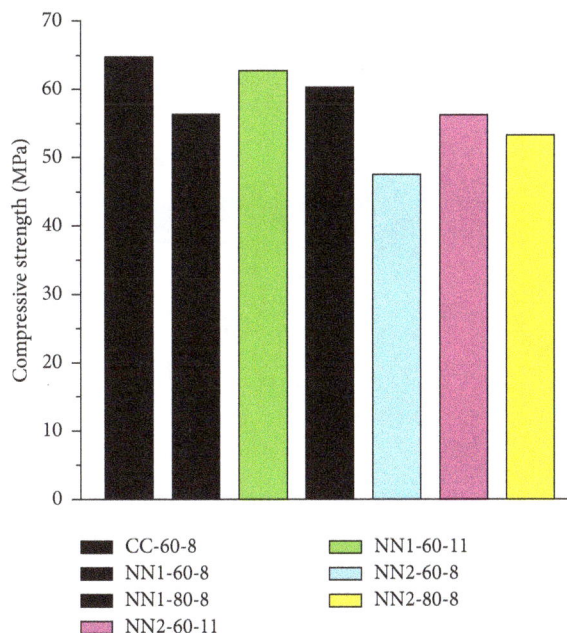

FIGURE 19: Demoulding strength of the steam-cured concrete with W/B of 0.32.

of 0.4 and 0.32, respectively. Due to the retarding effect of phosphorus slag on the early hydration of cement, the early strength of concrete decreases with the increase of phosphorus slag replacement. For concrete with W/B of 0.4, the compressive strengths of the concrete containing phosphorus slag achieve less than 4 MPa at the age of 18 h. Even at the age of 3 d, the compressive strengths of the concrete containing 15% and 30% phosphorus slag are 23.5% and 47.1% lower than that of plain cement concrete, respectively. However, the compressive strength of the concrete containing phosphorus slag is very close to that of the plain cement concrete at the late ages. Additionally, the growth rate of compressive strength of the concrete containing 30% phosphorus slag is significantly higher than that of plain cement concrete as well as the concrete containing 15% phosphorus slag at late ages. For concrete with W/B of 0.32, the trends are basically the same as those of the concrete with W/B of 0.4. The addition of phosphorus slag tends to decrease the early strength but enhance the strength development at the late ages. Note that the addition of phosphorus slag tends to improve the late-age pore structure of hardened paste (Figures 12 and 13). In summary, though phosphorus slag has an adverse effect on the early hardening rate of concrete, it makes considerable contribution to the late strength development: the reaction degree of phosphorus slag increases at late ages which produces secondary C-S-H and consumes $Ca(OH)_2$ [26], filling pores and improving the microstructure of interfacial transition zone of concrete.

Figures 18 and 19 show the demoulding strength of steam-cured concrete with the W/C of 0.4 and 0.32, respectively. The plain cement concrete cured at 60°C for 8 h is employed as the control group. It is clear that the demoulding strength of the concrete decreases significantly with the increase of the phosphorus slag replacement ratio if the temperature and the duration of the steam curing condition remain unchanged,

as a result of which the production of precast concrete cannot be carried out properly. As expected, the demoulding strength of the concrete containing phosphorus slag increases significantly with an extended steam curing duration to 11 h at 60°C or an elevated steam curing temperature at 80°C for 8 h. As shown in Figure 18, elevated steam curing temperature tends to enhance the demoulding strength of the concrete containing phosphorus slag more obviously than extended steam curing duration at the W/B of 0.4. This result is consistent with the w_n content result and hydration heat result: elevated steam curing temperature tends to increase the hydration degree of the binder containing phosphorus slag at demoulding time more obviously than extended steam curing duration at the W/B of 0.4. However, elevated steam curing temperature and extended steam curing duration have the similar enhancing effect on the demoulding strength of the concrete containing phosphorus slag at the W/B of 0.32 (Figure 19). This may be because the influence degree of a small increment of hydration products on the compressive strength of concrete is very limited at such a low W/B.

Figures 20 and 21 show the compressive strengths at the ages of 3, 7, and 90 d of steam-cured concrete with the W/C of 0.4 and 0.32, respectively. It is obvious that the compressive strength of the concrete containing phosphorus slag is lower than that of control concrete at the age of 3 d, no matter under the steam curing condition of 60°C for 11 h or 80°C for 8 h. It is an indication that though the demoulding strength of the concrete containing phosphorus slag is close to that of the control concrete, the hydration rate of the binder containing phosphorus slag is lower than that of plain cement during the period from demoulding time to 3 d. However, it is noteworthy that, with the increase of age, the compressive strength of the concrete containing phosphorus slag under the steam curing condition of 60°C for

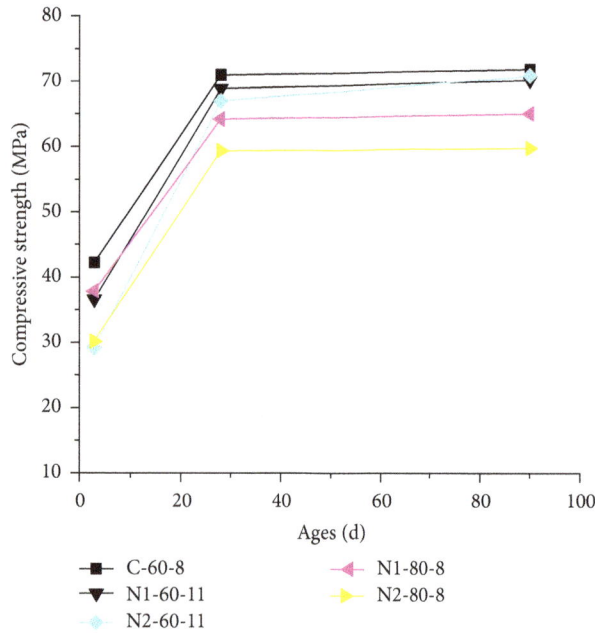

FIGURE 20: Compressive strength of the steam-cured concrete with W/B of 0.4.

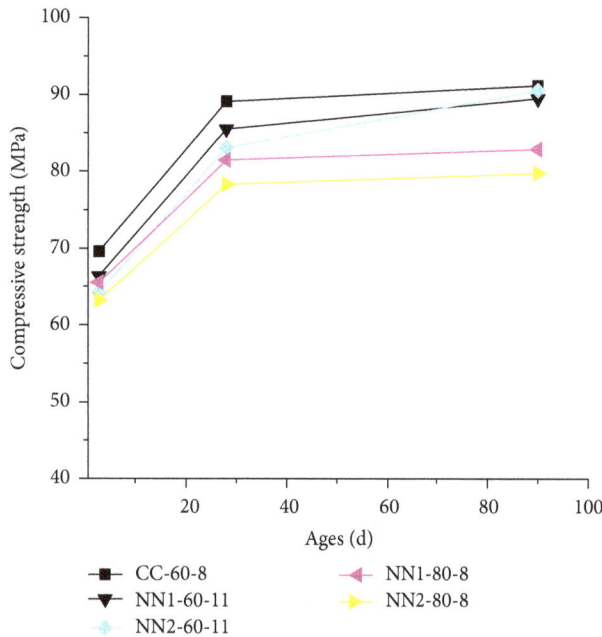

FIGURE 22: Chloride ion permeability of the standard-cured concrete with W/B of 0.40.

this difference is caused by the interfacial transition zone of concrete. Though elevated curing temperature improves the late-age pore structure of hardened paste, it has a significant adverse effect on the interfacial transition zone between hardened paste and aggregates, resulting in the decrease of compressive strength of concrete. Many researches have proved that elevated temperature curing at early ages has negative effects on the hydration of binder and the properties of concrete at late ages [18, 44]. The results of Figures 20 and 21 further confirm that the negative effect of elevated temperature curing on the late strength of concrete is still so obvious when the temperature is raised from 60°C to 80°C.

3.6. Chloride Ion Permeability. Figures 22 and 23 show the charge passed and the chloride ion permeability grade of standard-cured concrete at 28 d and 90 d with the W/C of 0.4 and 0.32, respectively. According to ASTM C1202, the chloride ion permeability grade of concrete is "very low" if the charge passed is between 100 and 1000 coulombs, and the grade is "low" or "moderate" if the charge passed is between 1000 and 2000 or between 2000 and 4000, respectively. As shown in Figure 22, the charge passed of the concrete with the W/C of 0.4 decreases with the increase of the phosphorus slag replacement ratio. The chloride ion permeability grade of concrete containing phosphorus slag at 28 d and 90 d decreases by one level compared with that of the plain concrete, which indicates that the resistance to chloride ion penetration of concrete is significantly enhanced by the addition of phosphorus slag. As shown in Figure 23, the charge passed of concrete with the W/C of 0.32 also decreases with the increase of the phosphorus slag replacement ratio. However, the matrix of the plain cement concrete with the W/C of 0.32 is already compact. Therefore, at such a low W/B, though the addition of phosphorus slag tends to reduce the

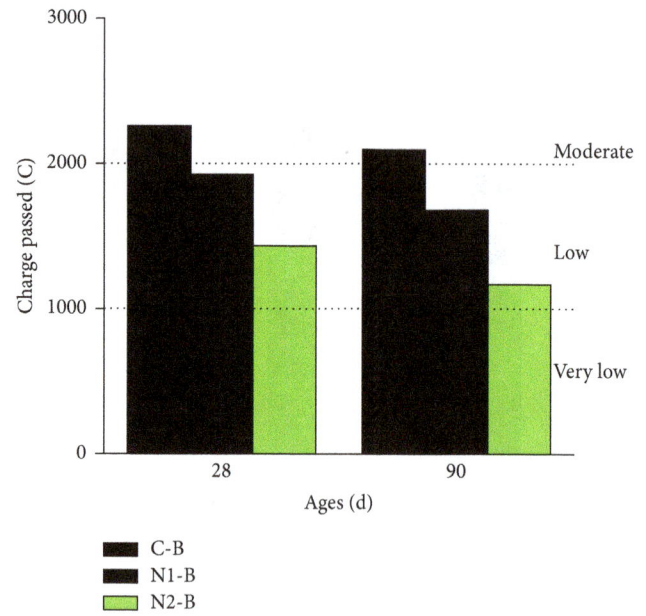

FIGURE 21: Compressive strength of the steam-cured concrete with W/B of 0.32.

11 h is close to that of control concrete at late ages, while the compressive strength of the concrete containing phosphorus slag under the steam curing condition of 80°C for 8 h is significantly lower than that of control concrete. This result is not consistent with the MIP result (Figures 14 and 15) which indicates that elevated curing temperature improves the late-age pore structure of hardened paste more obviously than extended steam curing duration. It is believed that

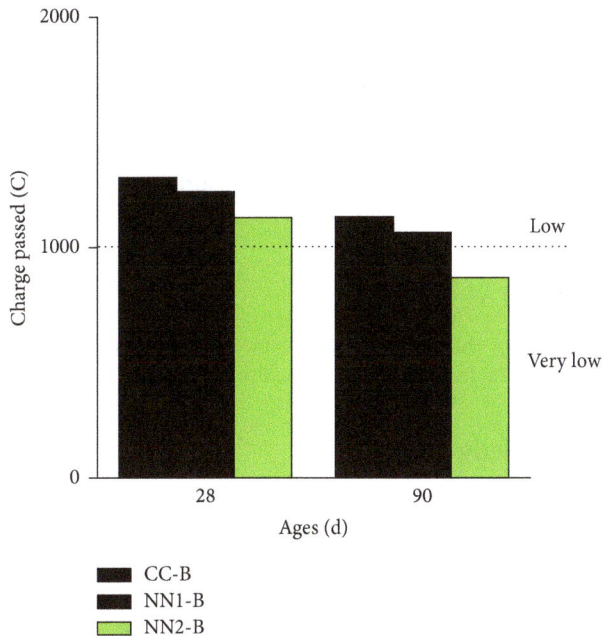

FIGURE 23: Chloride ion permeability of the standard-cured concrete with W/B of 0.32.

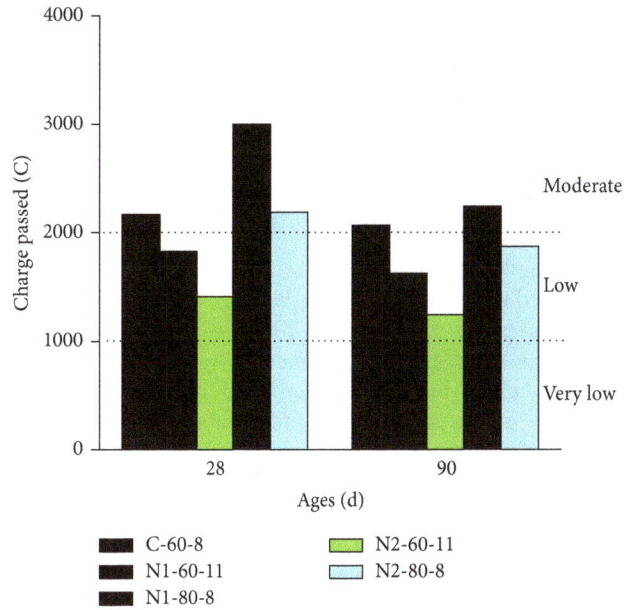

FIGURE 24: Chloride ion permeability of the steam-cured concrete with W/B of 0.40.

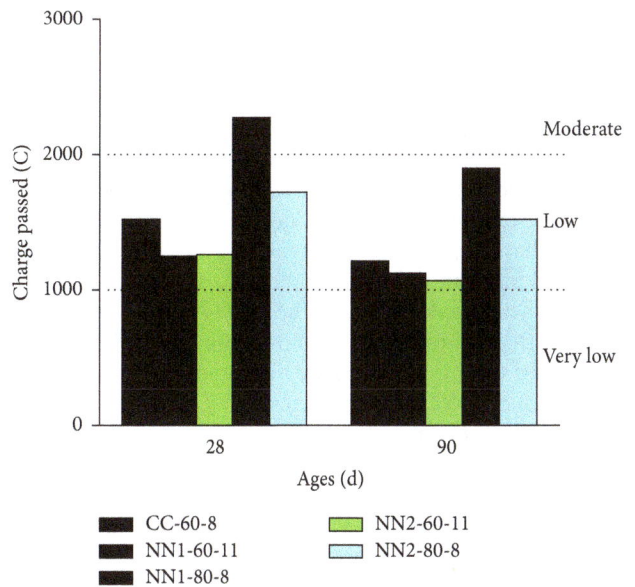

FIGURE 25: Chloride ion permeability of the steam-cured concrete with W/B of 0.32.

charge passed of the concrete, its influence on the chloride ion permeability grade is limited. The chloride ion permeability of concrete decreases by one level only when the phosphorus slag replacement ratio is 30% compared with that of the control concrete at the age of 90 d. On the whole, the addition of phosphorus slag can improve the microstructure and the resistance to chloride ion penetration of concrete in some cases.

Figure 24 shows the charge passed and the chloride ion permeability grade of steam-cured concrete at 28 d and 90 d with the W/C of 0.4. Whether at 28 d or 90 d, the chloride ion permeability grade of the control concrete is "moderate," and the chloride ion permeability grade of the concrete containing phosphorus slag cured at 60°C for 11 h is "low," which indicates that extending the duration of steam curing at 60°C can ensure that the concrete obtains a satisfactory resistance to chloride ion penetration. However, when the steam curing temperature is elevated to 80°C, the concrete containing phosphorus slag exhibits the same chloride ion permeability grade with the control concrete, except the concrete containing 30% phosphorus slag at the age of 90 d, which indicates that elevated steam curing temperature has a negative effect on the resistance to chloride ion penetration of concrete.

Figure 25 shows the charge passed and the chloride ion permeability grade of steam-cured concrete at 28 d and 90 d with the W/C of 0.32. Whether at 28 d or 90 d, the charge passed of the concrete containing phosphorus slag cured at 60°C for 11 h decreases compared with that of the control concrete, while the chloride ion permeability grade of the concrete containing phosphorus slag is the same with that of the control concrete. The charge passed of the concrete containing phosphorus slag cured at 80°C for 8 h

increases compared with that of the control concrete, and the chloride ion permeability grade of the concrete containing 15% phosphorus slag increases by one level at the age of 28 d. On the whole, when the curing temperature is increased from 60°C to 80°C, it tends to reduce the resistance of concrete to chloride ion penetration at certain circumstance.

In conclusion, the addition of phosphorus slag has positive effects on the late-age microstructure and the resistance to chloride ion penetration of concrete under the steam curing condition of 60°C. However, the resistance to chloride ion penetration of concrete containing phosphorus slag under the steam curing condition of 80°C is close to or even worse than the plain concrete, which might be due to the significant adverse effect of elevated curing temperature on the microstructure of interfacial transition zone of concrete.

4. Discussion

The demoulding strength of steam-cured concrete needs to be ensured in the first place. Under the condition of normal curing temperature, the addition of phosphorus slag would reduce the early strength of concrete significantly. Though elevated curing temperature tends to promote the early hydration of the composite binder containing phosphorus slag significantly, the concrete containing phosphorus slag cannot achieve similar demoulding strength with the plain cement concrete in the case of the same steam curing temperature and duration time. The demoulding strength results indicate that elevated steam curing temperature is more effective than extended steam curing duration, which is consistent with the trends of hydration heat results and nonevaporable water content results.

The MIP results indicate that the addition of phosphorus slag can improve the late-age pore structure of hardened paste whether cured at normal temperature or elevated temperature. Moreover, the late-age pore structure of hardened paste containing phosphorus slag cured at 80°C for 8 h is finer than that cured at 60°C for 11 h, and this trend is consistent with the trend of nonevaporable water content results. However, besides pore structure of hardened paste, the interfacial transition zone between matrix and aggregate is consistent with compressive strength and chloride ion permeability. It is notable that the effects of curing method on the properties of hardened paste and concrete might vary significantly due to the interfacial transition zone. The results of this paper show that the steam-cured concrete containing phosphorus slag cured at 60°C for 11 h can achieve higher late-age compressive strength and lower chloride permeability than that cured at 80°C for 8 h. Therefore, considering both demolding strength and late-age proprieties, extending the steam curing duration is a preferred method for the concrete containing phosphorus slag.

5. Conclusions

(1) Elevated steam curing temperature promotes the early hydration of the binder containing phosphorus slag more obviously than extended steam curing duration; moreover, the former method also results in higher late-age hydration degree of binder and finer pore structure of hardened paste.

(2) For steam-cured concrete containing phosphorus slag, the late-age strength decreases and the chloride permeability increases by increasing the curing temperature from 60°C to 80°C. The positive effects of phosphorus slag on the late-age properties of concrete are not well performed. The steam-cured temperature

of 80°C is not suitable for steam-cured concrete containing phosphorus.

(3) The steam-cured concrete containing phosphorus slag can achieve satisfied demoulding strength and late-age strength and chloride permeability by extending the steam curing duration at constant curing temperature of 60°C.

Conflicts of Interest

The authors declare that there are no conflicts of interest regarding the publication of this paper.

Acknowledgments

The support of China National Natural Science Foundation Project (no. 51572293) is acknowledged.

References

[1] M. Kim, Q. Wang, J. Park, J. C. Cheng, H. Sohn, and C. Chang, "Automated dimensional quality assurance of full-scale precast concrete elements using laser scanning and BIM," *Automation in Construction*, vol. 72, pp. 102–114, 2016.

[2] Q. Wang, M. Kim, S. Yoon, J. C. Cheng, and H. Sohn, "Corrigendum to "Automated quality assessment of precast concrete elements with geometry irregularities using terrestrial laser scanning" [Autom. Constr. 68 (2016) 170–182]," *Automation in Construction*, vol. 74, p. 1, 2016.

[3] M.-K. Kim, J. C. P. Cheng, H. Sohn, and C.-C. Chang, "A framework for dimensional and surface quality assessment of precast concrete elements using BIM and 3D laser scanning," *Automation in Construction*, vol. 49, pp. 225–238, 2015.

[4] J. Choi, S.-K. Park, H.-Y. Kim, and S. Hong, "Behavior of high-performance mortar and concrete connections in precast concrete elements: experimental investigation under static and cyclic loadings," *Engineering Structures*, vol. 100, pp. 633–644, 2015.

[5] Q. Wang, M. Li, and B. Zhang, "Influence of pre-curing time on the hydration of binder and the properties of concrete under steam curing condition," *Journal of Thermal Analysis and Calorimetry*, vol. 118, no. 3, pp. 1505–1512, 2014.

[6] G. Long, M. Wang, Y. Xie, and K. Ma, "Experimental investigation on dynamic mechanical characteristics and microstructure of steam-cured concrete," *Science China Technological Sciences*, vol. 57, no. 10, pp. 1902–1908, 2014.

[7] A. Gonzalez-Corominas, M. Etxeberria, and C. S. Poon, "Influence of steam curing on the pore structures and mechanical properties of fly-ash high performance concrete prepared with recycled aggregates," *Cement and Concrete Composites*, vol. 71, pp. 77–84, 2016.

[8] F. Cassagnabère, G. Escadeillas, and M. Mouret, "Study of the reactivity of cement/metakaolin binders at early age for specific use in steam cured precast concrete," *Construction and Building Materials*, vol. 23, no. 2, pp. 775–784, 2009.

[9] M. Gesoğlu, "Influence of steam curing on the properties of concretes incorporating metakaolin and silica fume," *Materials and Structures*, vol. 43, no. 8, pp. 1123–1134, 2010.

[10] D. W. S. Ho, C. W. Chua, and C. T. Tam, "Steam-cured concrete incorporating mineral admixtures," *Cement and Concrete Research*, vol. 33, no. 4, pp. 595–601, 2003.

[11] E. Gallucci, X. Zhang, and K. L. Scrivener, "Effect of temperature on the microstructure of calcium silicate hydrate (C-S-H)," *Cement and Concrete Research*, vol. 53, no. 2, pp. 185–195, 2013.

[12] A. M. Ramezanianpour, K. Esmaeili, S. A. Ghahari, and A. A. Ramezanianpour, "Influence of initial steam curing and different types of mineral additives on mechanical and durability properties of self-compacting concrete," *Construction and Building Materials*, vol. 73, pp. 187–194, 2014.

[13] C. Gu, W. Sun, L. Guo, and Q. Wang, "Effect of curing conditions on the durability of ultra-high performance concrete under flexural load," *Journal Wuhan University of Technology, Materials Science Edition*, vol. 31, no. 2, pp. 278–285, 2016.

[14] A. C. Aydin, A. Öz, R. Polat, and H. Mindivan, "Effects of the different atmospheric steam curing processes on the properties of self-compacting-concrete containing microsilica," *Sadhana - Academy Proceedings in Engineering Sciences*, vol. 40, no. 4, pp. 1361–1371, 2015.

[15] F. Sajedi and H. A. Razak, "Effects of curing regimes and cement fineness on the compressive strength of ordinary Portland cement mortars," *Construction and Building Materials*, vol. 25, no. 4, pp. 2036–2045, 2011.

[16] S. J. Barnett, M. N. Soutsos, S. G. Millard, and J. H. Bungey, "Strength development of mortars containing ground granulated blast-furnace slag: Effect of curing temperature and determination of apparent activation energies," *Cement and Concrete Research*, vol. 36, no. 3, pp. 434–440, 2006.

[17] S. Martínez-Ramírez and M. Frías, "The effect of curing temperature on white cement hydration," *Construction and Building Materials*, vol. 23, no. 3, pp. 1344–1348, 2009.

[18] Q. Wang, M. Miao, J. Feng, and P. Yan, "The influence of hightemperature curing on the hydration characteristics of a cement-GGBS binder," *Advances in Cement Research*, vol. 24, no. 1, pp. 33–40, 2012.

[19] S. Mengxiao, W. Qiang, and Z. Zhikai, "Comparison of the properties between high-volume fly ash concrete and high-volume steel slag concrete under temperature matching curing condition," *Construction and Building Materials*, vol. 98, pp. 649–655, 2015.

[20] H. Fanghui, W. Qiang, L. Mutian, and M. Yingjun, "Early hydration properties of composite binder containing limestone powder with different finenesses," *Journal of Thermal Analysis and Calorimetry*, vol. 123, no. 2, pp. 1141–1151, 2016.

[21] T. Zhang, Q. Yu, J. Wei, and J. Li, "Investigation on mechanical properties, durability and micro-structural development of steel slag blended cements," *Journal of Thermal Analysis and Calorimetry*, vol. 110, no. 2, pp. 633–639, 2012.

[22] G. Qian, S. Bai, S. Ju, and T. Huang, "Laboratory evaluation on recycling waste phosphorus slag as the mineral filler in hot-mix asphalt," *Journal of Materials in Civil Engineering*, vol. 25, no. 7, pp. 846–850, 2013.

[23] X.-W. Liu, L. Yang, and B. Zhang, "Utilization of phosphorus slag and fly ash for the preparation of ready-mixed mortar," *Applied Mechanics and Materials*, vol. 423–426, pp. 987–992, 2013.

[24] Y. Peng, J. Zhang, J. Liu, J. Ke, and F. Wang, "Properties and microstructure of reactive powder concrete having a high content of phosphorous slag powder and silica fume," *Construction and Building Materials*, vol. 101, pp. 482–487, 2015.

[25] C. Shi and J. Qian, "High performance cementing materials from industrial slags—a review," *Resources, Conservation and Recycling*, vol. 29, no. 3, pp. 195–207, 2000.

[26] P. Gao, X. Lu, C. Yang, X. Li, N. Shi, and S. Jin, "Microstructure and pore structure of concrete mixed with superfine phosphorous slag and superplasticizer," *Construction and Building Materials*, vol. 22, no. 5, pp. 837–840, 2008.

[27] X. Chen, L. Zeng, and K. Fang, "Anti-crack performance of phosphorus slag concrete," *Wuhan University Journal of Natural Sciences*, vol. 14, no. 1, pp. 80–86, 2009.

[28] L. Kalina, V. Bílek, R. Novotný, M. Mončeková, J. Másilko, and J. Koplík, "Effect of Na3PO4 on the hydration process of alkali-activated blast furnace slag," *Materials*, vol. 9, no. 5, article 395, 2016.

[29] X. Chen, K. H. Fang, H. Q. Yang, and H. Peng, "Hydration kinetics of phosphorus slag-cement paste," *Journal Wuhan University of Technology, Materials Science Edition*, vol. 26, no. 1, pp. 142–146, 2011.

[30] L. Dong-xu, C. Lin, X. Zhong-zi, and L. Zhi-min, "A blended cement containing blast furnace slag and phosphorous slag," *Journal of Wuhan University of Technology-Mater. Sci. Ed.*, vol. 17, no. 2, pp. 62–65, 2002.

[31] J. Feng, S. Liu, and Z. Wang, "Effects of ultrafine fly ash on the properties of high-strength concrete," *Journal of Thermal Analysis and Calorimetry*, vol. 121, no. 3, pp. 1213–1223, 2015.

[32] S. W. Tang, X. H. Cai, Z. He, H. Y. Shao, Z. J. Li, and E. Chen, "Hydration process of fly ash blended cement pastes by impedance measurement," *Construction and Building Materials*, vol. 113, pp. 939–950, 2016.

[33] S. K. Rao, P. Sravana, and T. C. Rao, "Abrasion resistance and mechanical properties of Roller Compacted Concrete with GGBS," *Construction and Building Materials*, vol. 114, pp. 925–933, 2016.

[34] G. D. Moon, S. Oh, and Y. C. Choi, "Effects of the physicochemical properties of fly ash on the compressive strength of high-volume fly ash mortar," *Construction and Building Materials*, vol. 124, pp. 1072–1080, 2016.

[35] M.-F. Ba, C.-X. Qian, X.-J. Guo, and X.-Y. Han, "Effects of steam curing on strength and porous structure of concrete with low water/binder ratio," *Construction and Building Materials*, vol. 25, no. 1, pp. 123–128, 2011.

[36] Y. Zhang, W. Zhang, W. She, L. Ma, and W. Zhu, "Ultrasound monitoring of setting and hardening process of ultra-high performance cementitious materials," *NDT and E International*, vol. 47, pp. 177–184, 2012.

[37] Q. Wang, M. Li, and G. Jiang, "The difference among the effects of high-temperature curing on the early hydration properties of different cementitious systems," *Journal of Thermal Analysis and Calorimetry*, vol. 118, no. 1, pp. 51–58, 2014.

[38] M.-F. Ba and C.-X. Qian, "Hydration evolution of pre-cast concrete with steam and water curing," *Journal of Central South University*, vol. 20, no. 10, pp. 2870–2878, 2013.

[39] Z. He, J. Liu, and K. Zhu, "Influence of mineral admixtures on the short and long-term performance of steam-cured concrete," in *Proceedings of the International Conference on Future Energy, Environment, and Materials, FEEM 2012*, pp. 836–841, China, April 2012.

[40] B. Lothenbach, F. Winnefeld, C. Alder, E. Wieland, and P. Lunk, "Effect of temperature on the pore solution, microstructure and hydration products of Portland cement pastes," *Cement and Concrete Research*, vol. 37, no. 4, pp. 483–491, 2007.

[41] J.-K. Kim, S. H. Han, and Y. C. Song, "Effect of temperature and aging on the mechanical properties of concrete: part I. Experimental results," *Cement and Concrete Research*, vol. 32, no. 7, pp. 1087–1094, 2002.

[42] R. Sarita, N. B. Singh, and N. P. Singh, "Interaction of tartaric acid during hydration of Portland cement," *Indian Journal of Chemical Technology*, vol. 13, no. 5, pp. 255–261, 2006.

[43] M. S. Morsy, "Effect of temperature on electrical conductivity of blended cement pastes," *Cement and Concrete Research*, vol. 29, no. 4, pp. 603–606, 1999.

[44] Q. Wang, J. J. Feng, and P. Y. Yan, "An explanation for the negative effect of elevated temperature at early ages on the late-age strength of concrete," *Journal of Materials Science*, vol. 46, no. 22, pp. 7279–7288, 2011.

Permissions

List of Contributors

Peng Gong, Zhanguo Ma and Xiaoyan Ni
State Key Laboratory for Geomechanics and Deep Underground Engineering, School of Mechanics and Civil Engineering, China University of Mining and Technology, Xuzhou, Jiangsu 221116, China

Ray Ruichong Zhang
Department of Mechanical Engineering, Colorado School of Mines, Golden, CO 80401, USA

Marcin Kozłowski
Department of Structural Engineering, Faculty of Civil Engineering, Silesian University of Technology, 5 Akademicka St., 44-100 Gliwice, Poland

Marta Kadela
Building Research Institute, 1 Filtrowa St., 00-611 Warszawa, Poland

Wenhui Zhao, Junjie Huang, Qian Su and Ting Liu
School of Civil Engineering, Southwest Jiaotong University, Chengdu, China
MOE Key Laboratory of High-Speed Railway Engineering, Southwest Jiaotong University, Chengdu, China

Daegeon Kim
Architecture Engineering, Dongseo University, Busan, Republic of Korea

Chengyao Liang, Chunxiang Qian, Huaicheng Chen and Wence Kang
School of Material Science and Engineering, Southeast University, Nanjing 211189, China
Research Institute of Green Construction Materials, Nanjing 211189, China

Tanakorn Phoo-ngernkham, Chattarika Phiangphimai and Jaksada Thumrongvut
Department of Civil Engineering, Faculty of Engineering and Architecture, Rajamangala University of Technology Isan, Nakhon Ratchasima 30000, Thailand

Nattapong Damrongwiriyanupap
Civil Engineering Program, School of Engineering, University of Phayao, Phayao 56000, Thailand

Sakonwan Hanjitsuwan
Program of Civil Technology, Faculty of Industrial Technology, Lampang Rajabhat University, Lampang 52100, Thailand

Prinya Chindaprasirt
Sustainable Infrastructure Research and Development Center, Department of Civil Engineering, Faculty of Engineering, Khon Kaen University, Khon Kaen 40002, Thailand
Academy of Science, The Royal Society of Thailand, Dusit, Bangkok 10300, Thailand

Lei Xu, Yongmiao Jin, Shuaizhao Jing, Yefei Huang and Changqiao Zhou
College of Water Conservancy and Hydropower Engineering, Hohai University, Nanjing 210098, China

Jie Liu
Chongqing Surveying and Design Institute of Water Resources, Electric Power and Architecture, Chongqing 400020, China

Xianggang Zhang and Jianhui Yang
Henan Province Engineering Laboratory of Eco-architecture and the Built Environment, Henan Polytechnic University, Jiaozuo 454000, China
School of Civil Engineering, Henan Polytechnic University, Jiaozuo 454000, China

Dapeng Deng
Henan Province Engineering Laboratory of Eco-architecture and the Built Environment, Henan Polytechnic University, Jiaozuo 454000, China

Ari Wibowo, Indradi Wijatmiko and Christin R. Nainggolan
Department of Civil Engineering, Faculty of Engineering, Brawijaya University, Malang 65149, Indonesia

Xiao-Yong Wang
College of Engineering, Department of Architectural Engineering, Kangwon National University, Chuncheon-si 200701, Republic of Korea

Ki-Bong Park and Xiao-Yong Wang
College of Engineering, Department of Architectural Engineering, Kangwon National University, Chuncheon-Si 200-701, Republic of Korea

Han-Seung Lee
Department of Architectural Engineering, Hanyang University, Ansan-Si 426-791, Republic of Korea

Qiang Du, Qiang Sun and Jing Lv
School of Civil Engineering, Chang'an University, Xi'an, Shaanxi 710061, China

Jian Yang
School of Civil Engineering, Birmingham University, Birmingham B15 2TT, UK

Wuman Zhang, Jingsong Zhang, Shuhang Chen and Sheng Gong
Department of Civil and Engineering, School of Transportation Science and Engineering, Beihang University, Beijing 100191, China

Min Sook Kim and Young Hak Lee
Department of Architectural Engineering, Kyung Hee University, 1732 Deogyeong-daero, Yongin, Republic of Korea

Joowon Kang
School of Architecture, Yeungnam University, 280 Daehak-ro, Gyeongsan, Republic of Korea

Hwasung Roh
Department of Civil Engineering, Chonbuk National University, Jeonju 561-756, Republic of Korea

Cheolwoo Park
Department of Civil Engineering, Kangwon National University, Samcheok 245-711, Republic of Korea

Do Young Moon
Department of Civil Engineering, Kyungsung University, Busan 608-736, Republic of Korea

Fang-Yuan Li, Cheng-Yuan Cao, Yun-Xuan Cui and Pei-Feng Wu
Department of Bridge Engineering, College of Civil Engineering, Tongji University, Shanghai 200092, China

Yubo Li and Shaobin Dai
Wuhan University of Technology, Wuhan 430070, China

Xingyang He and Ying Su
Hubei University of Technology, Wuhan 430070, China

Ruoyu Zhong, Ruichang Guo and Wen Deng
Missouri University of Science and Technology, Rolla, MO, USA

Jin Liu and Dongmin Wang
School of Chemical and Environmental Engineering, China University of Mining and Technology, Beijing 100083, China

Index

www.ingramcontent.com/pod-product-compliance
Lightning Source LLC
Chambersburg PA
CBHW080652200326
41458CB00013B/4831